Chemistry of
Sustainable
Energy

DATE DUE

Chemistry of Sustainable Energy

Nancy E. Carpenter

CRC Press
Taylor & Francis Group
Boca Raton   London   New York

CRC Press is an imprint of the
Taylor & Francis Group, an **informa** business

CRC Press
Taylor & Francis Group
6000 Broken Sound Parkway NW, Suite 300
Boca Raton, FL 33487-2742

Version Date: 20131227

International Standard Book Number-13: 978-1-4665-7532-5 (Paperback)

This book contains information obtained from authentic and highly regarded sources. Reasonable efforts have been made to publish reliable data and information, but the author and publisher cannot assume responsibility for the validity of all materials or the consequences of their use. The authors and publishers have attempted to trace the copyright holders of all material reproduced in this publication and apologize to copyright holders if permission to publish in this form has not been obtained. If any copyright material has not been acknowledged please write and let us know so we may rectify in any future reprint.

Library of Congress Cataloging-in-Publication Data

Carpenter, Nancy E., author.
  Chemistry of sustainable energy / Nancy E. Carpenter.
    pages cm
  Includes bibliographical references and index.
  ISBN 978-1-4665-7532-5 (paperback)
  1. Green chemistry. 2. Environmental chemistry. I. Title.

TP155.2.E58C37 2014
577'.14--dc23   2013049501

Visit the Taylor & Francis Web site at
http://www.taylorandfrancis.com

and the CRC Press Web site at
http://www.crcpress.com

This book is dedicated to my mother, Olga Y. Carpenter, who instilled in me love and respect for both the environment and for science, shaping my life's direction

and

to Jim Togeas and the late Jim Gremmels—two wonderful colleagues, mentors, and friends whose wisdom and humor have guided my professional career as an educator and writer.

Contents

Acknowledgments

This book would not have seen the light of day without the backing of Hilary Rowe, my acquisitions editor at Taylor & Francis, and its creation and completion rested on the input and assistance of many friends and colleagues. I gratefully acknowledge Julia Dabbs and Lorna Notch for their wise counsel in the initial phases of this project. Many colleagues provided helpful input: Andy Moses, Jim Cotter, and Stephen Crabtree for geological aspects; Ted Pappenfus for invaluable assistance with photovoltaics; Joe Alia, Jim Togeas, and Jenn Goodnough for help with thermodynamics and Jim Barbour for insight with respect to gasification and biomass energy. Special thanks to Susan Sutheimer for her enthusiastic encouragement at the onset and frequent and valuable feedback along the way. I truly could not have done this without their help.

I would particularly like to thank three others for their support and encouragement: Molly Carpenter, who gently asked how things were going when the question needed to be asked; Gretchen Gillis, who firmly encouraged me time and time again; and Cheryl Rempel for her unending patience, support, and faith in my ability.

Author

Professor Nancy E. Carpenter earned her PhD in organic chemistry from Northwestern University under the guidance of Professor Anthony G.M. Barrett. After a postdoctoral appointment with Professor Larry Overman at the University of California, Irvine, she came to the University of Minnesota, Morris, a four-year public liberal arts campus on the prairies of west-central Minnesota. Her research interests have spanned a diverse range of areas, from synthetic organometallic methodology to environmental remediation of chlorinated ethylenes and exploration of biodiesel from oilseeds and algae. She has been recognized with two teaching awards at the undergraduate level and was a co-recipient of the 2012 ACS-CEI Award for Incorporating Sustainability into Chemistry Education.

Introduction

Meeting global energy demand in a sustainable fashion will require not only increased energy efficiency and new methods of using existing carbon-based fuels but also a daunting amount of new carbon-neutral energy.

Nathan S. Lewis
Daniel G. Nocera
Henry Dreyfus Professor of Energy at the Massachusetts Institute of Technology
2006

The image on the cover is that of the solar-powered airplane Solar Impulse high over Belgium in 2011. The goal of Solar Impulse is grand: to fly both night and day relying solely upon solar energy. More broadly, however, Solar Impulse is meant to inspire: it is an innovation that has risen to a technological challenge to demonstrate that clean and sustainable energy can be achieved. It is an inspiration that is much needed when the impacts of global climate change are all around us. Based on the globally averaged temperature, 2012 was the tenth-warmest year since record-keeping began in 1880, and 2001–2012 rank among the 14 warmest years in this 133-year period. In the United States, 2012 was the warmest on record for the contiguous states and one of the most extreme with respect to temperature, precipitation, and tropical cyclones; 2013 continued with respect to extreme weather events (National Oceanic and Atmospheric Administration 2012).

Where is this climate change coming from? Overwhelming evidence points to the increasing amount of greenhouse gases—particularly carbon dioxide—in our atmosphere, a result of our insatiable consumption of fossil fuels (Bernstein et al. 2008). While writing this text, the alarming milestone of 400 ppm atmospheric CO_2 was surpassed—an ominous harbinger of climate change to come. Yet the world total primary energy consumption in 2011 increased to 531 quadrillion BTU (a quadrillion BTU is a *quad*); in 10 years that figure is expected to balloon to 638 and by 2032, to 741 quad (U.S. Energy Information Administration 2013). As the Earth warms and its population grows, our energy consumption puts us into a planetary Petri dish. The challenges before us are indeed daunting, and chemistry is a fundamental part of the solutions. This textbook was written to educate and prepare students of chemistry as they are called upon to help solve these global energy problems in the years to come.

Fossil fuels, of course, supply much more than energy for our society: from polymers to pharmaceuticals, most of our carbon comes from crude petroleum. While this is an important reality, this book is specifically focused on *energy*: processes that will reduce our reliance on fossil fuels, be it transportation-based, for electrical generation, or for heating our homes and businesses. It provides a survey of those areas of energy conversion that, arguably and at this point in time, show the most promise in terms of achieving some respectable level of sustainability: wind power, fuel cells, solar photovoltaics, and biomass conversion processes. In addition, the prospect of making use of next-generation nuclear power is an option that cannot be

ignored thus this topic is also included. Other aspects of energy and energy generation that are intimately tied to understanding the chemistry of sustainable energy are
also incorporated: fossil fuels, thermodynamics, hydrogen generation and storage,
and carbon capture. In this regard, this book provides a smorgasbord of sorts, with
some fascinating and relevant topics unfortunately omitted.

Chemistry of Sustainable Energy is, therefore, a survey text intended for use
in a one-semester advanced course for the student with a strong background in
chemistry. It is also a text that illustrates the increasingly interdisciplinary nature
of chemistry research and, in fact, scientific research as a whole. Examples from
the chemistry literature are inserted to provide relevant and interesting snapshots
of how solutions are developed and illustrate the living process of science. The
good news is that the progress in these fields is evolving rapidly, but as a result, the
examples in the text are unavoidably outdated as soon as they are put into print.
Nonetheless, they were chosen because they were interesting, recent at the time
of writing, seminal results, or promising. The aim of the text is to provide enough
background explanation for the reader to be able to peruse the most recent literature
and be able to read (and comprehend) scientific progress as it is published.

Texts on renewable energy and sustainability are being written and published at a
furious pace, and with good reason. Why this one? Simply stated, because it is focused
on *chemistry*, the so-called "central science," a discipline that holds the key to solving the pressing problems that face our planet (and our continued inhabitation of it).
Whether it be manufacturing new materials for solar or fuel cells or increasing the
efficiencies of agriculture production, wise implementation of chemical solutions is
required to meet the challenges that face us, from global climate change and dwindling
resources to population growth. Thus, first and foremost, this is a *chemistry* textbook.
In every topic, chapter, and paragraph I have striven to reinforce (and occasionally rein-
troduce) fundamental principles of chemistry as they relate to renewable or sustainable energy generation. My goal was to write a book from which advanced chemistry
students could learn of the breadth and depth of research being carried out to address
the problem of "meeting global energy demand in a sustainable fashion." As it has
been written by an organic chemist, there is a decidedly qualitative, structural bias.
There are glaring omissions—for example, thermodynamics is briefly covered but not
kinetics. The treatment is qualitative, not quantitative. Presentation of the analytical
methods key to research progress is, unfortunately, cursory. In addition, some attractive sustainable energy options are omitted, such as geothermal energy or tidal power.
Nevertheless, the overarching goal is to *teach chemistry* through the lens of several
sustainable energy options and provide a broad foundation for further exploration.

Energy is, of course, fundamental to our lives in all aspects and conversion of one
form of energy into another invariably involves chemistry. Ours is the era dominated
by the extraction and use of fossil fuels, a finite resource. Combustion of fossil fuels
generates the greenhouse gas carbon dioxide (CO_2) and increasing CO_2 levels have
inarguably contributed to global climate change. Chemistry is basic to these processes
and to the conversion of many other energy forms, including photoelectrochemical
devices, fuel cells, biochemical generation of fuel, energy storage and transmission—
the list goes on and on. Understanding the *chemistry* underlying sustainable energy is
the crux of any long-term solution to meeting our future energy needs.

Finally, what about *sustainability* in the context of energy generation? For all practical purposes the Earth is a closed system, and the scale of our energy demand is astronomical. Assessment of sustainability is inextricably tied to population, consumption, and environmental burden. Design of a "sustainable" energy process must take into account:

- The viability of the process (will it actually work?)
- The financial cost (will it survive in the current economic climate and vice versa?)
- The sustainability of the resources (e.g., rare earth metals) required to make it work
- The implications for global development (will all the Earth's inhabitants have equal opportunity?)
- The protection of our planetary home

This is a tall order indeed and one not wholly met by any of the arguably more sustainable energy conversion processes presented in this text. Although this text in no way professes to present a thorough, expert or complete analysis of the sustainability of the energy processes covered herein, the issues associated with sustainability will be raised throughout, with some hard questions undoubtedly resulting.

Will it work? Will the chemists, biochemists, and materials scientists of the coming years be able to compute, synthesize, and engineer our way to sustainable energy generation? That remains to be seen. Fundamentally, it is not a *science* problem, it is a *human* problem, and one that must be addressed by each one of us relying on this planet. But given (a) a problem that needs solving, (b) an incentive, and (c) enough ingenuity, science can, and must, contribute to the solutions. As the Solar Impulse soars, so must our hopes for the future.

SUGGESTED TEXTS FOR ADDITIONAL GENERAL READING

Armaroli, N. and V. Balzani. 2011. *Energy for a Sustainable World.* Weinheim, FRG: Wiley-VCH.

Bernstein, L., P. Bosch, O. Canziani et al. 2008. *Climate Change 2007 Synthesis Report. Intergovernmental Panel on Climate Change.* Geneva, Switzerland: Intergovernmental Panel on Climate Change.

Boyle, G., Ed. 2012. *Renewable Energy: Power for a Sustainable Future.* Oxford: Oxford University Press.

da Rosa, A. 2009. *Fundamentals of Renewable Energy Processes.* Burlington, MA: Academic Press/Elsevier.

Lewis, N.S. and D.G. Nocera. 2006. Powering the planet: Chemical challenges in solar energy utilization. *Proc. Natl. Acad. Sci.* 43:15729–15735.

MacKay, D.J.C. 2009. *Sustainable Energy—Without the Hot Air.* Cambridge: UIT Cambridge, Ltd.

National Oceanic and Atmospheric Administration. State of the Climate. 2-7-13. U.S. Government. 2012. http://www.ncdc.noaa.gov/sotc/

Sørensen, B. 2007. *Renewable Energy Conversion, Transmission and Storage.* Burlington, MA: Academic Press.

U.S. Energy Information Administration. 2013. *World Primary Energy Consumption by Region*. U.S. Department of Energy, Dec. 6, 2012 2013 [cited 14 February 2013]. Available from http://www.eia.gov/oiaf/aeo/tablebrowser/#release=IEO2011&subject =0-IEO2011&table=1-IEO2011&region=0-0&cases=Reference-0504a_1630.

Winterton, N. 2011. *Chemistry for Sustainable Technologies. A Foundation*. Cambridge: RSC Publishing.

1 Energy Basics

Every aspect of our existence rests upon energy conversion, but understanding energy is hampered by a morass of duplicative units and terms as well as confusion about just what energy is and where it comes from. The objective of this first chapter is to lay out the fundamentals common to all aspects of energy and to help unravel some of the confusion.

1.1 WHAT IS ENERGY?

We all "know" what energy is—or at least we know when we have it and when we have run out! We learn about *kinetic energy* and *potential energy* in elementary school and, as chemists, we are familiar with bond dissociation energy, free energy associated with chemical transformations, and electromagnetic radiation—energy in the form of photons, that is, quantized packets of energy. Several different *types* of energy are listed below. But what is *energy*, really?

Type of Energy	Example
Kinetic	Friction
Radiant	Electromagnetic spectrum
Thermal	Heat
Nuclear	Fusion, fission
Chemical	Chemical reactions
Electrical	Batteries/electricity
Gravitational	Gravity
Mechanical	Turbines

A dictionary will state that energy is the "capacity to do work," or "usable power," but these definitions can get fuzzy fairly quickly, especially when entering the realm of energy generation, conversion, and use. Energy and work are interchangeable and can both be measured in the units of joules, the SI unit of energy. But how do we define *work*? Again, we know what it is, but it is crucial to have a clear definition as we introduce the various concepts presented in this text. Work can be defined as *force* times *distance*, that is, using a force of some measured intensity to move something some distance—a classic example being raising an object against a gravitational field. The key word in our definition of energy is *capacity:* some *amount* of energy is required to carry out some task ("doing work"). Thus, energy is not *force* but it is clearly related to it. Similarly, energy is not *power*—power is the *rate* at which energy is expended (or generated) *over time*. Thus, for example, a *watt* is a unit of power (equivalent to 1 J/s), but a *watt-hour* (more familiarly, kilowatt hour, kWh) is the unit that reflects a quantity of energy. The types of energy, the amount

1

of work that can be done with a specified amount of energy, and the rate at which energy is generated or consumed are all important and relevant concepts under the umbrella of "energy."

In order to understand the role that chemistry must play in our energy future, we have to lay the foundation: what forms of energy do we have as potential resources and where did they come from? It all started with the big bang some 15–20 billion years ago. The hydrogen and helium that resulted from this event coalesced into stars, where nuclear fusion ultimately resulted in the formation of other elements. Galaxies, solar systems, and our own planetary home were formed. Initially, Earth was just matter condensed from a spinning mass of gas and dust. As the interior became hotter, heat was released as radioactive elements decayed and the planetary bulk accumulated due to gravity and compression. Ultimately, this led to the separation of the matter into layers—the familiar core, inner and outer mantle, crust, ocean, and atmosphere. (It is worth noting that the distribution was uneven—including the distribution of *elements*, which are ultimately the source of all of our materials—hence the stage was set for global geopolitics based on resource availability.) Eventually, photosynthesis took root and plant and animal life evolved.

There is general agreement that life on our planet began 3.5–4 billion years ago. At this time, the atmosphere consisted mostly of carbon dioxide and nitrogen with some ammonia and methane. Since the atmosphere was devoid of oxygen, life consisted of one-celled anaerobic bacteria (*prokaryotes*). Over time, some bacteria evolved with the ability to use photosynthesis to generate oxygen as a waste product (Equation 1.1).

$$2H_2O \rightarrow 4H^+ + O_2 + 4e^- \qquad (1.1)$$

As a result, about 2.5 billion years ago, there was a marked increase in the oxygen levels of Earth's atmosphere. As these oxygen-producing cyanobacteria thrived, anaerobic bacteria died off.

During the Paleozoic era (some 250–550 million years ago; Table 1.1), life began to diversify, with microorganisms, ferns, fishes, insects, and reptiles all coming into being. Three mass extinctions occurred during this era, including a mass extinction at the end of the Permian period in which it is estimated that about 96% of species known at that time disappeared (Purves et al. 2004). As these plant and animals died, the dead and decaying plant and animal matter were compressed in an oxygen-free environment, eventually leading to *fossil fuels* (Chapter 2).

Humans appeared at the beginning of the Quaternary period (see Table 1.1), a mere 1.8 million years ago. Initially, we took advantage of the sun and simply used solar energy for providing the heat we needed to survive. Solar energy dried materials and allowed foodstuffs to grow. Eating chemicals (in the form of plants and animals) led to the capacity for humans and animals to do work (consume energy!) and exist as hunter-gatherers. Then, of course, came fire. Combustion of cellulosic materials gave mankind a controllable, portable, and somewhat reliable way of heating. About 10,000 years ago, humankind developed a less nomadic lifestyle, domesticating animals for use in agriculture. This led to energy needs beyond that required

TABLE 1.1
Geologic Time Scale

Eon	Era	Period	Millions of Years Ago[a]
Phanerozoic	Cenozoic	Quaternary	1.6
		Tertiary	66
	Mesozoic	Cretaceous	138
		Jurassic	205
		Triassic	240
	Paleozoic	Permian	290
		Pennsylvanian	330
		Mississippian	360
		Devonian	410
		Silurian	435
		Ordovician	500
		Cambrian	570
Proterozoic Archean	Precambrian		2500
	Pre-Archean		3800?

Source: Adapted from U.S. Geological Survey. 1997. The Numeric Time Scale. http://pubs.usgs.gov/gip/fossils/numeric.html.
[a] Approximations.

for individual survival—energy to plow, energy to irrigate, energy to transport crops, energy to build for crop storage—energy to support a new way of life. The use of wind as a source of energy blew onto the scene approximately 9000 years ago in the form of sailing. Windmills were first seen in India about 2400 years ago, with water-driven mills eventually being built in settled agricultural communities.

Figure 1.1 summarizes the historical rate of growth of our energy use per capita. The big leap in energy usage came during the Industrial Revolution (late 1700s to roughly 1850). Our reliance on the energy of wind and water and the limited power of livestock was suddenly and dramatically displaced by the development of the steam engine. Energy use increased steadily as wood was eventually replaced as a primary fuel source with coal, then oil and natural gas. The ability to achieve higher temperatures led to improvements in the manufacture of iron, which led to better/ stronger/faster machinery and transportation. Oil consumption began to climb dramatically in the 1930s. The Allies' control of oil supplies in World War II cemented oil's use as a primary fuel source, and the explosion of research in nuclear chemistry and physics led to not only the atomic bomb but also nuclear power plants for the generation of electricity. Around 1950, the use of natural gas began to climb steadily.

With the increased use of fossil fuels came an increase in anthropomorphically generated carbon emissions (Figure 1.2) and the concomitant initiation of human-aided global climate change. As Figure 1.3 illustrates, our heavy reliance upon fossil fuels to supply our primary energy needs continues with minimal abatement.

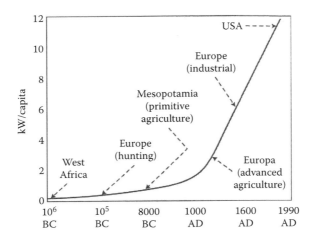

FIGURE 1.1 Per capita energy use over history. (Reprinted from *Fundamentals of Renewable Energy Processes*, da Rosa, A. V. Generalities, p. 7, 2009. With permission from Academic Press/Elsevier, Burlington, MA.)

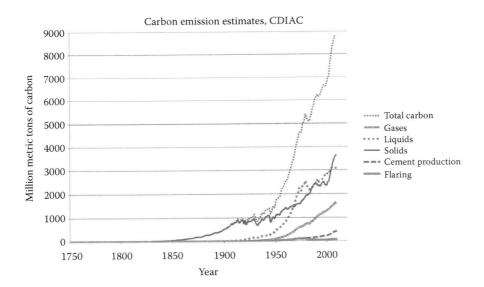

FIGURE 1.2 Global carbon emissions from cement manufacture and fossil fuel use. (Adapted from Boden, T.A., G. Marland, and R.J. Andres. 2010. Global, Regional, and National Fossil-Fuel CO_2 Emissions. Carbon Dioxide Information Analysis Center, Oak Ridge National Laboratory, U.S. Department of Energy, Oak Ridge, TN, USA.)

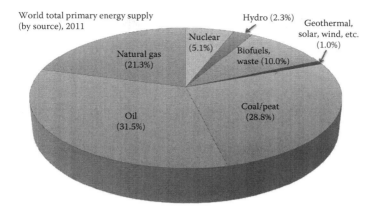

FIGURE 1.3 World primary energy supply (by source), 2009. (Adapted from IEA/ International Energy Agency 2013. *2013 Key World Energy Statistics.* IEA/International Energy Agency Paris, France, © OECD/IEA 2013.)

Furthermore, all of the global challenges that have resulted from this change and growth have been exacerbated by population growth. In prehistory, there were far fewer than 10 million human beings on the Earth at any one time using at most hundreds of watts of power per day. Compare that to the projected world population of almost 9 billion by 2050. From Neolithic times to the beginning of the Industrial Revolution, our per capita energy use was fairly constant and low, but as Figure 1.4 shows, the *rate* of increase of energy use exploded upon the implementation of the machinery of the Industrial Revolution. The per capita energy use in the United States in 2010 was 316 million BTU, a value that boggles the mind (over 91 million

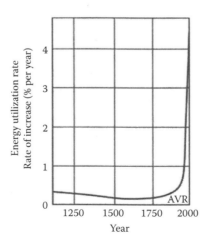

FIGURE 1.4 Average rate of increase in energy utilization. (Reprinted from *Fundamentals of Renewable Energy Processes*, da Rosa, A. V. Generalities, p. 7, 2009. With permission from Academic Press/Elsevier, Burlington, MA.)

watt-hours *per individual*). Given our energy use and population growth, we are in the midst of "progress" that is anything but sustainable (United Nations Department of Economic and Social Affairs Population Division 2004; U.S. Energy Information Administration 2012).

1.2 ENERGY, TECHNOLOGY, AND SUSTAINABILITY

1.2.1 WHAT DOES SUSTAINABILITY MEAN?

Energy use revolves around energy *conversion*, that is, transforming *primary* energy (raw energy) to a more useful form—*final* energy—that is functional for the need at hand. We can eat a candy bar to convert chemical energy (in the form of carbohydrates) into mechanical energy (in the form of muscle power), but that will not power our car. Primary energy sources such as solar energy or crude oil must be converted into electricity or heat, for example, for convenient use. Energy conversions are all around us: stars convert the nuclear energy of their core into light and heat. Plants take solar energy and convert it, via photosynthesis, into chemicals that we can burn for thermal energy or eat for mechanical (muscle) energy. The palette of energy transformations available to humans is summarized in Table 1.2; it is from this palette we must work to provide "sustainable" energy.

But just what do we mean by sustainable anyway? The word is often bantered about with little thought to its concrete connotation. Is this *ecologically* sustainable? *Economically* sustainable? *Sustainable for what and for whom?* One of the more frequently cited definitions of sustainability is "meet[ing] the needs of the present

TABLE 1.2
Energy Transformations

Initial Energy Form	Converted Energy Form[a]				
	Chemical	Radiant	Electrical	Mechanical	Thermal
Nuclear					Nuclear reactor
Chemical			Fuel cell/ battery	Ignition (internal combustion engine)	Burner/boiler
Radiant	Photolysis		Photovoltaic cell		Solar collector
Electrical	Electrolysis	Lamp		Electric motor	Heat pump
Mechanical			Piezoelectric generator	Turbine	Friction
Thermal			Thermoelectric generator	Thermodynamic engine	Radiator

[a] Note that often one form of converted energy accompanies another, for example, light and heat from a chemical reaction.

without compromising the ability of future generations to meet their own needs," a decidedly vague definition with a clear human focus (U.N. World Commission on Environment and Development 1987). The appropriate question may not be "is this sustainable?" but "will this leave our planet in better or worse shape for future generations?" As humans, of course, we are interested in sustaining our way of life, but as we have come to increasingly dominate the planet, we have sustained *our* way of life at the expense of others—the environment for future generations has been irreversibly altered because of loss of habitat, decreased biodiversity, and irreparable damage to the environment.

Not only is sustainability (and sustainable development) hard to define, it is also relative. When it comes to consumption of energy, sustainability for an urban household in the United States is immensely different from, say, sustainability for an aboriginal clan in Australia. Support of the standard of living in North America consumes drastically more energy than the rest of the world, such that if everyone on the globe were to attain this same level of comfort, at least twice the Earth's natural resources would be required (Simms et al. 2010). Science and technology have helped to create a wildly disparate distribution of wealth, health, and well-being across the globe. As we look toward offering "sustainable energy solutions," we must consider the Earth and all its inhabitants.

Finally, we must ask the question "is this level of sustainability even achievable?" An in-depth treatment of the planet's "carrying capacity" is beyond the scope of this text, but an oversimplified definition would be that *carrying capacity* is the maximum number of individuals (humans, for example) that can be sustained indefinitely by an ecosystem without causing irreparable damage. One way to gauge Earth's carrying capacity and our impact on our ecosystem is to look at our *ecological footprint*, a measure of our demands on nature. These representations compare the resource use (in this case, by humans) to the resource capacity (the planet Earth). As can be seen from the graph in Figure 1.5, we have been overextending the carrying capacity of our planet since 1976 (a value of 1 means that what the Earth can sustainably provide in 1 year was completely consumed). It is now estimated that we are consuming the resources of over one and a half Earths (Roney 2010). Ultimately, "nature will decide what is sustainable; it always has and always will" (Zencey May/June 2010).

1.2.2 CARBON CYCLE

One of the biggest concerns associated with energy use is the increasing amount of carbon in the atmosphere and its impact on global climate change. Clearly not all energy is about carbon, but because of our reliance on fossil fuels, quite a lot of it is. It is therefore important to review what we know about carbon's fate in our environment. The *carbon cycle* describes this exchange of carbon through four main reservoirs: the atmosphere, the terrestrial biosphere, the oceans, and the sediments (which include fossil fuels). Carbon in various forms moves through these sinks by chemical, physical, biological, and geological processes in a cycle and is graphically depicted in Figure 1.6 (National Aeronautic and Space Administration 2013; Olah et al. 2011). Carbon dioxide is the major source of carbon in our environment. It is found in the atmosphere (0.825×10^{15} kg) and has appreciable solubility in the

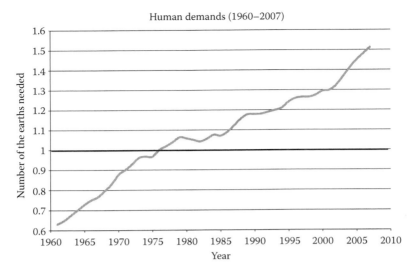

FIGURE 1.5 Human demands on the carrying capacity of the Earth. (Reprinted with permission from Roney, J.M. 2010. *Humanity's Ecological Footprint, 1961–2007*, edited by B. Barbeau: Global Footprint Network, www.earth-policy.org.)

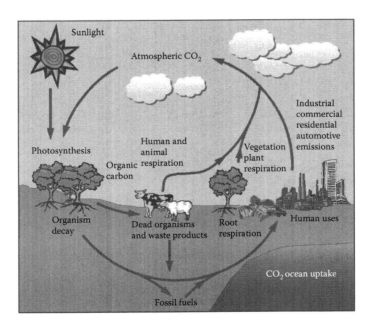

FIGURE 1.6 The carbon cycle. (Reprinted with permission from Olah, G. A. et al. 2011. Anthropogenic chemical carbon cycle for a sustainable future. *J. Am. Chem. Soc.* 133:12881–12898. Copyright 2011, American Chemical Society.)

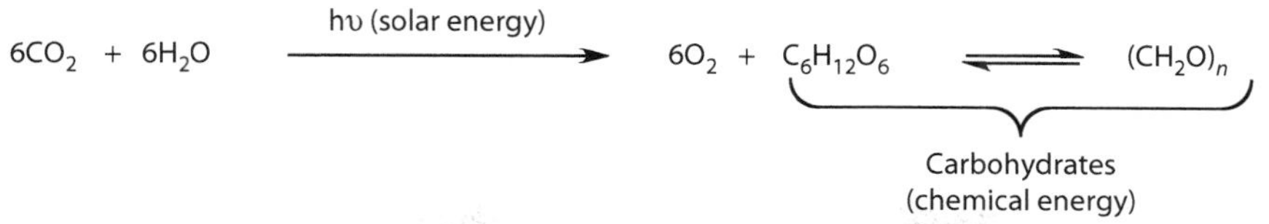

$$6CO_2 + 6H_2O \xrightarrow{\text{h}\upsilon \text{ (solar energy)}} 6O_2 + C_6H_{12}O_6 \rightleftharpoons (CH_2O)_n$$

Carbohydrates
(chemical energy)

FIGURE 1.7   Chemical conversions in photosynthesis.

oceans (45×10^{15} kg), but there is little exchange of $CO_2^{\cdot}$ from the oceans. Other significant amounts of carbon are present in carbonate minerals (e.g., limestone) and, obviously, in the dwindling supplies of hydrocarbon fossil fuels (10×10^{15} kg) (Sørensen et al. 2008). Although it is hard to imagine, the carbon in your sandwich (be it peanut butter or pastrami) may once have been the carbon in a coal seam, carbonate in the ocean, or CO_2 in the atmosphere thanks to the carbon cycle. The processes in this cycle take anywhere from hours to millions of years and have occurred many times over the course of Earth's history (McElroy 2010).

Plants and animals play a central role in the carbon cycle. Atmospheric CO_2 is taken up by plants, which utilize photosynthesis (Figure 1.7) to generate glucose plus oxygen. Carbon dioxide, in turn, is produced from the metabolism of glucose in cellular respiration (Equation 1.2). Thus, cellular respiration is the reverse of photosynthesis in terms of the overall production of CO_2, although the actual chemical mechanisms of the two processes are very different.

$$(CH_2O)_n + O_2 \rightarrow 6H_2O + 6CO_2 + \text{energy} \tag{1.2}$$

A close look at Figure 1.6 clearly illustrates the problem of fossil fuels in the overall balance of CO_2 in the carbon cycle. A huge amount of carbon was safely ensconced in the Earth in the form of coal, oil, and natural gas. By combusting this sequestered carbon pool in our vehicles and power plants, the balance of carbon in the environment has been significantly altered, producing increasing levels of CO_2 in the atmosphere (from 270 ppm in late 1800 to over 400 ppm today) and contributing to global climate change (Equation 1.3, Tans and Keeling 2013).

$$C_nH_{2n+2} + O_2 \rightarrow CO_2 + H_2O + \text{energy} \tag{1.3}$$

Atmospheric levels of CO_2 continue to climb with little effective effort by humanity to reduce or reverse this disturbing trend.

1.2.3   RESOURCE AVAILABILITY

As we further consider the sustainability of our energy solutions, resource availability is a concern that is amplified by the problem of scale. Several ostensibly renewable, clean, or sustainable energy solutions rely upon materials whose long-term existence is uncertain. For example, the scarce rare earth elements neodymium and dysprosium are used in the generators in wind turbines. Lanthanum, another rare earth, is found in the catalysts used in cracking hydrocarbons to make fuels from crude oil, and all of these rare earth metals are supplied by only one country—China (Knowledge Transfer Network 2010). These elements are considered "critical" or "near-critical" in terms of supply risk (Bauer et al. 2011). We will see that the platinum group metals (Ni, Pd, Pt) are exceedingly important metals in the catalytic conversion of all kinds of materials in energy production; tons of ore are required to make just one troy ounce of platinum, palladium, or rhodium. Furthermore, almost all of the world's supply of platinum is from two mines in South Africa. Many other transition metals (ruthenium, osmium, iridium, silver, etc.) are considered "endangered elements" in waning supply due to rapidly increasing use (Stanier and Hutchinson 2011). Several materials used in solar photovoltaics, too, are of special concern. Figure 1.8 presents some of the elements that are of particular value in the manufacture of photovoltaics (Chapter 7). The occurrence in Earth's crust is shown in ppm in the upper half of the graph; the cost below. Clearly, indium (used extensively in photovoltaics and electronics) and tellurium (a particularly effective material in thin-film photovoltaics) are especially scarce; their sustainable use is highly questionable.

But rare earth and transition metals are not the only elements at risk: with increasing population comes the need for additional biomass (for food and, potentially, for energy production), and phosphorus is essential for biomass production. The current consumption of phosphorus is not sustainable: like fossil fuels, phosphorus-containing

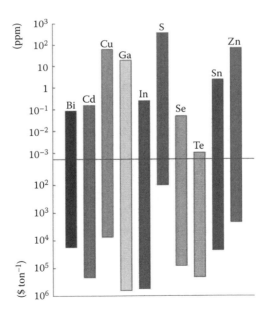

FIGURE 1.8 (**See color insert.**) Cost versus availability of materials important in the solar photovoltaic industry. (Peter, L. M. 2011. Towards Sustainable Photovoltaics: The Search for New Materials. *Philos. Trans. R. Soc. Lond. A*, 369(1942): 1840–1856. By permission of the Royal Society.)

rock is mined from nonrenewable deposits and used to manufacture, among other things, fertilizer. As with the other elements mentioned above, as the supplies become more limited, the cost of extracting these materials increases dramatically. The global production of phosphorus has been predicted to peak as early as 2035 (Schröder et al. 2010). Recycling or recovery of waste phosphorus is currently limited at best, and large amounts of phosphorus are lost in runoff from agricultural fields, contributing to eutrification in reservoirs and the infamous "dead zone" of the Gulf of Mexico (Massefski and Capelli 2012). Given the importance of phosphorus in agriculture, it is imperative that sustainable phosphorus production and use be achieved (Schröder et al. 2010).

Of course (as we will see in Chapter 2), a huge resource that is of limited and dwindling availability is fossil fuel. While new methods for the extraction of natural gas and petroleum are unlocking previously inaccessible reserves, fossil fuels are, ultimately, a *nonrenewable* resource and their continued use contributes to global climate change. Furthermore, all of these materials pose a dilemma when considering their implementation on the global scale that is required to meet our current energy requirements. However, one substance perhaps more than any other is especially important when it comes to sustainable development of energy solutions: water. Water is used in enormous volumes in manufacturing, agriculture, and energy production. A tempting source of clean energy is hydrogen gas, which can be produced from the "splitting" of water (Chapter 5). However, as the planet warms, the population grows and water becomes increasingly scarce; there is real concern that water will become the "new oil," with ample strife already resulting from competition for this vital resource in arid areas across the globe (Wachman 2007).

In the end analysis, our place in this closed system that is the Earth requires that we not only attempt to find replacements for at-risk elements and materials but also *reduce* our use and develop methods to *recycle* what we do use. The value of resources in our waste stream is often higher than what it costs to obtain them in the first place. But in order to efficiently recycle valuable materials, products must be designed with recovery in mind, a feature that is not prevalent in today's disposable culture.

1.3   ENERGY UNITS, TERMS, AND ABBREVIATIONS

Energy value. The amount of energy that can be delivered per unit mass or volume is an important facet of fuels. *Fuels* deliver their energy specifically by combustion, in contrast to other energy sources such as batteries or solar cells. It is relatively straightforward to directly compare the amount of energy contained, say, in a ton of corn cobs to that in a barrel of oil if we focus on their heat of combustion. As Table 1.3 illustrates, the heat of combustion per gram or mole of substance increases with increasing molecular weight (c.f. butane, hexane, heptane, and octane), and hydrocarbons release much more heat per gram or milliliter than alcohols (c.f. ethane/ethanol or butane/butanol). Thus, octane (a good approximation for gasoline) has a higher heat of combustion—and more energy per gram or milliliter—than ethanol or butanol.

When the energy content of a fuel is reported on a per mass or per volume basis, it is often referred to as the material's *energy value*. Energy value comes in two forms: the *lower heating value* (LHV; also known as *net calorific value*) and *higher*

TABLE 1.3
Heat of Combustion Data for Various Substances

Substance	Heat of Combustion (kJ/mol)	Heat of Combustion (kg·cal/g) (Temperature, °C)	Density (g/mL, 20°C)
Ethanol	−1368	326.7 (25)	0.7893
Ethane	−1560 (gas phase)	372.8 (25)	—
1-Butanol	−2670	639.5 (25)	0.8098
n-Butane	−2876 (gas phase)		—
n-Hexane	−4163	995.0 (25)	0.6603
n-Heptane	−4817	1150.0 (20)	0.6838
n-Octane	−5430	1302.7 (20)	0.7025

Source: Data from Weast, R., Ed. 1974. *CRC Handbook of Chemistry and Physics.* Cleveland, OH: CRC Press, D243–248.

heating value (HHV or *gross calorific value*). HHV is defined as the amount of heat released by combusting a specific quantity (in mass or volume) at an initial temperature of 25°C until it is completely combusted and the products have returned to a temperature of 25°C. The HHV includes the heat released in bringing the vaporized water back to the liquid state. LHV, on the other hand, is defined as the amount of heat released (as above) when the products cool to only 150°C. In this case, there is no heat of vaporization to capture; hence a smaller heating value is obtained. This is the case, for example, when running a system at high temperatures, as in a boiler. A comparison of HHV and LHV for various fuels is presented in Table 1.4.

TABLE 1.4
Comparison of Higher (HHV) and Lower Heating Values (LHV)

	HHV	LHV	HHV	LHV
	MJ/kg		MJ/L	
Natural gas (32°F/1 atm)	52.2	47.1	39.0	35.2
Propane	50.2	46.3	99.8	91.9
Gasoline	46.5	43.4	34.8	32.7
Hydrogen (70 MPa)	142.2	120.2	5.63	4.76
Crude oil	45.5	42.7	38.8	36.8
Ethanol	29.8	27.0	23.5	21.3

Source: Data from Boundy, B. et al. 2011. *Biomass Energy Data Book.* U.S. Department of Energy/Office of the Biomass Program/Energy Efficiency and Renewable Energy; Staffell, I. 2011. *The Energy and Fuel Data Sheet.* http://works.bepress.com (accessed 13 April 2013).

TABLE 1.5
Comparative Data for the Energy Density of Various Substances

Substance	Energy Density	
	MJ/kg	MJ/L
Natural gas	47.1	35.2
Gasoline	43.4	32.7
Hydrogen gas	120	0.01 (LHV)
Crude oil	42	36.8 (LHV)
Ethanol	28	21.3
Coal	32	42.0
Dry wood	12.5	10.0

Source: Data from Sørensen, B. 2007. *Renewable Energy Conversion, Transmission & Storage.* Burlington MA: Academic Press, 262; Staffell, I. 2011. *The Energy and Fuel Data Sheet.* http://works.bepress.com (accessed 13 April 2013); Waldheim, L. and T. Nilsson. 2001. *Heating Value of Gases from Biomass Gasification,* Nyköping, Sweden: IEA Bioenergy Agreement Subcommittee on Thermal Gasification of Biomass.

Energy density values (Table 1.5) can be used in certain situations to provide an approximate comparison (Waldheim and Nilsson 2001; Sørensen 2007; Staffell 2011). In general, the value reported as "energy density" is the same as the material's LHV. These data have been collected from a very wide variety of sources with the result that there is almost invariably some disparity in reported values; therefore, the information presented in these tables is relative but provides some interesting insights. For example, while H_2 has the highest energy density by mass, its energy density by volume is (as expected) abysmal. Another application of energy density is in the realm of energy storage (Section 6.10), where it refers to the ratio of the mass of energy stored to the volume of the storage device.

Energy units and abbreviations. One of the more exasperating aspects of working in an energy-related field is the panoply of units, abbreviations, and equations. The SI units we are familiar with (joules, watts, etc.) are not necessarily the units used in industrial applications, and of course the use of SI units is, unfortunately, not universal. What is worse is that many units are specific to a certain kind of industry. Furthermore, a calorie is not necessarily a calorie: its measurement is temperature dependent and "1 calorie" may be 4.18674 J, 4.19002 J, or 4.18580 J depending upon what *particular* calorie one is referring to (the *international steam table calorie,* the *mean calorie,* or the *thermochemical calorie*) or even the particular country that is reporting the value. The familiar BTU is similarly afflicted. The International Energy Agency maintains a unit converter (http://www.iea.org/stats/unit.asp) as does the U.S. Energy Information Agency (http://www.eia.gov/energy-explained/index.cfm?page=about_energy_conversion_calculator) and Appendices

I through IV provide most of the important conversion factors, SI prefixes, and energy-related equations. Most global energy statistics are given in units of *toe*— tonnes of oil equivalent. As one might expect, a tonne of oil equivalent is the amount of thermal energy released when a metric tonne (1000 kg) of oil is combusted, a value that is standardized at 41.868 GJ. The world consumption of final energy has increased from roughly 4000 to over 12,000 Mtoe from 1971 to 2011—a truly staggering 5×10^8 terajoules (TJ) of energy (British Petroleum 2012; IEA/International Energy Agency 2012).

1.4 ELECTRICITY GENERATION AND STORAGE

Energy use can be arbitrarily broken down into major uses: fuels for transportation, energy for manufacturing, energy for heating/cooling systems, and finally energy for our smartphones, computers, televisions, and so on. It is in the latter category that electricity plays a large role with at least 40% of the energy in the United States being used to generate electricity (Girard 2010). Globally, about 40% of electricity use is industrial (IEA/International Energy Agency 2012) with the world's total final consumption of electricity in 2010 reaching 1536 Mtoe—well over 17,800 terawatt-hours. Fossil fuels figure prominently in the production of electricity, with coal alone providing about 40% of the world's electricity (Figure 1.9) (Reisch 2012). As a result, roughly one-third of global CO_2 emissions comes just from electricity generation (Milne and Field 2013). Because this most convenient of final energies plays such a large role in humanity's energy use, it is important to understand electricity and how it is made.

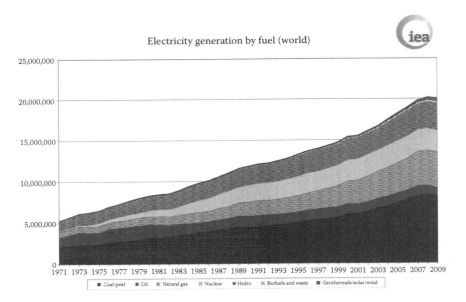

FIGURE 1.9 World electricity generation by fuel (GWh). (Adapted from IEA/International Energy Agency. 2012. *2012 Key World Energy Statistics*. Paris, France, International Energy Agency.)

It was the French physicist Charles Augustin Coulomb (1736–1806) who recognized the nature of the electrostatic force between charged particles and showed that the force between them is proportional to the product of their charges (and inversely proportional to the distance between them), as shown by Coulomb's law (Equation 1.4, where q designates charge in the units of coulombs, F is the force in newtons, r is the distance between the two charges in meters, and k is a constant equal to 8.99×10^9 N · m²/C²).

$$F = \frac{kq_1q_2}{r^2} \qquad (1.4)$$

Just as a mass at some height has a potential energy due to gravitational forces, so does a charged particle in an electromagnetic field: a charged particle (e.g., an electron) will be forced to move by being either attracted or repulsed by the field, depending upon the nature of the charges involved. This electrical potential energy (or *electromotive force*, EMF) has the units of volts, where a volt is equivalent to joule/coulomb. Thus, when some material (e.g., a coil of wire in a generator) is moved through a magnetic field, the electrons in the material can be forced to move, generating electricity. Obviously, moving the coil of wire through the magnetic field is work that requires some source of energy, as in the raw energy of a fossil fuel. In a typical coal-fired power plant, combustion of coal provides the thermal energy to generate high-pressure steam, which turns a steam turbine coupled to a generator: voilá—electricity!

In the United States, the percentage use of natural gas for electrical generation is approaching that of coal (see Figure 1.10), with hydroelectric and renewable electricity

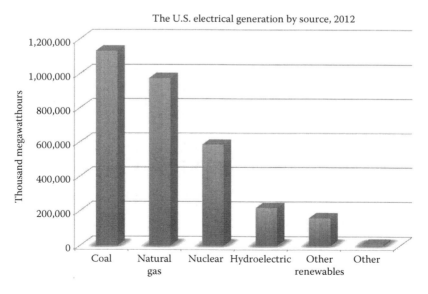

FIGURE 1.10 U.S. electrical generation by fuel type (%), 2011. (Adapted from U.S. Energy Information Administration, 2013b. *Net Generation by Energy Source: Total (All Sectors).* Retrieved 19 April 2013, from http://www.eia.gov/electricity/data.cfm#generation.)

generation making up almost 12% of the total (U.S. Energy Information Administration 2013b). Many U.S. natural gas and coal-fired power plants have generating capacities of well over 1000 MW, with operating efficiencies limited to 30–40% as a result of the laws of thermodynamics, which we will turn to in Chapter 3. The largest nuclear power plants in the United States are rated at a capacity of around 1300 MW (the net total MWh used in the United States in 2011, for comparison, was 4,100,656 *thousand* MWh (U.S. Energy Information Administration 2011)). There are other cleaner ways of generating electricity: wind turbines (Chapter 4), fuel cells (Chapter 6), and solar photovoltaics (Chapter 7). While these are very promising sources of electrical energy, recall the terawatt level of consumption noted above. The comparisons listed below demonstrate quite bluntly why the matter of scale will be a constant reminder as we contemplate the chemistry of sustainable energy (Kirubakaran et al. 2009).

	Turbine Generator	Photovoltaics	Wind Turbine	Fuel Cells
Capacity (range)	500 kW to 25 MW	1 kW to 1 MW	10 kW to 1 MW	200 kW to 2 MW
Efficiency (%)	29–42	6–19	25	40–60

The huge scale required for electricity generation and the variability of sustainable sources of electricity such as solar and wind raise additional concerns. How useful is wind energy when the wind is not blowing? How can photovoltaics provide power for us when the sun is not shining? These concerns are the driving force for the development of a so-called *smart grid* and efforts to develop effective, large-scale modes of storing electricity. While weaving together a smart grid is all about digital technology (and thus not the purview of this text), electrical energy storage (EES) is strongly embedded in the area of chemistry and materials science. Storage of electricity as some other form of energy that can be converted back into electricity, possibly at a moment's notice, can even out the variability of a wind- or light-driven electricity supply. There are several approaches to stationary EES systems, including supercapacitors that store electricity in electrical charges or flywheels that convert kinetic energy back into electrical energy. Potential energy can be used to store electrical energy, for example, by pumping large volumes of water into a raised reservoir where it can reside until it is freed to pass through a hydroelectric generating plant. The EES most relevant to this text is electrochemical energy storage, a.k.a. *batteries*. Because the research progress associated with the development of large-scale EES is closely related to that of fuel cells, this topic will be covered in some detail in Chapter 6.

OTHER RESOURCES

BOOKS

Letcher, T.M. 2008. *Future Energy*. Oxford, U.K.: Elsevier.
McElroy, M.B. 2010. *Energy: Perspectives, Problems & Prospects*. New York: Oxford University Press.
National Research Council. 2010. *Electricity from Renewable Resources: Status, Prospects, and Impediments*. Washington, D.C.: National Academies Press.

ONLINE RESOURCES

Energy Resources Conservation Board (Canada): http://ercb.ca/
International Energy Agency: http://www.iea.org/
United States Energy Information Administration: http://www.eia.gov/
United States Department of Energy: http://www.doe.gov/—National Energy Technology Laboratory: http://www.netl.doe.gov/index.html—Energy Efficiency and Renewable Energy: http://www.eere.energy.gov/
United Nations Development Programme, Environment & Energy: http://www.undp.org/content/undp/en/home/librarypage/environment-energy.html
U.S. National Renewable Energy Laboratory: http://www.nrel.gov/
World Energy Council: http://www.worldenergy.org/

REFERENCES

Bauer, D., D. Diamond, J. Li et al. 2011. Critical Materials Strategy U.S. Department of Energy. U.S. Department of Energy, http://energy.gov/sites/prod/files/DOE_CMS2011_FINAL_Full.pdf.

British Petroleum. 2012. *World Primary Energy Consumption Grew by 2.5% in 2011, Roughly in Line with the 10-Year Average.* British Petroleum 2012 [cited 1 May 2013]. Available from http://www.bp.com/extendedsectiongenericarticle.do?categoryId=9041234&contentId=7075077.

Boden, T.A., G. Marland, and R.J. Andres. 2010. Global, Regional, and National Fossil-Fuel CO_2 Emissions. Carbon Dioxide Information Analysis Center, Oak Ridge National Laboratory, U.S. Department of Energy, Oak Ridge, TN, USA.

Boundy, B., S. W. Diegel et al. 2011. *Biomass Energy Data Book.* U.S. Department of Energy/Office of the Biomass Program/Energy Efficiency and Renewable Energy.

da Rosa, A.V. 2009. *Generalities, Fundamentals of Renewable Energy Processes.* Burlington, MA: Academic Press/Elsevier, p. 7.

Girard, J.E. 2010. *Principles of Environmental Chemistry*, 2nd ed. Sudbury, MA: Jones & Bartlett.

IEA/International Energy Agency. 2013. *2013 Key World Energy Statistics.* International Energy Agency. Paris: France.

Kirubakaran, A., S. Jain, and R.K. Nema. 2009. A review on fuel cell technologies and power electronic interface. *Renew. Sustain. Energy Rev.* 13:2430–2440.

Knowledge Transfer Network. 2010. *Minerals and Elements Review.* Chemistry Innovation Ltd. 2010 [cited 8 May 2013]. Available from http://www.chemistryinnovation.co.uk/stroadmap/files/dox/MineralsandElementspages.pdf.

Massefski, A. and K. Capelli. 2012. Dead zone: The source of the Gulf of Mexico's hypoxia. In *Science Features*: USGS.

McElroy, M.B. 2010. *Energy: Perspectives, Problems & Prospects.* New York: Oxford University Press.

Milne, J.L. and C.B. Field. 2013. Assessment report from the GCEP workshop on energy supply with negative carbon emissions. Stanford University Global Climate and Energy Project.

National Aeronautic and Space Administration. 2013. *Earth Observatory Glossary* 2013 [cited 2 April 2013].

Olah, G.A., G.K.S. Prakash, and A. Goeppert. 2011. Anthropogenic chemical carbon cycle for a sustainable future. *J. Am. Chem. Soc.* 133 (33):12881–12898.

Peter, L. M. 2011. Towards Sustainable Photovoltaics: The Search for New Materials. *Philos. Trans. R. Soc. Lond. A*, 369(1942):1840–1856.

Purves, W.K., D. Sadava, G.H. Orians et al. 2004. *Life: The Science of Biology.* 7th ed. Sunderland, MA: Sinauer Associates, Inc.

Reisch, M.C. 2012. Coal's enduring power. *C. & E. News* 90 (47):12–17.

Roney, J.M. 2010. *Humanity's Ecological Footprint, 1961–2007,* edited by B. Barbeau. Global Footprint Network. www.earth-policy.org/datacenter/xls/book_wote_ch1_9.xls

Schröder, J.J., D. Cordell, A.L. Smit et al. 2010. Sustainable Use of Phosphorus. Report 357. EU Tender ENV.B.1/ETU/2009/0025. Plant Research International/Business Unit Agrosystems and Stockholm Environment Institute (SEI). Wageningen, The Netherlands, DLO Foundation.

Simms, A., V. Johnson, and P. Chowla. 2010. *Growth Isn't Possible. Why We Need a New Economic Direction. The Great Transition.* London, UK: New Economics Foundation.

Sørensen, B. 2007. *Renewable Energy Conversion, Transmission & Storage.* Burlington, MA: Academic Press.

Sørensen, B., P. Breeze, T. Storvik et al. 2008. *Renewable Energy Focus Handbook.* Amsterdam: Academic Press.

Staffell, I. 2011. *The Energy and Fuel Data Sheet.* works.bepress.com 2011 [cited 16 April 2013]. Available from http://www.academia.edu/1073990/The_Energy_and_Fuel_Data_Sheet.

Stanier, C. and J. Hutchinson. 2011. A sustainable global society: How can materials chemistry help? A white paper from the chemical sciences and society Summit (CS3) 2010. Royal Society of Chemistry.

Tans, P. and R. Keeling. 2013. *Trends in Atmospheric Carbon Dioxide.* National Oceanic & Atmospheric Administration, Earth System Research Laboratory 2013 [cited 2 April 2013]. Available from http://www.esrl.noaa.gov/gmd/ccgg/trends/mlo.html.

U.N. World Commission on Environment and Development. 1987. *Our Common Future.* New York: United Nations.

U.S. Energy Information Administration. 2012. *Annual Energy Review, 2011.* U.S. Energy Information Administration. Available from http://www.eia.gov/totalenergy/data/annual.

U.S. Energy Information Administration. 2013a. *Electric Power Annual.* U.S. Energy Information Administration, January 30, 2013 [cited 1 May 2013]. Available from http://www.eia.gov/electricity/annual/.

U.S. Energy Information Administration. 2013b. *Net Generation by Energy Source: Total (All Sectors).* U.S. Energy Information Administration [cited 19 April 2013]. Available from http://www.eia.gov/electricity/data.cfm-generation.

U.S. Geological Survey. 1997. The Numeric Time Scale. http://pubs.usgs.gov/gip/fossils/numeric.html.

United Nations Department of Economic and Social Affairs Population Division. 2004. *World Population to 2300.* New York: United Nations. http://www.unpopulation.org.

Wachman, R. 2007. Water becomes the new oil as world runs dry. *The Observer,* 8 December 2007.

Waldheim, L. and T. Nilsson. 2001. *Heating Value of Gases from Biomass Gasification.* TPS-01/16. IEA Bioenergy Agreement Subcommittee on Thermal Gasification of Biomass. TPS Termiska Processer AB, Studsvik, 611 82 Nyköing, Sweden.

Weast, R., Ed. 1974. *CRC Handbook of Chemistry and Physics.* Cleveland, OH: CRC Press, D243–248.

Zencey, E. May/June 2010. Sustainability. *Orion* May/June: 34–37.

2 Fossil Fuels

Before we can make any progress toward a sustainable energy future, we need to learn from our past and present. Although proven reserves of fossil fuels may continue to meet our energy needs for some years to come and nonconventional sources of fossil fuels are pushing back the "end" of oil, there will come an end, particularly because our reliance on fossil fuels remains great (recall Figure 1.2). The story of fossil fuels is a fascinating mix of history, politics, and science, with global security, economics, and most certainly the environment demanding that we begin to phase out our dependence on fossil fuels. The aim of this chapter is to understand what fossil fuels are, where they came from, how we got to this point, and where we are going with respect to finding fossil fuels for the foreseeable future. (N.B. Appendix IV provides some additional information with respect to units and conversions peculiar to the oil and gas industry.)

2.1 FORMATION OF OIL AND GAS

Contrary to popular belief, deceased and decomposed dinosaurs are not the primary source of fossil fuels that we use to power our planet. But before we can fully understand the source of *petroleum* (literally "rock oil"), we need to review a little geology. Scientists have devised many cycles to explain the processes and conversions of our planet, such as the carbon cycle to explain the chemical fate of carbon in the environment. The analogous *rock cycle* explains how rocks (far from sedentary objects) are moved about by internal and external processes on and in our planet. In this cycle, igneous, metamorphic, and sedimentary rocks are transformed and transported. We are most concerned with sedimentary rocks for they are the source of oil and gas. Produced from sediments deposited either on the land or at the bottom of a body of water, sedimentary rocks represent only about 5% of Earth's crust (Skinner and Porter 2000). Sediment—and hence, sedimentary rocks—comes from everywhere and everything, including rock fragments, chemical precipitates (e.g., calcium carbonate from bones and shells), long-dead marine organisms, and detritus in general. In the sedimentation process, various kinds of materials are stratified such that sandy sediments (sandstone) overlay clay sediments (shales), with calcium carbonate-containing sediments (limestones) at the bottom of the strata. Sediment is buried and compacted to become sedimentary rock, typically at or near the site at which it was originally deposited. As a result, oil (which ultimately comes from sedimentary rock, *vide infra*) is not widely distributed on the planet, with obvious geopolitical consequences.

Given that sediment comes from everywhere and everything, the fact that a small portion of sedimentary rocks consists of organic matter is self-evident. Roughly 18% of the total amount of carbon in Earth's crust is organic carbon in the form of oil, gas, and coal (Schidlowski et al. 1974). Living things die and the carbon, hydrogen, nitrogen, and oxygen that were once a part of living biota become part of the land-scape. In an aerobic environment, this organic matter is oxidized, but in an oxygen-free environment (as when trapped in a layer of sediment), oxidation is negligible. Numerous types of organic compounds, then, are trapped in sedimentary rocks and converted into crude petroleum and other carbon-rich materials under conditions of heat and pressure over a geologic timescale.

Fossil fuels, then, are mixtures of hydrocarbons in a low oxidation state, formed over a period of millions of years and trapped in sediment that makes them fairly inac-cessible. What is the chemical composition of these mixtures and what is their ori-gin? Given that ancient crude oil primarily consists of even-numbered carbon chains and that the material is both optically active and levorotatory, it is widely accepted that the vast majority of oil is biological in origin (Selley 1998). Decomposition of this once-living matter into fossil fuels is a process whereby the highly oxygenated organic matter is reduced to hydrocarbon species.

Lipids and *lignins* from the original plant or animal material are the primary organic materials that decay to hydrocarbon fossil fuels. (Ultimately, of course, these carbon atoms came from carbon dioxide in the process of photosynthesis.) *Lignins*, found in all woody plant materials as a very complex polymeric mixture of phenols and glycerol, are the source of solid fossil fuels such as peat and coal (Figure 2.1).

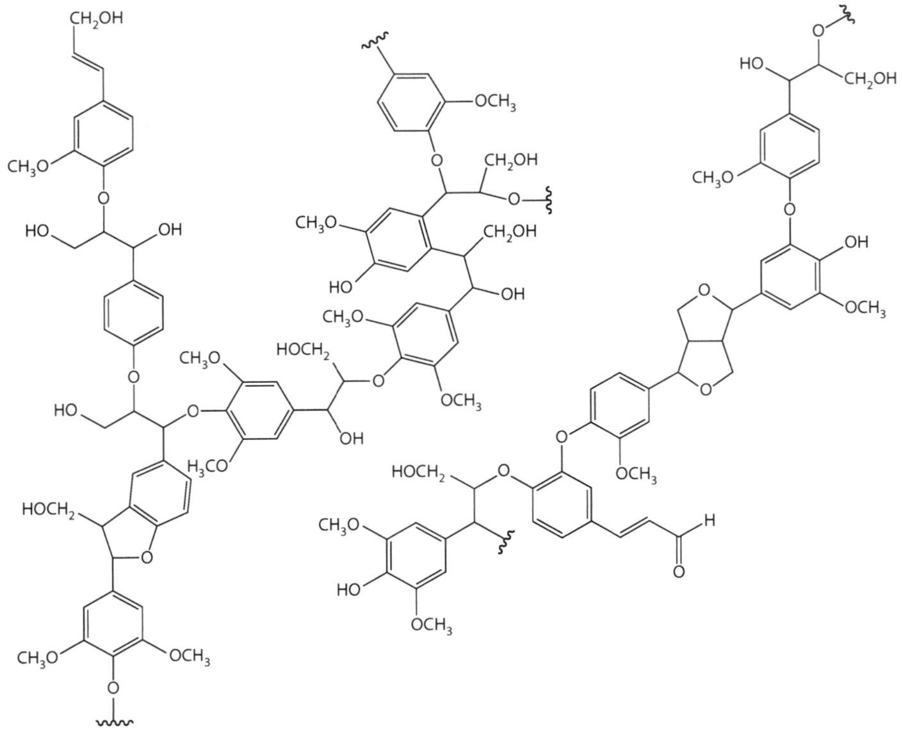

FIGURE 2.1 Lignin structure.

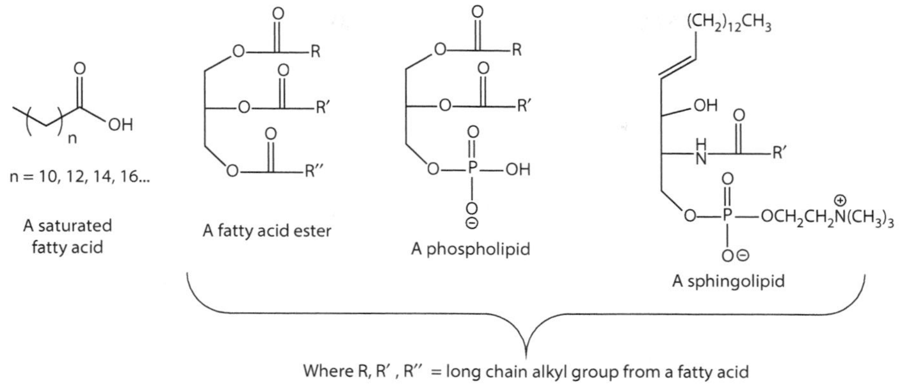

FIGURE 2.2   Representative lipid families.

Lipids are organic compounds characterized by their solubility behavior: they are of low polarity and are thus highly hydrophobic. Terpenes, fatty acids, fatty acid esters, and phospho- and sphingolipids (Figure 2.2) containing long (even-numbered) hydrocarbon chains are all considered lipids.

Lipids are the primary source of crude oil. Several million years ago, a multitude of marine phytoplankton and bacteria lived, died, and were deposited in sediment. Through reduction by anaerobic bacteria, burial, and compression in the rock cycle, the material decomposed with the loss of small molecules such as water, methane, and carbon dioxide. In this process known as *diagenesis*, an enriched organic material is formed. Dark layers of this material are found primarily in oil shale and are called *kerogen*, a very complex solid mixture containing primarily carbon and hydrogen with lesser percentages of nitrogen, oxygen, and sulfur.

There are three basic types of kerogen, distinguished by both their origin and their general structures. Kerogen that is algal in origin is known as "Type I" kerogen and is heavily aliphatic (see Figure 2.3), being formed from lipids. This type of kerogen ultimately becomes oil and gas. Types II and III kerogens are considerably more aromatic in nature and derive from marine microorganisms (Type II) and woody plants (Type III). Kerogen from the Green River Formation oil shale deposit in the western United States, for example, is rich in algal kerogen and has an approximate composition of $C_{215}H_{330}O_{12}N_5S$ (Cane 1976). It is the thermal decomposition of kerogen under heat and compression at depths well below 1000 m—a process known as *maturation*—that leads to oil, gas, and coal. If liquid, the petroleum is expelled and migrates to nearby deposits.

While lipid-based kerogens lead primarily to petroleum, lignin-based kerogen (from woody plants) leads to solid fossil fuels such as peat and coal. Coal, by definition, is a biogenic sedimentary rock composed of at least 50% decomposed plant matter. Millions of years ago, tropical flora flourished in the swamps, bogs, and forests of the Earth. Layers of dead plant matter became incorporated into parts of Earth's crust. Peat, the precursor to coal, is one example. Having a high moisture content, as the peat is compressed, small molecules (e.g., water and methane) are

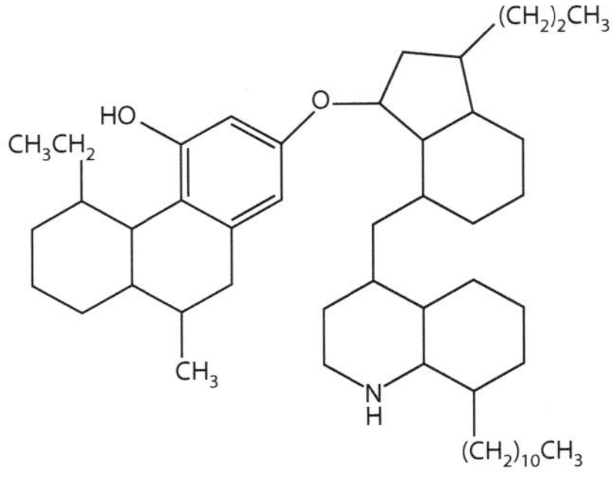

FIGURE 2.3   Type I kerogen.

expressed out, increasing the proportion of carbon and leading to varying grades of coal. The overall approximate chemical composition of coal is $CH_{0.8}S_xN_yO_z$ (where x, y, and z are each <0.1). Coal is classified into various grades (Tester et al. 2005):

Anthracite is a hard, lustrous coal containing a high percentage (86–97%) of carbon and few volatile components. It typically has the highest heating value of coal types (see Table 2.1) and is largely used for residential and commercial space heating because of its limited availability. No new anthracite is being mined in the United States.

Bituminous coal is 45–86% carbon and accounts for almost one-half the coal produced in the United States. Bituminous and subbituminous coals (see below) are used primarily for steam-electric power generation.

Subbituminous coal consists of 35–45% carbon and is midrange between bituminous coal and lignite in terms of properties. Similarly, its appearance ranges from the lustrous, hard appearance of the higher grades of coal to a soft brown-to-black form.

Lignite contains 25–35% carbon and has a high moisture content. Lignite is also known as "brown coal." One step removed from peat, lignite (unlike peat) is free

TABLE 2.1
Types and Properties of Coal

Coal Type	Typical Heat Content Range (mmBtu/ton)	Typical Sulfur Content (%)	Average Ash Content (%)
Bituminous	22–26	≈2	11
Subbituminous	16–18	≈0.3	13.8
Lignite	12–13	≈0.9	5.3
Anthracite	22–28		

Source: Data from U.S. Department of Energy Information Administration, EIA-923 Monthly Time Series File, Fuel Receipts and Cost, Schedules 2 (March 2012).

of cellulose. A 50-m layer of peat will be compressed over time to a 10-m layer of lignite (Skinner and Porter 2000).

While the carbon and hydrogen in coal generate heat by combustion, the other trace elements in coal make it a significant source of airborne pollution. Particulate matter, SO_2, NO_x, as well as mercury (both in its elemental and oxidized form) are released upon combustion, making coal-fired plants the largest single source of anthropogenic mercury pollution (United Nations Environment Programme 2012). Several approaches have been developed for mercury reduction, from injection of activated carbon to trapping particulates by a fabric filter or electrostatic precipitation. *Gasification* of coal (Chapter 5) is the basis for "clean coal technology," wherein coal is converted into hydrogen gas with cogeneration of CO_2 (which can be sequestered) in what is known as "IGCC" (integrated gasification combined cycle) technology (Armaroli and Balzani 2011). By integrating coal gasification with heat-driven turbines, waste heat that would otherwise be lost in the exhaust stream is captured, leading to higher efficiencies. More details on CO_2 sequestration are provided in Section 2.5.

The hydrocarbon fuels upon which most of our way of life is based are aptly labeled as fossil fuels: it has taken unimaginably staggering volumes of marine microorganisms with fantastic pressures and temperatures over millions of years to yield sedimentary rock sources from which we can (with some difficulty) extract oil. Conversion of decaying plants through the same processes forms seams of coal that must be mined at significant financial, human, and environmental cost. "Hubbert's peak" describes the point at which the demand for these fossil fuels outpaces the global production capacity—a peak that may have already been reached for oil, and is predicted for the latter part of the twenty-first century for natural gas (Deffeyes 2005). These are in no sense sustainable fuels.

2.2 EXTRACTION OF FOSSIL FUELS

Fossil fuels, then, are finite forms of fuel that humankind has taken advantage of over the past century to support our way of life. It is problematic enough that the formation of these fuels took eons, but getting these sources of energy out of the ground presents additional challenges and takes a severe toll. The more scarce the fuel, the more drastic our methods for extraction become as new *nonconventional* methods for extracting petroleum resources illustrate (*vide infra*).

2.2.1 CONVENTIONAL PETROLEUM

In the early years of oil exploration and discovery, petroleum geologists were able to tap into reservoirs of light, free-flowing oil trapped underground. As can be seen in Figure 2.4, a *structural oil trap* stratifies water, oil, and gas on the basis of different densities. The oil reserves in the Middle East are a good example of oil-rich structural oil traps. Relatively straightforward technology is used to vertically drill through the impervious rock trapping the oil, releasing and then capturing the oil and gas. The petroleum industry has been producing oil and gas in this manner— the "conventional" manner—since the late nineteenth century. As readily accessible

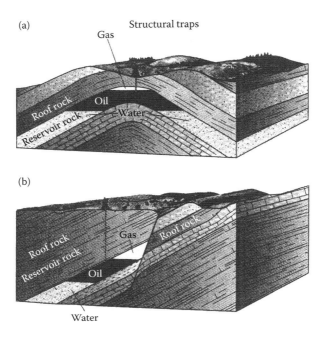

FIGURE 2.4 Types of structural oil traps: (a) anticline and (b) fault. (Skinner, B.J. and S.C. Porter: *The Dynamic Earth. An Introduction to Physical Geology.* 4th ed. 2000. Copyright Wiley-VCH Verlag GmbH & Co. KGaA. Reprinted with permission.)

onshore supplies have dwindled, exploration and recovery have moved offshore, with concomitant increase in risk and cost. The BP *Deepwater Horizon* explosion and oil spill of April 2010 has the distinction of being the largest unintentional oil spill in our planet's history. Having killed 11 oil rig workers and spewed an estimated 4.9 million barrels of crude oil into the Gulf of Mexico, the costs of this event are incalculable.

2.2.2 NONCONVENTIONAL SOURCES

As supplies of conventional oil and gas have declined, new methods for accessing the more elusive hydrocarbons trapped onshore have evolved. Once deemed nonproducible, these nonconventional sources of oil and gas (also called *tight* oil or gas) include shale gas, oil shale, tar sands, methane coalbeds, and methane clathrates. In other words, as the relatively accessible supplies dry up, the petroleum industry must go to further and further lengths to find and extract petroleum from the planet, using breakthrough technology in physics, chemistry, and engineering to meet our demands.

2.2.2.1 Shale Oil and Gas

Shale is a fine-grained sedimentary rock that can trap natural gas in natural fractures and pores, or adsorb it onto organic matter and minerals in the shale (Gregory et al. 2011). The relatively recent development of horizontal drilling and hydraulic fracturing ("fracking") techniques have grown exponentially in the extraction of tight fossil fuels, largely supplanting vertical well drilling (Figure 2.5). The Barnett Shale in Texas, the Marcellus formation in the northeastern United States, the Bakken formation in western North Dakota and Eastern Wyoming, the natural gas and "light tight crude" locked in shale are now recoverable in what were once considered poor risks for petroleum and natural gas recovery. In fact, shale gas is the fastest-growing source of natural gas (U.S. Energy Information Administration 2011).

Horizontal drilling is self-explanatory: the vertical drill is sunk to the point at which a layer of gas-infused shale is reached. The drill is then curved to enter the shale stratum horizontally and boring is continued. After some length of horizontal bore has been reached, hydraulic fracturing of the shale is used to liberate the trapped hydrocarbons (Figure 2.6). *Slick water*—a fracturing fluid made up of sand (the *proppant*), water, and a small amount of numerous other chemicals (Table 2.2)— is pumped at high pressure into the horizontal wellbore to fracture the shale. Natural gas and light crude trapped in the shale is released and percolates through the sand back to the surface, where it is captured at the wellhead.

While fracking has proven to be successful at recovering what was once considered to be inaccessible petroleum reserves, controversy abounds. Water use and water quality are huge issues—millions of gallons of fracking fluid are required for each well (Manning 2013). Although these fracking fluids are greater than 98% water, the friction reducers, biocides, and corrosion inhibitors are many (see Table 2.2 with the caveat that not all of these chemicals are necessarily used simultaneously; the actual mixture depends on the site). The water that percolates back to the surface (*flowback*) has a very high total dissolved solids concentration and consists of salts, sediment, bacteria, and some heavy and light hydrocarbons. Extensive treatment of the recovered water is required for reuse or recycling, but only about 35% of the fracking fluid returns to the surface—the remainder remains

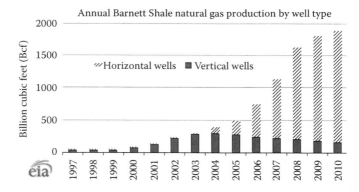

FIGURE 2.5 Growth in horizontal well drilling in the Barnett Shale region. (Adapted from U.S. Energy Information Administration. 2011. Technology drives natural gas production growth from shale gas formations. *Today in Energy*, July 12, 2011. http://www.eia. gov/todayinenergy/images/2011.07.12/barnettbarpII.png. Accessed 11 June 2013.)

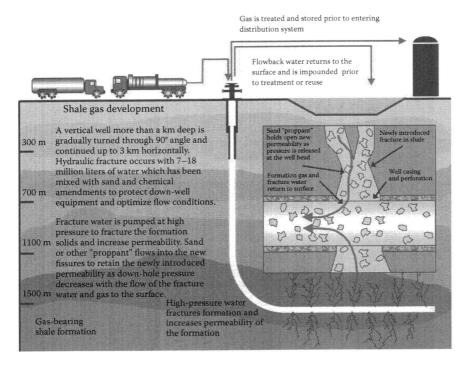

Gas is treated and stored prior to entering distribution system

Flowback water returns to the surface and is impounded prior to treatment or reuse

Shale gas development

A vertical well more than a km deep is gradually turned through 90° angle and continued up to 3 km horizontally. Hydraulic fracture occurs with 7–18 million liters of water which has been mixed with sand and chemical amendments to protect down-well equipment and optimize flow conditions.

Fracture water is pumped at high pressure to fracture the formation solids and increase permeability. Sand or other "proppant" flows into the new fissures to retain the newly introduced permeability as down-hole pressure decreases with the flow of the fracture water and gas to the surface.

300 m
700 m
1100 m
1500 m

Gas-bearing shale formation

High-pressure water fractures formation and increases permeability of the formation

Sand "proppant" holds open new permeability as pressure is released at the well head

Formation gas and fracture water return to surface

Newly introduced fracture in shale

Well casing and perforation

FIGURE 2.6 (See color insert.) Hydraulic fracturing in the extraction of shale gas. (Reprinted with permission from Gregory, K.B., R.D. Vidic, and D.A. Dzombak. 2011. Water management challenges associated with the production of shale gas by hydraulic fracturing. *Elements* 7 (3):181–186. Copyright 2011, Mineralogical Society of America.)

a potential threat to groundwater contamination and there have been reports of water supplies being literally flammable (Bomgardner 2012). Fracturing shale has led to increased seismic activity; discharge of nitrogen oxides and volatile organic compounds at the sites results in decreased air quality (Schmidt 2011), and studies have shown that leakage of methane from fracking sites has led to a greater carbon footprint than the extraction of coal or conventional natural gas wells (Howarth et al. 2011; Weber and Clavin 2012). Furthermore, the mining of enormous amounts of fracking sand has led to considerable community controversy across the United States. Overall, the ecological and health effects of fracturing warrant serious concern.

2.2.2.2 Heavy Oil

Heavy oil is a complex hydrocarbon mixture that is too viscous to flow at ambient temperatures due to loss of the light hydrocarbon fraction during migration or microbial decay. *Asphaltenes* (Figure 2.7) exemplify the type of compounds found in heavy oil. Given their high molecular weight and ample opportunity for intramolecular attractions and conformational entanglement, it is clear why these are highly viscous materials. As a result, either heating or dilution with a solvent is required

TABLE 2.2
Example Composition of Fracking Fluid

Ingredient Function	Chemical	Maximum Ingredient Concentration, % by Mass
Carrier/base fluid	Water	85.48
Proppant	Crystalline silica (fracking sand)	12.66
Acid	Hydrochloric acid	1.30
Gelling agent	Petroleum distillate blend	0.14
	Polysaccharide blend	0.14
Cross-linker	Methanol	0.05
	Boric acid	0.01
Breaker	Sodium chloride	0.04
Friction reducer	Petroleum distillate, light	0.01
pH-adjusting agent	Potassium hydroxide	0.01
Scale inhibitor	Ethylene glycol	0.005
	Diethylene glycol	0.001
Iron control agent	Citric acid	0.004
Antibacterial agent	Glutaraldehyde	0.002
	Dimethyl benzyl ammonium chloride	0.001
Corrosion inhibitor	Methanol	0.001
	Propargyl alcohol	<0.001

Source: FracFocus Chemical Disclosure Registry. http://fracfocus.org/chemical-use/what-chemicals-are-used; accessed 8 July 2013.

to get this material to flow, with most *enhanced oil recovery* (EOR) methods using steam injection to force the material to the surface. (In fact, one use of captured CO_2 is for enhanced oil recovery.) In the United States, the largest reservoirs of heavy oil are located in Alaska and California.

Oil shale contains trapped organic matter that has not "cooked" enough at the needed temperature and pressure. It is generally found all over the world at depths of <900 m and has a high *bitumen* (essentially tar) content. In oil shale, the bitumen–kerogen mixture trapped in the shale is, like heavy oil, immobile. This material is typically mined and crushed and the pulverized material then heated to 450–500°C in the absence of air (a process known as retorting) to obtain crude oil. Other methods for extracting oil from oil shale (including *in situ* retorting while the material is still underground) are under development, but the environmental impact and costs remain very high with commercial-scale viability still uncertain.

2.2.2.3 Oil Sands

Heavy oil is, in essence, tar—more accurately referred to as bitumen or asphalt. Tar pits, tar pools, and oil sands (often less accurately referred to as "tar sands") are scattered about the globe at shallow depths; in some areas, the pitch has seeped

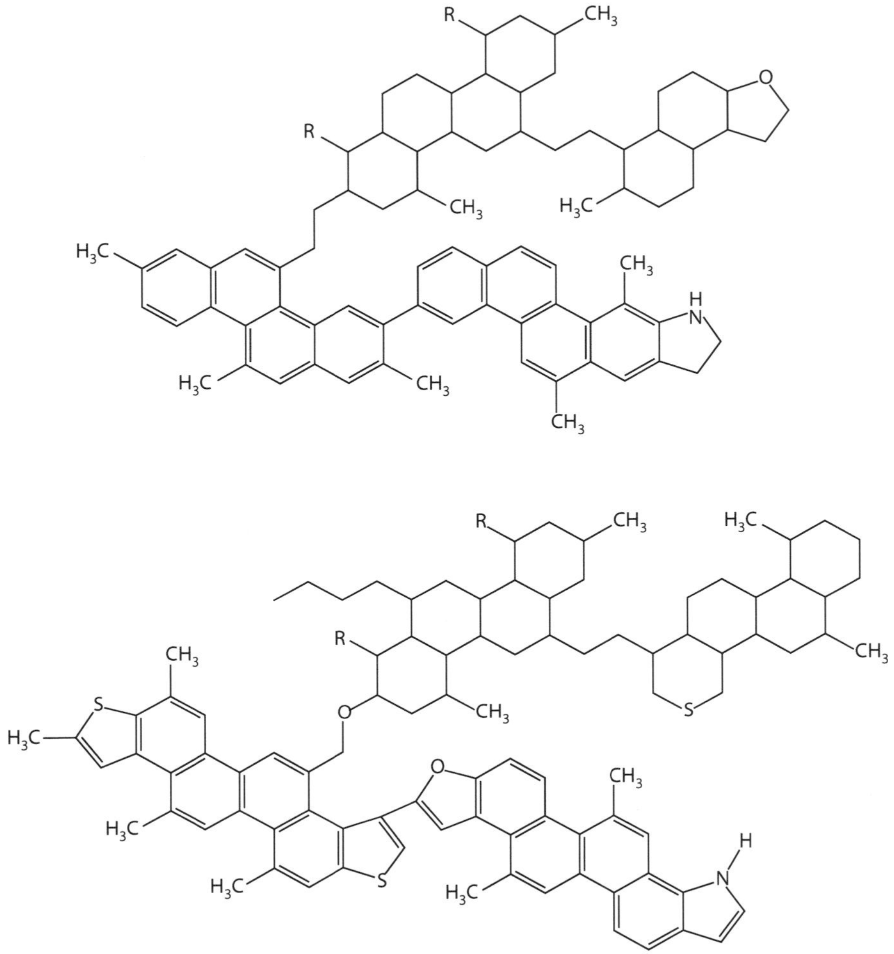

FIGURE 2.7 Some representative asphaltenes.

to the surface. Evaporation of the more volatile components and oxidation of the residue yields the tarry, carbon-rich fuel source. The Athabasca Valley in Alberta, Canada, is one of the largest oil sands reserves in the world, with an estimated 170 billion barrels of oil locked in these bitumen-soaked sands (Figure 2.8; Energy Resources Conservation Board 2011). But bitumen is notoriously difficult to handle and correspondingly hard to extract. The methods whereby the oil sands are converted into a useable fuel—open pit mining or forcing superheated steam into a well to force the bitumen to flow to the surface—again have serious environmental consequences. Recent findings show that levels of toxic polycyclic aromatic hydrocarbons (thiophenes, anthracenes, pyrenes, etc.) are increasing in the environment of northern Alberta, in parallel with the increase in oil sands production. Levels at one site are now as much as 23 times higher than the 1960 levels (Kurek et al. 2013).

FIGURE 2.8 (**See color insert.**) An oil sands development in northern Alberta, Canada. (Shutterstock Image id 48011344.)

Furthermore, the transport of the crude product from Alberta to refineries in the U.S. Gulf of Mexico has been proposed via the "Keystone XL" pipeline, an extremely controversial project due to environmental and safety concerns. As is evident, non-conventional fossil fuel sources are laden with difficult choices.

2.2.2.4 Coal Bed Methane and Methane Hydrates

2.2.2.4.1 Coal Bed Methane

It is only too well known that underground coal mines contain potentially hazardous amounts of methane gas, occasionally exploding with tragic results. Extensive safety precautions are taken—including ventilation—in order to prevent such tragedies. But as fossil fuel supplies decrease, capturing this *firedamp* becomes an attractive alternative as a source of natural gas. Because of its highly porous nature, coal is an ideal substrate for adsorption of large volumes of methane. In order to obtain methane from coal beds (primarily bituminous), the water that permeates the coal beds must first be pumped out. The coal bed methane is then freed from the coal bed by desorption and the gas flows to the wellhead. (Note that a new approach to coal bed methane production is under investigation in which pressurized CO_2, ideally captured from combustion of fossil fuel, is used to displace the methane gas as a carbon capture and sequestration strategy; see Section 2.4.2.1.) The U.S. production of coal bed methane in 2010 was about 1886 billion cubic feet (roughly 7.5% of all U.S. natural gas marketed in 2010; U.S. Energy Information Administration 2012).

2.2.2.4.2 Methane Hydrates

All it takes is water and the microbial decay of organic matter, plus low temperature (4°C) and high pressure (60 bar), to trap methane in an ice cage (the methane

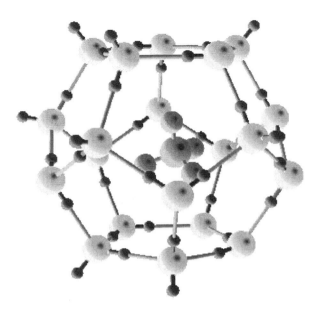

FIGURE 2.9   Methane hydrate. (Adapted from United States Geological Survey. The U.S. Geological Survey Gas Hydrates Project Gas Hydrates Primer. Retrieved 14 May 2013, from http://woodshole.er.usgs.gov/project-pages/hydrates/primer.html.)

hydrate; see Figure 2.9). The concept of harvesting significant quantities of methane from methane hydrates (also known as methane clathrates) is relatively recent. The permafrost and continental margins (e.g., the Gulf of Mexico in the United States) are locations where large quantities of methane hydrates are found. Methane hydrate recovery is a very active area of research.

Methane hydrates lead a schizophrenic life: they are hazards when it comes to conventional natural gas exploration and recovery in that accidental release of methane from the hydrate (degassing) can result in explosion or environmental catastrophe. At the same time, methane hydrates clearly pose great potential as a production target. A 2010 report of the National Research Council suggested that methane hydrate remains a feasible source for commercial production of methane (Committee on Assessment of the Department of Energy's Methane Hydrate Research and Development Program: Evaluating Methane Hydrate as a Future Energy Resource 2010). As with every other nonconventional (and conventional!) source of fossil fuel, the trick is figuring out a way to extract the fuel from the planetary resource safely and with minimal environmental impact.

2.3   REFINING

2.3.1   Crude Petroleum

Regardless of the source or type of material extracted from the earth, crude petroleum is a mixture of many hundreds of compounds that must be upgraded in order to obtain marketable products. The plethora of molecules that make up crude oil can be put into four basic categories (N.B. Bear in mind that these relative amounts vary depending upon the source of the oil.):

- *Saturated hydrocarbons.* Approximately 30% (based on molecular weight) of oil is made up of acyclic aliphatics (alkanes), also known as *paraffins.*
- *Cycloalkanes.* Roughly 49% consists of cyclopentanes, cyclohexanes, and cycloheptanes, many with short-chain substituents. These are also known as the *napthenes.*
- *Aromatics.* Benzene, toluene, and xylenes ("BTX"), among other aromatics, make up ca. 15%.
- *Asphaltics.* Residual material with more than 38–40 carbon atoms constitutes about 6% of crude petroleum by molecular weight.

Nitrogen heterocycles, sulfur (as thiols and thiophenes), and oxygen contaminants (from water to phenols) are also present in small (trace up to approximately 4%) amounts. Alkenes are present in only very limited amounts due to their natural reactivity.

It is futile to attempt to identify and purify every molecule making up crude oil; what is important is separating the mess into cleaner fractions based on similar physical properties and (as a result) similar end uses—this is the aim of petroleum refining. Petroleum refining is quite different from isolating a single, identifiable compound, but the separation techniques for refining petroleum are, for the most part, identical to those used in any undergraduate organic chemistry laboratory: distillation, extraction, chromatography—all techniques that separate the crude material into fractions (or *cuts*) based on similar physical properties. Further conversion of the separated materials by chemical means takes place to convert these fractions into more useful products. The pathway from crude oil to finished product via the petroleum refinery is very complex (and very big business indeed), as can be seen in Figure 2.10. Our focus will be on four main processes: distillation, extraction, cracking, and reforming.

2.3.1.1 Distillation

Distillation of crude petroleum is the first step in the process of refining the material. Given the complexity of the mixture, simple distillation is insufficient to separate the fractions. Instead, fractional distillation in a distillation tower (Figure 2.11) is required. As in simple distillation, fractional distillation separates on the basis of volatility (vapor pressure/boiling point) but the efficiency of separation is greatly improved by allowing repeated vaporization–condensation cycles on the tower's internal surfaces as the crude distillate flows upward through the *multiplate* fractionating column. Multiplate refers to the fact that the increased surface area—a result of packing material inside the tower and represented by the squiggly lines in Figure 2.11—increases the number of theoretical plates upon which the vaporization–condensation equilibrium can be reestablished many times over. The fractions are split off at varying heights in the tower as the distillation continues, separating the

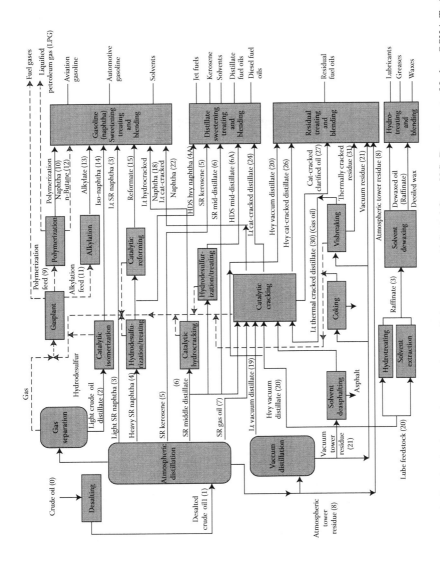

FIGURE 2.10 The complexity that is the modern petroleum refining process. (Adapted from U.S. Department of Labor OSHA Technical Manual, Section IV: Chapter 2. Petroleum Refining Processes. http://www.osha.gov/dts/osta/otm/otm_iv/otm_iv_2.html#3. Accessed 29 August 2012.)

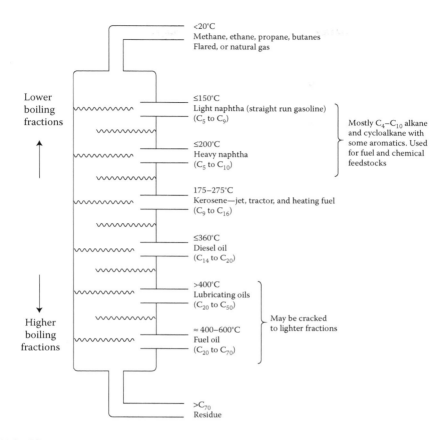

FIGURE 2.11   Schematic of a distillation tower.

crude oil into several cuts, from gases (butane and lower-molecular-weight alkanes) to residual heavy oils (see Figure 2.11). It is important to note that the fractions thus obtained are not discrete products with a sharp boiling point. Instead, each fraction consists of an array of compounds exhibiting a band of boiling point ranges. Ultimately, the higher-boiling fractions are transferred to a new tower where a vacuum distillation is carried out to separate even more material from less-volatile asphaltic residues.

2.3.1.2   Extraction

The asphaltic residues can also be further fractionated by solvent extraction with liquified propane or butane, solubilizing the nonpolar solutes. Even more separation can take place by adsorbing the solubilized material onto some adsorbent (e.g., clay, alumina, or silica) and extracting different fractions based on the partitioning ability of an added solvent. Thus, dichloromethane will extract a different type of material than methyl ethyl ketone, for example. The extracted materials can be used for heavy lubricating oils, greases, and so on once the solvent is removed (Speight 2007).

2.3.1.3 Cracking

Once the crude petroleum has been separated into its various fractions (so-called *straight-run* fractions), the next step consists of chemical conversion into materials that exhibit more desirable properties: primarily lower boiling range fractions that make for good transportation fuels. This conversion is known as *cracking*, of which there are three main types: (a) *thermal cracking*, (b) *catalytic cracking*, and (c) *hydrocracking*. Cracking is intense chemistry: while the principles of basic organic chemistry apply, cracking begins with mixtures, treats them to harsh conditions, and ends with different mixtures. A complete understanding of these processes at the molecular level can only be based on simpler models.

2.3.1.3.1 Thermal Cracking

Thermal cracking is precisely what it sounds like: using high temperature (450–450°C) and pressure (690–6900 kPa) to crack molecules into smaller fragments. The high temperatures and pressures under neutral conditions used in thermal cracking lead to fragmentation and product reforming by homolytic (radical) mechanisms. Thus, for example, hexanes can be converted into ethene and the *n*-butyl radical under these extreme conditions (see Figure 2.12). Control of the cracking products is an obvious issue. As the temperature is increased, more and more low-molecular-weight alkenes are formed. The pressure at which the cracking takes place as well as the length of time the feedstock is heated can also influence the product distribution. Thermal cracking was the historical method for petroleum conversion but has now been largely replaced by catalytic cracking.

2.3.1.3.2 Catalytic Cracking

Catalytic cracking (also known as *fluidized catalytic cracking*, or FCC) is milder and more selective than thermal cracking. FCC takes higher-boiling fractions and converts them into higher-value saturated compounds (branched paraffins and naphthenes) and aromatics. Most catalysts used for FCC are based on natural or synthetic cage-like aluminosilicates known as *zeolites*, the "crown jewels of catalysis" (Bartholomew, 2006). Zeolites have the general formula $M_y(AlO_2)_x(SiO_2)_y \cdot zH_2O$, with the aluminum and silicon oxide species sharing oxygen atoms in tetrahedral AlO_4 and SiO_4 building blocks for the zeolite unit cell and the metal cation residing in an exchangeable cationic site in the center of the cage (see Figure 2.13) (Bartholomew and Farrauto 2006). Activation of the zeolite catalyst into an acidic form facilitates the conversion of higher-molecular-weight compounds into lower ones via carbocation intermediates. Replacement of the aluminum with a rare earth element (e.g., La, Ce, Pr, Nd, Sm, Eu, or Gd) imparts greater stability to the catalyst,

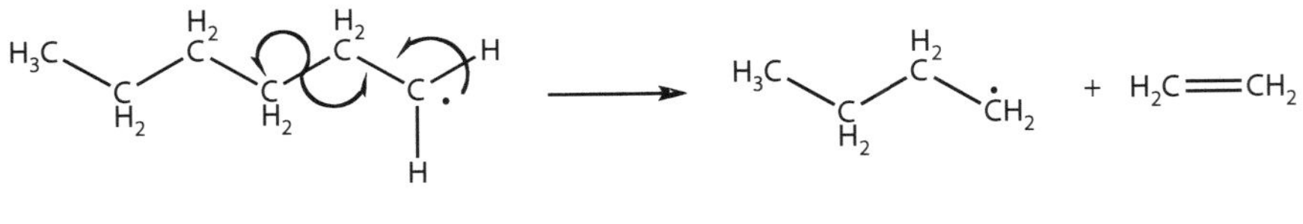

FIGURE 2.12 Hypothetical homolytic thermal cracking mechanism.

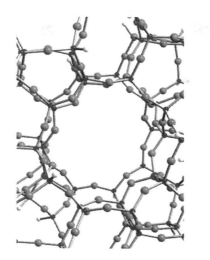

FIGURE 2.13   A portion of zeolite ZSM-5. (Adapted from Yumura, T. et al. 2011. *Inorganic Chemistry* 50(14): 6533–6542. With permission.)

but the use of these rare-earth elements makes the petroleum industry vulnerable to resource availability issues (Bauer et al. 2011).

While zeolites are naturally occurring, they can be synthetically prepared as well, such that catalysts can be designed with a variety of pore sizes and geometries. By controlling the pore size of the catalyst and varying the reaction conditions (temperature and pressure), the product distribution can be controlled (Rahimi and Karimzadeh 2011). The process is typically run at slight pressure (70–140 kPa) and at 450–550°C. Under these conditions and in the presence of the solid acid catalyst, protonation or abstraction of a hydride from a saturated hydrocarbon yields a cation intermediate that can undergo rearrangement, beta-elimination, fragmentation, or hydrogen transfer to give the gasoline-range products (Figure 2.14).

2.3.1.3.3   Hydrocracking
Hydrocracking is catalytic cracking in the presence of hydrogen. Under these conditions, no alkenes or alkynes are present in the product mixture and heteroatom

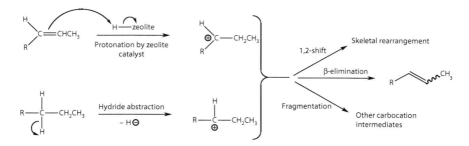

FIGURE 2.14   Cationic fragmentation mechanisms.

(N, O, S) impurities are removed as ammonia, hydrogen sulfide, and water. Hydrocracking is carried out at a lower temperature than conventional catalytic cracking (ca. 250–350°C) and requires a different catalyst, typically a metallic hydrogenation–dehydrogenation couple. Palladium or platinum on an alumina support, or a cobalt–molybdenum catalyst, can be used in the presence of hydrogen gas at 3500–6900 kPa to convert heavier oils into lower-molecular-weight hydrocarbon distillates.

2.3.1.4 Reforming

A subsequent stage for further refining the cracked petroleum fractions is known as reforming. The difference between reforming and cracking is simply the feedstock and what you do with it. In cracking, the objective is to crack the higher-boiling components into smaller molecules with a lower boiling point range. Reforming takes that lower-boiling cut and rearranges the molecular skeleton to give more branched compounds and thus a higher octane rating. Thus, for example, the gasoline cut can be refined into a higher-octane product by a reforming process that converts the low-octane straight-chain alkanes into higher-octane branched alkanes. As with cracking, the process of reforming can be carried out thermally or catalytically, with the thermal methods being replaced by catalytic processes due to their greater efficiency and economy.

Catalytic reforming takes advantage of organometallic chemistry to isomerize the straight-chain alkanes found in distilled gasoline into structural isomers that give a higher-quality product. Typical reforming catalysts are based on molybdenum, platinum, or chromium on an inert inorganic support such as alumina or silica. If straight-chain paraffins such as *n*-butane, *n*-pentane, or *n*-hexane are part of the feedstock, catalytic reforming isomerizes these normal alkanes to their iso counterparts (e.g., isobutane and isopentane). If run in the presence of hydrogen, any alkenes formed are hydrogenated to give only saturated products.

2.3.2 NATURAL GAS

The conventional source of natural gas is a well that produces a relatively pure gas that is mostly methane—typically well over 85%—that contains small amounts of N, O, and S contaminants. Purification of natural gas removes moisture, hydrogen sulfide, carbon dioxide, and other impurities. Passage of impure natural gas through an aqueous solution of an amine (usually mono- or diethanolamine) removes carbon dioxide by carbamate formation (Figure 2.15), acidic impurities by protonation, and H_2S by absorption into the alkaline solution in a process known as *sweetening* (Kohl and Nielsen 1997). Compared to petroleum refining, the purification of natural gas is relatively simple.

Because of the surprising increase in the availability of natural gas due to hydraulic fracturing, it is rapidly being viewed as the fossil fuel of choice for many applications. It burns more cleanly than coal and possesses fewer contaminants than petroleum, but it is still a carbon-based fuel. As a result, its combustion, of course, generates the greenhouse gas CO_2. *Natural gas combined cycle* (NGCC) power plants utilize natural gas as a fuel source and combine both a gas turbine and a steam turbine to generate electricity with higher thermal efficiency: instead of wasting the

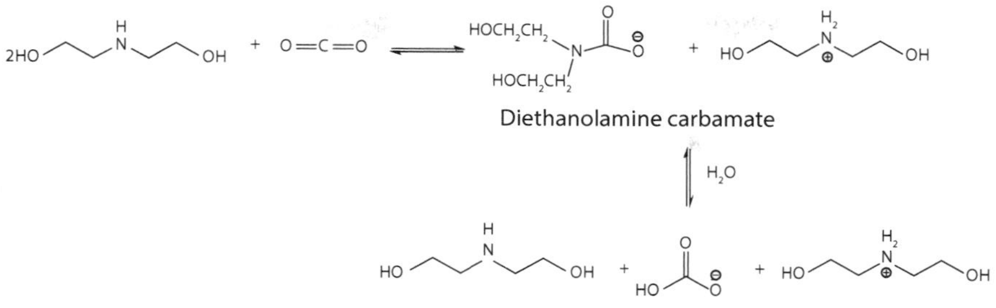

FIGURE 2.15 CO_2 scrubbing by carbamate formation.

heat from the gas turbine it is used to generate the steam, which in turn drives the electricity generation. While NGCC plants may be more efficient, the production of CO_2 from the combustion of any fossil fuel cannot be ignored; what to do with this carbon waste is the topic of the next section.

2.4   CARBON CAPTURE AND STORAGE

An alarming milestone was reached on May 14, 2013: the average daily CO_2 measurement as determined at the U.S. NOAA Mauna Loa Observatory reached 400.03 ppm. While the level of atmospheric CO_2 has been increasing dramatically since the time of the Industrial Revolution, this value is symbolic in its significance—this level of atmospheric CO_2 has not been known for at least 800,000 years based on ice core data (and more likely 2–4 million years) (Montaigne 2013). Gigatons of CO_2 are emitted into the atmosphere every year and there is currently in excess of one teraton in Earth's atmosphere (Mikkelsen et al. 2010). As a potent greenhouse gas, its reduction is of crucial importance, but humanity has yet to show the needed resolve to make significant progress on this front. In fact, a record high of CO_2 emissions was reached in 2012; 31.6 billion tons of CO_2 were spewed into the atmosphere, up 1.4% from the previous year (Chestney 2013). Since, thus far, we have failed to significantly curtail the main sources of CO_2 emissions, and natural processes in the carbon cycle can recycle or sequester only about one-half of the total amount of global annual emissions of CO_2, pressure exists to lower the amount of postproduction CO_2 (Olah et al. 2011).

In this section, an overview of the vast chemistry of separation and capture of CO_2 from various gas streams is presented. There are three major separation problems associated with CO_2:

- *Separation of the CO_2 impurity from natural gas (CH_4)*. This encompasses the well-developed amine-treatment methods already discussed in Section 2.3.2.
- *Separation of CO_2 from nitrogen*. The combustion of a carbon-containing fuel in air results in a dilute CO_2 waste stream in nitrogen gas. For example, a coal-fired power plant is a large point source of CO_2 emissions.

- *Separation of CO_2 from hydrogen.* We will see in Chapter 5 that *syngas*, a mixture of CO and hydrogen, can be produced as a useful fuel. Some CO_2 invariably contaminates this gas stream.

Separation of CO_2 from these other gases is a huge challenge that will require the development of a method that somehow exhibits excellent selectivity for one gas over the other. Perhaps the more important question is "what do we do with the separated CO_2?" The two main approaches for dealing with CO_2 postcapture are *sequestration/storage* (CCS)—hiding away the captured CO_2 where it can do no harm, or *utilization* (CCU)—using the CO_2 as a carbon source for value-added products or as an aid in the extraction of yet more fossil fuels from the Earth. Chemistry plays a vital role in both separation and utilization technologies.

2.4.1 CAPTURE AND SEPARATION

The idea behind sustainable energy sources is to use carbon-neutral or carbon-free alternatives to fossil fuels that do not generate CO_2 in the first place, but complete cessation of CO_2 emissions is impossible—not to mention that there is already far too much of it in our atmosphere. While widely dispersed sources like automobiles, airplanes, and even individual organisms contribute to the problem, most CO_2 is produced from large-scale point sources such as

- Power plants—from the combustion of fossil fuels.
- Cement manufacture—from the calcining of limestone ($CaCO_3$) to give lime (CaO) and CO_2.
- Aluminum manufacture—both in the huge amount of electricity required (thus generating more CO_2 from the combustion of fossil fuels in a power plant) and from the oxidation of carbon at the anode in the electrochemical process whereby Al_2O_3 is reduced to aluminum metal.

Certainly, plant matter captures CO_2 by photosynthesis and, in fact, algae-based biodiesel (Section 8.6.3.2) presents an attractive "closed-loop" method of energy production with built-in carbon capture, as illustrated in Figure 2.16: combusting the biodiesel that was made from algae that grew on the CO_2 generated from the combustion of biodiesel, and so on. However, this can contribute only minimally to

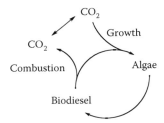

FIGURE 2.16   Illustration of the CO_2 "closed loop" in algae-based biodiesel.

the capture of CO_2. We must find ways to capture significant quantities of CO_2 from our largest point sources.

To that end, what are the options for reducing the level of this climate-changing greenhouse gas by capturing it as part of the combustion of carbonaceous fuels? Several different approaches have been developed, including some that attempt to capture carbon precombustion via *gasification*. Because CO_2 is generated during the gasification process, efforts are well underway to carry out coal gasification with simultaneous carbon capture of the cogenerated CO_2. Addition of CaO as an *in situ* CO_2 sorbent was shown to increase the percent of hydrogen gas in the product stream by nearly 53% (Equation 2.1) (An et al. 2012).

$$CaO(s) + C(s) + 2H_2O(g) \rightleftarrows 2H_2(g) + Ca(CO_3)(s) \qquad (2.1)$$

An alternative approach to CO_2 capture is seen in the concept of underground coal gasification, where the entire process is actually carried out in the coal seam and the generated CO_2 ostensibly trapped in subsurface cavities (Shafirovich and Varma 2009).

We will examine gasification in detail in Section 5.2.3. Our focus in this section will be on *postcombustion capture* that refers to the technology that removes the CO_2 from flue gases after combustion. The typical composition of postcombustion flue gas is 73–77% (by volume) nitrogen, 15–16% CO_2, 5–7% water vapor, 3–4% oxygen, and trace amounts of several other contaminants (Granite and Pennline 2002). The low concentration of CO_2 makes capture more difficult and presents a large technological challenge for postcombustion methods. *Oxycombustion* strives to address this issue by using pure oxygen to combust coal. This results in a higher concentration of CO_2 in the waste stream leading to a simpler and lower cost separation of the gas. A technology known as *chemical looping combustion* (CLC) has also been developed to produce a higher concentration of CO_2 in the gas stream. For example, in CLC, a solid metal oxide (instead of air) is used to oxidize the fuel in a separate fuel reactor, generating CO_2 and H_2O flue gases that are free of N_2, the main diluent when air is used as the combustion agent. It is a simple matter to condense out the water in the flue gas mixture to yield relatively pure CO_2. The reduced metal oxide is sent to a separate reactor where it is reoxidized with air and the process is repeated continuously.

Much research has been devoted to the CLC process, including optimization of the metal oxide. Ideally, it should be a sustainable resource, be inexpensive, have high mechanical and thermal stability, and have high activity with respect to both oxidation and reduction. Several transition metal oxides (Ni, Cu, Co, Fe, and Mn) have all been used with some success in CLC systems (Hossain and deLasa 2008). At this stage, CLC has been primarily applied to gaseous fuels (e.g., natural gas) due to the increased reactivity of the metal oxide with a gaseous fuel. However, application to the combustion of solid fuels is clearly a very high priority.

While CLC presents an attractive possibility, it is far from large-scale implementation. As noted in Section 2.4, it is well-established technology to remove CO_2 from natural gas by scrubbing with amines (see Figure 2.15); thus, why not apply this technology to postcombustion capture? Unfortunately, this method has serious draw-backs: the amines are highly corrosive, they decompose with time, and the energy penalty for regeneration of the amine adds too much to the cost of electricity generation. Other well-established methods take advantage of low-temperature adsorption of CO_2, including the Rectisol® method (which separates CO_2 by absorption into chilled methanol), and the Selexol® process (which takes advantage of absorption into chilled ethylene glycol). But these methods also add too much to the cost of electricity for large-scale application (Merkel et al. 2012). Thus, new methods are under development for the cost-effective, large-scale separation of CO_2 in postcombustion processes. We will look at a few examples of promising technologies in the following sections.

2.4.1.1 Membrane Technology

The development of membranes that can effectively separate CO_2 from H_2 or N_2 is a very active research area. All gas separation membranes should, ideally, be easily fabricated into thin films, have good mechanical, thermal, and chemical stability, be relatively inexpensive, and, of course, be made from sustainable or reusable materials. Inorganic membranes that include palladium are exceedingly effective at capturing hydrogen, but the cost and resource scarcity of Pd, along with the difficulty in fabricating large-scale inorganic membranes, make this a poor choice for industrial application. In terms of cost and ease of fabrication, organic polymer membranes are a much better option, but the effectiveness of the membrane is considerably lower. The area of "mixed matrix" membranes—incorporating active inorganic fillers into a polymeric membrane—is also an area of vigorous research.

The selectivity (α) of the membrane for a particular gas is a balance of diffusion (through the membrane) and solubility (in the membrane) and is represented by Equation 2.2, where P_A is the permeability coefficient for gas A (as measured in units of Barrer), a product of the diffusion coefficient and the solubility coefficient.

$$\alpha = \frac{P_A}{P_B} \tag{2.2}$$

Selectivities as high as 55 have been reported for $\alpha(CO_2/N_2)$, but the permeability coefficient for hydrogen is very similar to that of CO_2, making their separation particularly difficult (Yampolskii 2012). Hydrogen, a smaller molecule than CO_2, will always have a higher diffusion coefficient, but CO_2, the more easily condensed molecule, generally has a higher solubility in the organic membrane. As we will see in Section 4.4, these competing attributes can be adjusted advantageously by wise design of the polymer material (Merkel et al. 2012). Several examples of polymeric and mixed matrix gas separation membranes will be presented in Section 4.2.3.

2.4.1.2 Ionic Liquids

Ionic liquids (ILs) are organic salts. That are liquids by virtue of their low melting point. They contain an organic cation (see Figure 2.17) with noncoordinating anions like PF_6^-, BF_4^-, or triflate ($CF_3SO_3^-$). While ILs are by no means a new discovery,

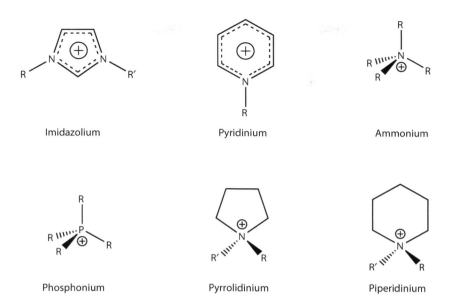

Imidazolium   Pyridinium   Ammonium

Phosphonium   Pyrrolidinium   Piperidinium

FIGURE 2.17 Representative cations for ionic liquids.

the manifold uses of these unique solvents have been an area of increasing interest, particularly in the area of *green chemistry* (Section 4.6). They are essentially non-volatile, their properties can be tuned by judicious choice of cation (there are tens of thousands of possible ILs), they are thermally stable and nonflammable, and they have a high solubility for CO_2. In addition, the regeneration of the IL and release of CO_2 should take place at a lower temperature with a smaller contribution to the cost of electricity generation (Figueroa et al. 2008). Hence their use in the separation and capture of CO_2—particularly as a replacement for the amine-based capture process—is especially encouraging.

The physical absorption of CO_2 by ILs has been thoroughly studied. The solubility of the gas in the IL is primarily related to the structure of the anion in terms of both van der Waals interactions and its impact on the molar volume of the IL material. The greater the IL molecular weight, molar volume, and free volume (the empty space between molecules), the higher the solubility of CO_2 in the IL. Since the solubility of CO_2 in ILs is much higher than that of either H_2 or N_2, selective separation of CO_2 from these gases by dissolution in ILs would seem promising. However, at this stage, the use of an IL to selectively absorb CO_2 in postcombustion applications is not feasible. A significant problem is the large increase in viscosity of the IL upon absorption of CO_2, turning the liquid into a thick, hard-to-handle gel. Additionally, the low concentration of CO_2 in postcombustion applications renders the use of ILs for this purpose out of the question with respect to capacity. An interesting alternative couples the use of ILs with the chemical capture of CO_2 by amine functionality, just as in the amine capture technology of natural gas purification (Figure 2.18). While this approach strongly enhanced the CO_2 solubility, the increase in viscosity

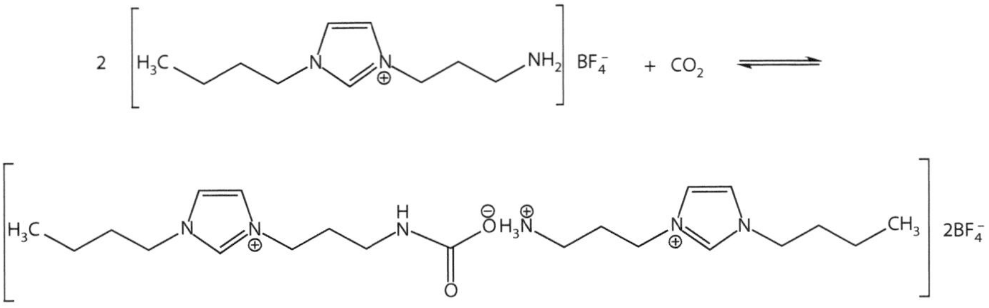

FIGURE 2.18 A functionalized ionic liquid for CO_2 capture.

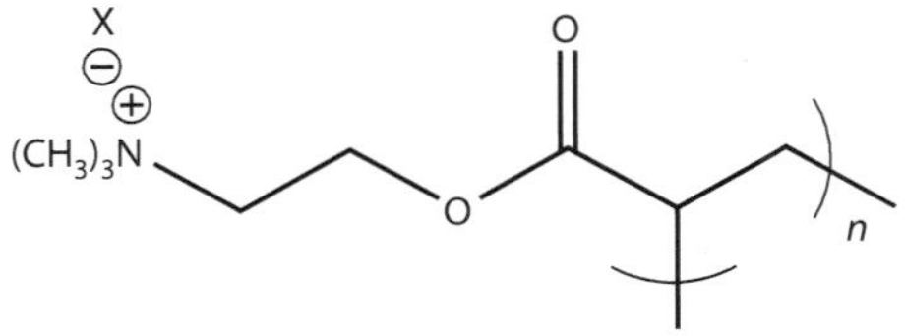

FIGURE 2.19 Poly[2-(methylacryloyloxy)ethyl]trimethylammonium ionic liquid.

was still a problem. Another attempt to make use of ILs for CO_2 separation coupled ILs with polymer membranes by synthesizing poly([2-(methylacryloyloxy)ethyl] trimethylammonium chloride) (Figure 2.19) and carrying out an ion exchange to replace the chloride ion with the more typical noncoordinating ions found in ILs. Again, excellent selectivity (CO_2/N_2) was seen due to the lack of absorption of N_2 by the material, but the capacity remains low (Samadi et al. 2010). At this point, the potential for use of ILs in postcombustion CO_2 capture remains unfulfilled, especially in light of the much higher cost of ILs compared to the solvents currently in use (Ramdin et al. 2012).

2.4.1.3 Solid Sorbents

Solid materials can be used to remove CO_2 from postcombustion flue gases by either *physisorption* or *chemisorption*. Physisorption is simply the adsorption of the molecule onto the surface of the adsorbent by virtue of a huge number of surface interactions (e.g., van der Waals forces). A wide variety of solid materials can adsorb CO_2 and, as such, presents an attractive alternative to liquid absorption technology for CO_2 separation and capture, especially since no solvent is required and the subsequent regeneration of CO_2 uses less energy. Furthermore, solid sorbents are commonly less prone to thermal degradation and, unlike amine scrubbers, noncorrosive. Chemisorption occurs as a result of chemical modification of the sorbent surface to add functionality that will actually react with the CO_2, as in amines. In either case, these materials have the same requirements of other CO_2 capture techniques:

- High selectivity for CO_2 over N_2 or H_2 and other gaseous impurities
- High capacity

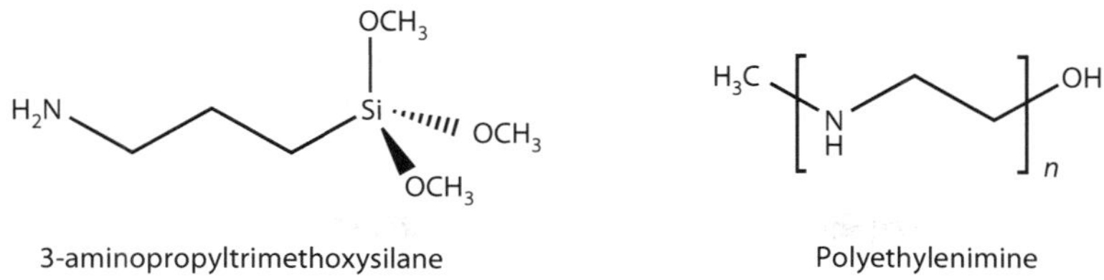

3-aminopropyltrimethoxysilane

Polyethylenimine

FIGURE 2.20 Amine carriers for functionalization of montmorillonite nanoclay.

- Good mechanical, thermal, and chemical stability
- Ease of regeneration/release of CO_2

Sorbents that have been studied include carbon (as in activated carbon, a form of solid carbon that is quite porous and therefore has a high surface area), zeolites, clays, silica gels, and metal–organic frameworks, among others. Discussion of metal–organic frameworks is deferred to Chapter 5 where their use as a material for hydrogen storage is presented in Section 5.3.1.

All of these materials rely upon porosity and surface area for the physical separation of gases. The specific surface area can be estimated by measuring the adsorption of gas molecules and making use of the Brunauer–Emmett–Teller (BET) isotherm to give the *BET surface area*. For example, the BET surface area of activated carbon is 1300 m²/g (Na et al. 2001) and that of Zeolite 13x is 515 m²/g (Samanta et al. 2011). Both carbon sorbents and zeolites readily absorb CO_2, and regeneration (unlike amine-based capture) does not require high-energy inputs. However, the capacity of CO_2 adsorption by these materials is highly dependent upon temperature (decreasing at high temperatures) and the presence of other impurities, including water vapor (a likely contaminant of flue gases). Thus, the use of solid sorbents for the physisorption of postcombustion CO_2 is not yet particularly promising.

Modified sorbents for chemisorption currently appears to be the more promising technology. In one example, researchers modified a natural montmorillonite clay consisting of a central alumina (aluminum oxide) or magnesia (magnesium oxide) octahedral layer sandwiched between two layers of silicate (silicon oxide) tetrahedra. The material was prepared so that the particle sizes were nanoscale with a surface area of about 750 m²/g. Treatment with 3-aminopropyltrimethoxysilane and polyethyleneimine (Figure 2.20) was then undertaken in order to load the clay with the desired amine functionality. Fourier transform infrared spectroscopy was used to confirm the presence of both amine-containing materials on the nanoclay, which was then tested with respect to CO_2 adsorption. The treated nanoclay was able to capture 7.5 wt.% CO_2 at 85°C and 1 atm pressure from a nitrogen-diluted CO_2 stream and regeneration could take place either at 100°C over multiple cycles or by vacuum regeneration. Thus this relatively inexpensive material and simple process for modification gave good initial results (Roth et al. 2013).

2.4.2 Conversion and Utilization

While progress is being made in the area of potentially separating and selectively capturing CO_2 from large point sources, the important concern, as noted before, is what do we *do* with the huge volumes of CO_2 so captured? The primary options are to nullify the CO_2 by storing it in some inert manner or to use the CO_2 as the starting material for useful products, including fuels. From the perspective of sustainability, CO_2 utilization is the more attractive option provided that the technology can be developed to carry out such transformations in a cost-effective and sustainable manner. However, there is far too much CO_2 for utilization methods to consume, thus some form of sequestration must be employed.

2.4.2.1   Sequestration

In an ironic or innovative twist on the solid sorbent technology described earlier, one approach to CO_2 storage is to use pressurized CO_2 to push coal bed methane out of coal seams, instead of recovering the methane by depressurizing the coal bed (see Section 2.2.2.4). The residual coal should then act as a porous sorbent for retaining the CO_2, particularly as the rate of adsorption for CO_2 is roughly twice that of methane (U.S. Department of Energy). However, the investigation of this concept is at the most preliminary stage. In contrast, geological storage of CO_2 is already in operation at several sites across the globe with the potential to store billions of tons of CO_2. These CCS sites inject captured CO_2 in various geological receptacles, for example

- Under the North Sea
- At a natural gas extraction site in the Sahara Desert
- In a sandstone formation under the Barents Sea

Yet another form of geologic storage is to use captured CO_2 for the purpose of enhanced oil recovery. Two sites, one in the Rocky Mountains (USA) and one in Canada, make use of this form of storage. For example, CO_2 captured at the Great Plains Synfuels plant in North Dakota is piped to Canada and injected into depleted oilfields in Saskatchewan (International Energy Agency 2010). However, geologic storage of CO_2 by entrapment or adsorption requires long-term monitoring of the site to ensure that CO_2 leakage is not occurring. (N.B. While the leakage of CO_2 may at first glance seem like a minimal threat, the sudden release of CO_2 can be lethal: in 1986, a catastrophic release of gas (primarily CO_2) from Lake Nyos in northwest Cameroon, West Africa, killed at least 1700 people by carbon dioxide asphyxiation (Kling et al. 1987).)

A method that eliminates the need for long-term monitoring of the site for escaping CO_2 (g) is its storage by mineral carbonation. This technology has the potential to mitigate billions of tonnes of CO_2 per year by reacting CO_2 with magnesium, calcium, or iron silicates to form inert and effectively permanent CO_2 storage in the form of the various carbonates (Equations 2.3 through 2.6):

$$Mg_2SiO_4 + 2CO_2 + 2H_2O \rightarrow 2MgCO_3 + Si(OH)_4 \qquad (2.3)$$

$$Mg_3Si_2O_5(OH)_4 + 3CO_2 + 2H_2O \rightarrow 3MgCO_3 + 2Si(OH)_4 \qquad (2.4)$$

$$Fe_2SiO_4 + 2CO_2 + 2H_2O \rightarrow 2FeCO_3 + Si(OH)_4 \qquad (2.5)$$

$$CaSiO_3 + CO_2 + 2H_2O \rightarrow CaCO_3 + Si(OH)_4 \tag{2.6}$$

Thus, unlike the injection of CO_2 into coal seams or depleted oil and gas reserves, in mineralization, the CO_2 is trapped as the stable solid carbonate. The problem with mineralization, as usual, is cost as it relates to energy input. The starting silicates must be ground into fine particles and both high temperature (150–600°C) and pressure (1–1.5×10^5 kPa) are required for the carbonation reaction (Sanna et al. 2012).

2.4.2.2 Utilization

Deep saline reservoirs are also considered good prospects for long-term storage of CO_2, as there are many such formations available for CO_2 injection and the capacity is huge. Perhaps more applicable to this text, however, is the *utilization* of CO_2. Figure 2.21 illustrates the many uses of CO_2—some of which are industrial applications of the unmodified molecule and some (the boxes in **bold**) that require a chemical transformation. Several of these uses are well established and familiar (e.g., the use of supercritical CO_2 as an extractant to decaffeinate coffee, or use in the beverage industry for the production of carbonated drinks), but our current consumption of CO_2 uses less than 1% of the amount of CO_2 generated annually (Mikkelsen et al. 2010).

The chemical transformation of CO_2 to other value-added compounds is the focus of this section, but it should be noted that the *biochemical* fixation of CO_2 happens, of course, every day in that plants convert around 560 million tonnes of CO_2 into carbohydrates via the process of photosynthesis (Kember et al. 2011). Unfortunately,

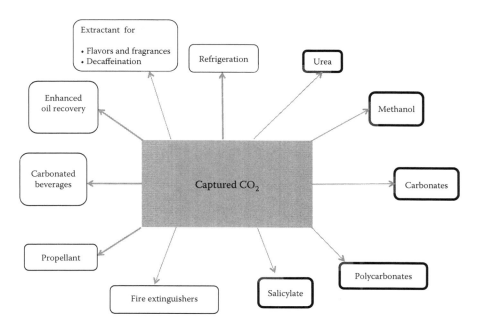

FIGURE 2.21 Uses of carbon dioxide.

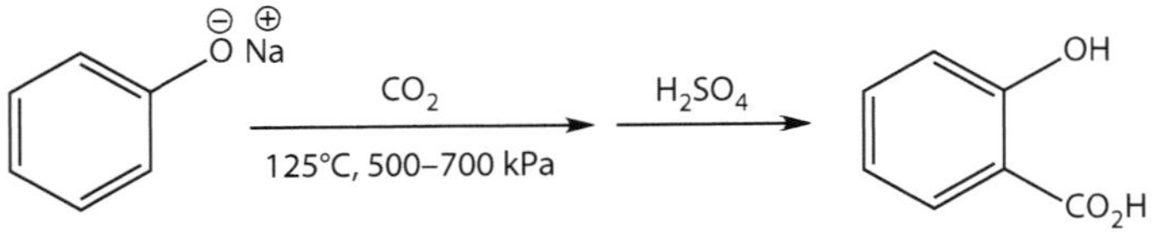

FIGURE 2.22   The Kolbe–Schmitt synthesis of salicylic acid.

humans have not mastered this level of volume or chemical expertise. For us, CO_2 is a stubborn molecule that requires activation in order to undergo a reaction. It is carbon at its most highly oxidized and it is thermodynamically very stable. In order to make the reaction be thermodynamically favorable, the following approaches are chemically logical:

- Use a high-energy reaction partner (e.g., epoxides).
- Drive toward a lower-energy reaction product (e.g., a *cyclic* carbonate).
- Force the reaction to the product side by taking advantage of Le Chatelier's principle.
- Take advantage of catalysis to enhance the reactivity of CO_2.

The use of CO_2 as an industrial raw material has existed for many years; for example, the carboxylation of salicylic acid by the Kolbe–Schmitt process (Figure 2.22) and the industrial process for the synthesis of urea (Figure 2.23) are both over 100 years old. Salicylic acid is converted into acetylsalicylic acid, and urea has many applications: it is a fertilizer, can be trimerized to make melamine (Figure 2.23), and is a key component in urea–formaldehyde resins. But the reduction of CO_2 to methanol is the holy grail of CO_2 transformation, as methanol is an important industrial feedstock and holds great potential for use as a fuel.

The catalytic reduction of CO_2 to methanol with hydrogen gas has long been known (Equation 2.7), although the mechanism is not necessarily well understood (Olah et al. 2011).

$$CO_2 + 3H_2 \xrightarrow{\text{metal oxide catalyst}} CH_3OH + H_2O \quad (\Delta H_{298} = -49.4\,\text{kJ/mol}) \quad (2.7)$$

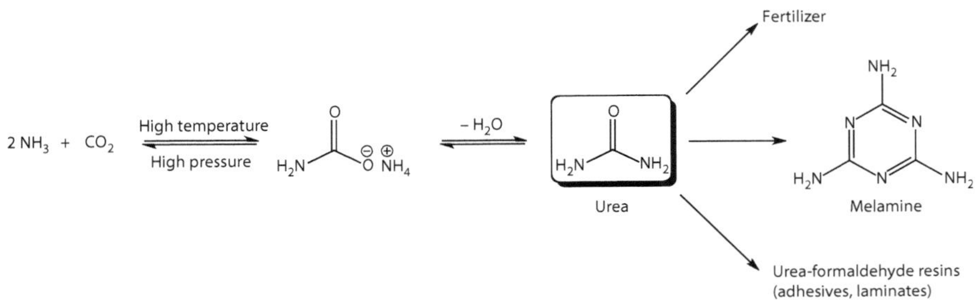

FIGURE 2.23   Synthesis of urea from CO_2 and ammonia, and its applications.

This transformation can be carried out with or without prior reduction of CO_2 to carbon monoxide, a compound that is also readily reduced to methanol (*vide infra*). Finding a catalyst to reduce the energy costs associated with the reduction of CO_2 is an active area of research. For example, the photoelectrochemical reduction of CO_2 holds great promise in harnessing sunlight to overcome the energy barrier for this reduction. Use of a gallium phosphide electrode to convert CO_2 to methanol via the intermediacy of a homogeneous pyridinium catalyst was found to take place selectively and with good efficiency; other photoelectrode materials are continually being investigated (Barton et al. 2008; Gu et al. 2013).

The energy-intensive reduction of carbon dioxide also requires a sustainable source of hydrogen, which is a challenge in its own right (see Section 5.2). The primary route to methanol currently is from *synthesis gas* (also known as *syngas*, a mixture of primarily CO and H_2 with some CO_2 contamination), which is ultimately made from fossil fuels (Equation 2.8). As can be seen, syngas conversion to methanol also consumes a small amount of CO_2.

$$\text{Coal or natural gas} \rightarrow \text{syngas} \equiv$$
$$3CO + 9H_2 + CO_2 \xrightarrow{\text{catalyst}} 4CH_3OH + H_2O \tag{2.8}$$

In terms of its potential as a fuel, methanol can be used in direct methanol fuel cells (Section 6.6.4) and it can be directly dehydrated to dimethyl ether (DME, Equation 2.9), a good prospect for use in diesel engines due to its high octane number, high vapor pressure, and lack of particulate pollutants (Ji et al. 2011).

$$2CH_3OH \xrightarrow{\text{catalyst}} CH_3OCH_3 + H_2O \quad (\Delta H_{298} = -21.0\,\text{kJ/mol}) \tag{2.9}$$

(Rahimpour et al. 2013).

More recently, CO_2 has been explored for conversion into cyclic carbonates or polycarbonates (Figure 2.24). The use of CO_2 in the synthesis of polycarbonates is covered in the chapter on polymers (Section 4.2.1). The nonpolymer cousins—simple carbonates—are useful as nonvolatile solvents or as precursors to other valuable compounds. They are especially valuable as a nontoxic alternative to the C_1 reagent phosgene (ClC(O)Cl).

The key to the preparation of carbonates is the activation of the CO_2 molecule by some sort of catalyst. As we will see in Chapter 5, finding an effective catalyst is a monumental undertaking. Catalytic activation during carbonate formation can proceed by a wide variety of possible mechanisms. Ultimately, any transformation utilizing CO_2 takes advantage of its electrophilic carbon and its nucleophilic oxygens (Figure 2.25). For example, in a polar mechanism reacting CO_2 with an epoxide partner, a Lewis acid may coordinate to the nucleophilic oxygen on either the CO_2 or the epoxide, leading to either of the plausible intermediates shown in Figure 2.26. Alternatively, in the case of catalysis by a transition metal, a reasonable expectation is that the CO_2 would insert into a metal–oxygen bond, as shown in Figure 2.27. In any case, a wide variety of catalysts have been used to effect this transformation: for

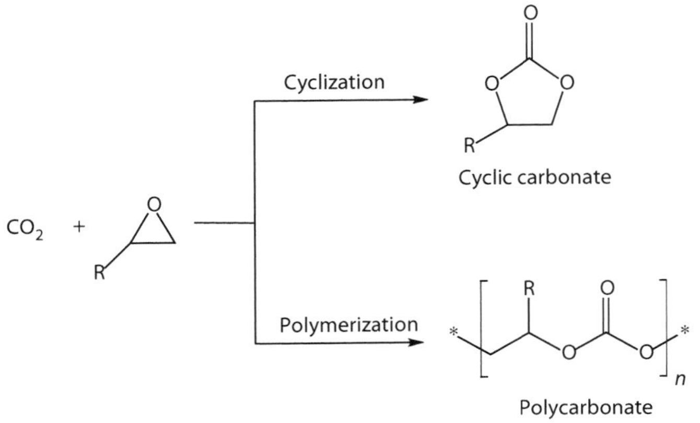

FIGURE 2.24 Carbonates from CO_2.

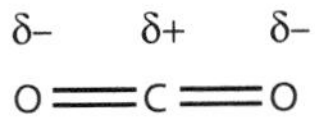

$$\overset{\delta-}{O}\!=\!\!=\!\!\overset{\delta+}{C}\!=\!\!=\!\overset{\delta-}{O}$$

FIGURE 2.25 The nucleophilic and electrophilic sites of CO_2.

example, a catalytic amount of the Lewis acid di-*n*-butyldimethoxy tin was used to convert methanol and supercritical CO_2 into dimethyl carbonate (DMC) by using molecular sieves to remove water as it is formed, shifting the equilibrium to favor the carbonate formation (Figure 2.28). In this case, it is believed that the reaction proceeds via CO_2 insertion into the tin–methoxy bond as shown (Figure 2.28) followed by thermolysis to release the DMC product (Choi et al. 2002).

The formation of cyclic carbonates by reaction of CO_2 with epoxides is a much more favorable enterprise due to the high energy of the epoxide starting material and

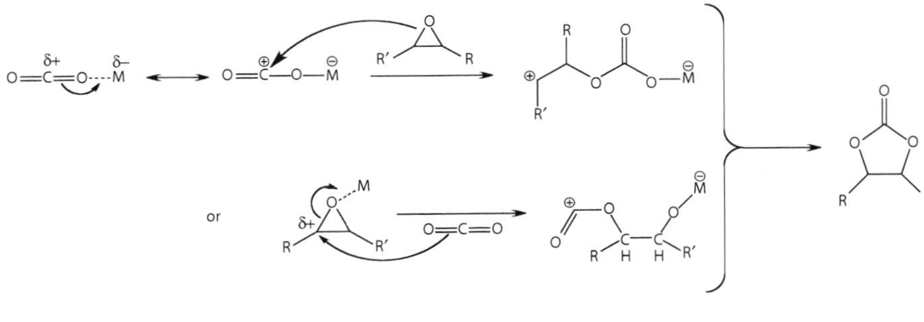

FIGURE 2.26 Plausible intermediates in the formation of a cyclic carbonate via a polar mechanism.

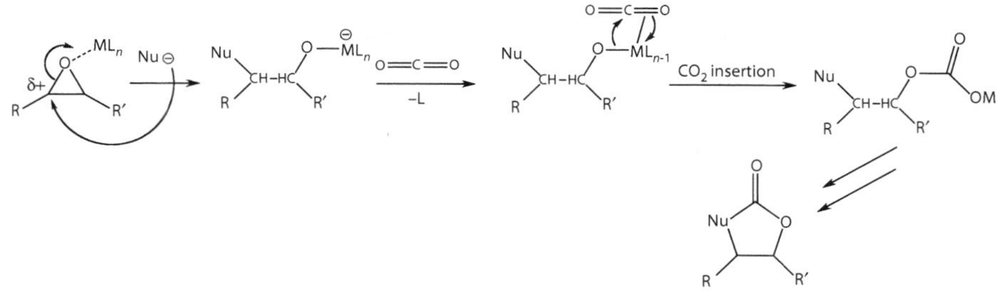

FIGURE 2.27 Transition metal-catalyzed carbonate formation with CO_2 insertion.

the thermodynamic stability of the cyclic carbonate product. The cyclic carbonate can then be used as a precursor to DMC (and other compounds) or used as a non-volatile solvent in its own right. In a one-pot synthesis of DMC, researchers used catalytic n-Bu_4N^+ Br–/n-Bu_3N with methanol, styrene oxide, and supercritical CO_2 at 150°C and 15 MPa pressure to obtain up to 84% DMC directly (Figure 2.29) (Tian et al. 2006). This pathway illustrates the activation of the epoxide by the Lewis acid n-Bu_4N^+ Br^- as shown in the catalytic cycle in Figure 2.30. Fortuitously, reaction of aziridines with CO_2 can also be achieved to form oxazolidinones, as shown in Figure 2.31. Chiral oxazolidinones such as the product in Figure 2.31 are valuable chiral auxiliaries for organic synthesis. While these CO_2 transformations do not contribute significantly to mitigation of the level of atmospheric CO_2, they do make a small contribution and further illustrate the potential utility of CO_2 as an alternative source of carbon for the synthesis of value-added chemicals. In the end analysis, however, CO_2 can neither be captured on a large scale due to the expense associated with the present technologies, nor be utilized on a large enough scale to address the grave impact its growing level has on our environment.

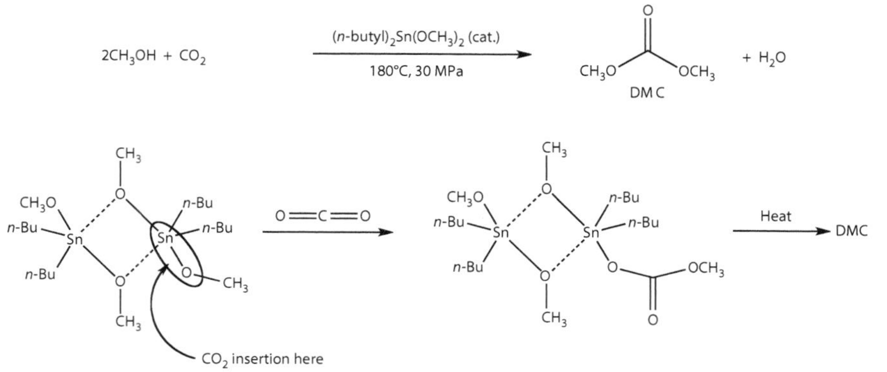

FIGURE 2.28 Tin-catalyzed synthesis of dimethyl carbonate.

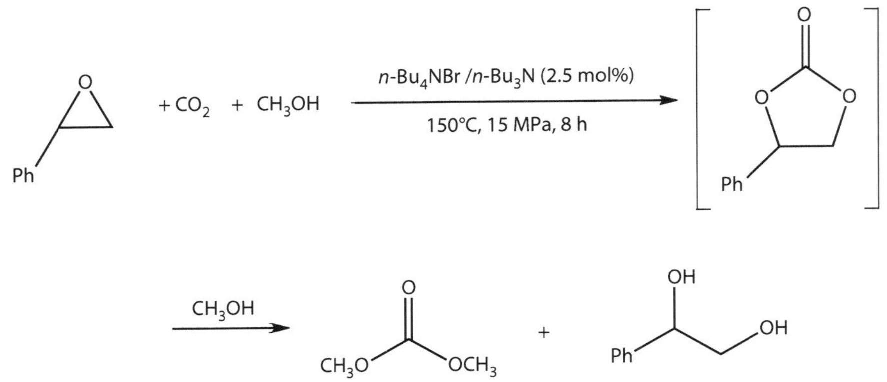

FIGURE 2.29   Synthesis of dimethyl carbonate via a cyclic carbonate.

2.5   SUMMARY

The history and evolution of fossil fuels on our planet is a fascinating illustration of the overlap between chemistry, physics, geology, and biology. Life became rocks, which became oil over millenia; now the cycle of carbon into our atmosphere has led to difficult and dangerous decisions for our ecosystem. The messy mixtures of fossil fuels that are made from lipids and lignins have been transformed into electricity and heat through extraordinary processes that require enormous amounts of ingenuity and energy.

The problems of extraction, environmental impact, and sustainability are by no means limited to fossil fuels. We depend on phosphate fertilizers to support the heavy demand to feed seven billion people on this planet. Similarly, the ever-developing array of technological gadgets demands the mining of a multitude of metals. The problems exhibited by the extraction of fossil fuels (and lessons learned) extend to all manner of environmental issues.

In this chapter, we have focused on the use of fossil fuels as our primary source of energy for transportation, heating our homes, businesses, manufacturing, and so on. But there is an additional crucial aspect of hydrocarbons that deserves mention: fossil fuels are ultimately a major source of *all* our carbon-containing materials. About 10% of these nonrenewable hydrocarbon feedstocks are made into alcohols, alkenes, carboxylic acids, and aromatics—the basic building blocks for much of our material existence, from pharmaceuticals to polymers. Thus, fossil "fuels" are of critical importance well beyond their use for energy. The capture and utilization of the harmful by-product of fossil fuel combustion, CO_2, may someday be the carbon feedstock for a variety of chemicals, mitigating the impact of this greenhouse gas in the atmosphere. However, many years of research progress are needed before carbon capture and utilization technologies can contribute significantly to solving the fossil fuel problem.

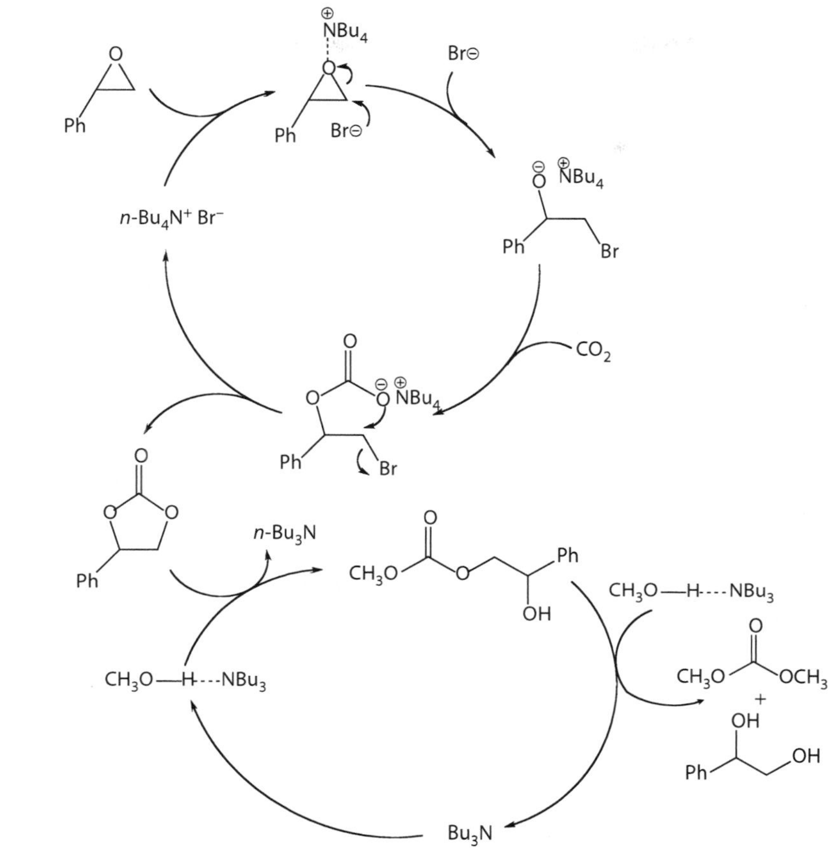

FIGURE 2.30 Proposed mechanism for the synthesis of dimethyl carbonate. (Reprinted from *Appl. Catal. A Gen.*, 301 (2), Tian, J.-S. et al. One-pot synthesis of dimethyl carbonate catalyzed by N-Bu$_4$n-Br/N-Bu$_3$n from methanol, epoxides, and supercritical CO$_2$, 215–221. Copyright 2006, with permission from Elsevier.)

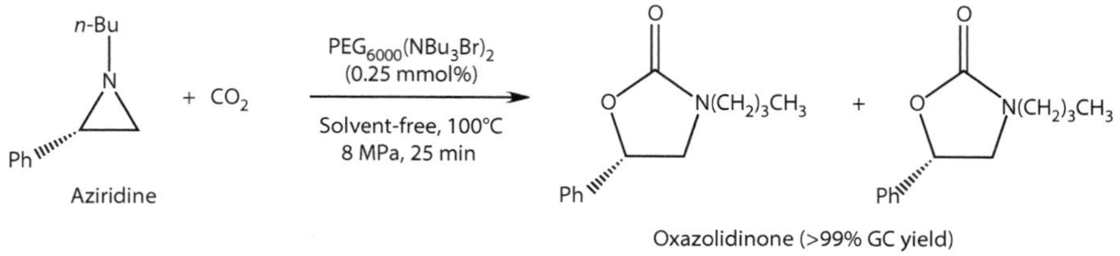

FIGURE 2.31 Transformation of CO$_2$ to oxazolidinones.

OTHER RESOURCES

BOOKS

Legault, A. 2008. *Oil, Gas & Other Energies. A Primer*. Paris: Editions Technip.
Selley, R.C. 1998. *Elements of Petroleum Geology*. San Diego: Academic Press.
Speight, J.G. 2007. *The Chemistry and Technology of Petroleum*. Boca Raton, FL: CRC Press.

ONLINE RESOURCES

International Energy Agency—Coal: http://www.iea.org/topics/coal/; Oil: http://www.iea.org/topics/oil/; Natural gas: http://www.iea.org/topics/naturalgas/.
U.S. Energy Information Administration—petroleum: http://www.eia.gov/petroleum/; Coal: http://www.eia.gov/coal/; Natural gas: http://www.eia.gov/naturalgas/
U.S. Geological Survey Energy Resources Program: http://energy.usgs.gov/
U.S. Task Force on Strategic Unconventional Fuels: http://www.unconventionalfuels.org/home.html

ONLINE RESOURCES RELATED TO CARBON CAPTURE AND SEQUESTRATION

International Energy Agency/CCS: http://www.iea.org/topics/ccs/
U.S. Department of Energy Clean Coal Technologies: http://www.fossil.energy.gov/programs/powersystems/index.html

REFERENCES

An, H., T. Song, L. Shen et al. 2012. Hydrogen production from a Victorian brown coal with *in situ* CO_2 capture in a 1 KWth dual fluidized-bed gasification reactor. *Ind. Eng. Chem. Res.* 51:13046–13053.
Armaroli, N. and V. Balzani. 2011. The hydrogen issue. *ChemSusChem* 4:21–36.
Bartholomew, C.H. and R.J. Farrauto. 2006. *Fundamentals of Industrial Catalytic Processes*. 2nd ed. Hoboken, NJ: Wiley & Sons.
Barton, E.E., D.M. Rampulla, and A.B. Bocarsly. 2008. Selective solar-driven reduction of CO_2 to methanol using a catalyzed P-Gap based photoelectrochemical cell. *J. Am. Chem. Soc.* 130 (20):6342–6344.
Bauer, D., D. Diamond, J. Li et al. Critical Materials Strategy 2011. U.S. Department of Energy,
Bomgardner, M.M. 2012. Cleaner fracking. *C. & E. News* 13:13–16.
Cane, R.F. 1976. The origin and formation of oil shale. In *Developments in Petroleum Science*, edited by Y.T. Fu and V.C. George, 27–60. Amsterdam, Oxford, New York: Elsevier.
Chestney, N. 2013. Global carbon emissions hit record high in 2012. *Reuters News Agency*, June 10, 2013.
Choi, J.-C., L.-N. He, H. Yasuda et al. 2002. Selective and high yield synthesis of dimethyl carbonate directly from carbon dioxide and methanol. *Green Chem.* 4 (3):230–234.
Committee on Assessment of the Department of Energy's Methane Hydrate Research and Development Program: Evaluating Methane Hydrate as a Future Energy Resource. 2010. Realizing the Energy Potential of Methane Hydrate for the United States. Washington, DC: National Academies Press.
Deffeyes, K.S. 2005. *Beyond Oil, the View from Hubbert's Peak*. New York: Hill and Wang.
Energy Resources Conservation Board. 2011. *Big Reserves, Big Responsibilities. Developing Alberta's Oil Sands. Oil Sands Regulatory Case Study 201104*. [cited 26 June 2012]. Available from http://www.ercb.ca/

Figueroa, J.D., T. Fout, S. Plasynski et al. 2008. Advances in CO_2 capture technology—The U.S. Department of Energy's carbon sequestration program. *Int. J. Greenhouse Gas Control* 2:9–20.

FracFocus Chemical Disclosure Registry. http://fracfocus.org/chemical-use/what-chemicals-are-used; accessed 8 July 2013.

Granite, E.J. and H.W. Pennline. 2002. Photochemical removal of mercury from flue gas. *Ind. Eng. Chem. Res.* 41 (22):5470–5476.

Gregory, K.B., R.D. Vidic, and D.A. Dzombak. 2011. Water management challenges associated with the production of shale gas by hydraulic fracturing. *Elements* 7 (3):181–186.

Gu, J., A. Wuttig, J.W. Krizan et al. 2013. Mg-Doped $CuFeO_2$ photocathodes for photoelectrochemical reduction of carbon dioxide. *J. Phys. Chem. C* 117 (24):12415–12422.

Hossain, M.M. and H.I. de Lasa. 2008. Chemical-looping combustion (CLC) for inherent separations—A review. *Chem. Eng. Sci.* 63 (18):4433–4451.

Howarth, R.W., R. Santoro, and A. Ingraffea. 2011. Methane and the greenhouse-gas footprint of natural gas from shale formations. *Clim. Change* 106 (4):679–690.

International Energy Agency. Carbon Capture and Storage: Progress and Next Steps. 2010. Paris, France.

Ji, C., C. Liang, and S. Wang. 2011. Investigation on combustion and emissions of DME/Gasoline mixtures in a spark-ignition engine. *Fuel* 90 (3):1133–1138.

Kember, M.R., A. Buchard, and C.K. Williams. 2011. Catalysts for CO_2/epoxide copolymerisation. *Chem. Commun.* 47 (1):141–163.

Kling, G.W., M.A. Clark, G.N. Wagner et al. 1987. The 1986 Lake Nyos gas disaster in Cameroon, West Africa. *Science* 236:169–175.

Kohl, A. and R.B. Nielsen. 1997. *Gas Purification, 5th Ed.* Houston: Gulf Publishing Co. (an imprint of Elsevier).

Kurek, J., J.L. Kirk, D.C.G. Muir et al. 2013. Legacy of a half century of Athabasca oil sands development recorded by lake ecosystems. *Proc. Natl. Acad. Sci. U.S.A.* Published ahead of print January, 2013, doi:10.1073/pnas.1217675110.

Manning, R. 2013. Letter from Elkhorn Ranch. Bakken business. The price of North Dakota's fracking boom. *Harper's*, March 2013, 29–38.

Merkel, T.C., M. Zhou, and R.W. Baker. 2012. Carbon dioxide capture with membranes at an IGCC power plant. *J. Memb. Sci.* 389 (0):441–450.

Mikkelsen, M., M. Jørgensen, and F.C. Krebs. 2010. The teraton challenge. A review of fixation and transformation of carbon dioxide. *Energy Environ. Sci.* 3:43–81.

Montaigne, F. 2013. Record 400 ppm CO_2 milestone 'feels like we're moving into another era'. *The Guardian*, 14 May 2013.

Na, B.-K., K.-K. Koo, H.-M. Eum et al. 2001. CO_2 recovery from flue gas by PSA process using activated carbon. *Korean J. Chem. Eng.* 18 (2):220–227.

Olah, G.A., G.K. Surya Prakash, and A. Goeppert. 2011. Anthropogenic chemical carbon cycle for a sustainable future. *J. Am. Chem. Soc.* 133 (33):12881–12898.

Rahimi, N. and R. Karimzadeh. 2011. Catalytic cracking of hydrocarbons over modified ZSM-5 zeolites to produce light olefins: A review. *Appl. Catal. A Gen.* 398 (1–2):1–17.

Rahimpour, M.R., M. Farniaei, M. Abbasi et al. 2013. Comparative study on simultaneous production of methanol, hydrogen, and DME using a novel integrated thermally double-coupled reactor. *Energy Fuels* 27 (4):1982–1993.

Ramdin, M., T.W. de Loos, and T.J.H. Vlugt. 2012. State-of-the-art of CO_2 capture with ionic liquids. *Ind. Eng. Chem. Res.* 51 (24):8149–8177.

Roth, E.A., S. Agarwal, and R.K. Gupta. 2013. Nanoclay-based solid sorbents for CO_2 Capture. *Energy Fuels* 27 (8):4129–4136.

Samadi, A., R.K. Kemmerlin, and S.M. Husson. 2010. Polymerized ionic liquid sorbents for CO_2 separation. *Energy Fuels* 24 (10):5797–5804.

Samanta, A., A. Zhao, G.K.H. Shimizu et al. 2011. Post-combustion CO_2 capture using solid sorbents: A review. *Ind. Eng. Chem. Res.* 51 (4):1438–1463.

Sanna, A., M.R. Hall, and M. Maroto-Valer. 2012. Post-processing pathways in carbon capture and storage by mineral carbonation (CCSM) towards the introduction of carbon neutral materials. *Energy Environ. Sci.* 5 (7):7781–7796.

Schidlowski, M., R. Eichmann, and C.E. Junge. 1974. Evolution des irdischen sauerstoff—Budgets und Entwicklung Der Erdatmosphäre. *Umschau* 22:703–707.

Schmidt, C.W. 2011. Blind rush? Shale gas boom proceeds amid human health questions. *Environ. Health Perspect.* 119 (8):a348–a353.

Selley, R.C. 1998. *Elements of Petroleum Geology.* 2nd ed. San Diego: Academic Press.

Shafirovich, E. and A. Varma. 2009. Underground coal gasification: A brief review of current status. *Ind. Eng. Chem. Res.* 48:7865–7875.

Skinner, B.J. and S.C. Porter. 2000. *The Dynamic Earth. An Introduction to Physical Geology.* 4th ed. New York: John Wiley & Sons.

Speight, J.G. 2007. *The Chemistry and Technology of Petroleum.* 4th ed. Boca Raton, FL: CRC Press.

Tester, J.W., E.M. Drake, M.J. Driscoll et al. 2005. *Sustainable Energy. Choosing among Options.* Cambridge, MA: MIT Press.

Tian, J.-S., J.-Q. Wang, J.-Y. Chen et al. 2006. One-pot synthesis of dimethyl carbonate catalyzed by N-Bu_4n-Br/N-Bu_3n from methanol, epoxides, and supercritical CO_2. *Appl. Catal. A Gen.* 301 (2):215–221.

U.S. Department of Energy, Office of Fossil Energy. 2013. *Carbon Storage R&D* [cited 20 May 2013]. Available from http://energy.gov/fe/science-innovation/carbon-capture-and-storage/carbon-storage-rd.

U.S. Department of Labor OSHA Technical Manual, Section IV: Chapter 2. Petroleum Refining Processes. http://www.osha.gov/dts/osta/otm/otm_iv/otm_iv_2.html#3. Accessed 29 August 2012.

U.S. Energy Information Administration. U.S. Coalbed Methane Production. U.S. EIA 2012. Available from http://www.eia.gov/dnav/ng/hist/rngr52nus_1a.htm.

U.S. Energy Information Administration. EIA-923 Monthly Time Series File, Fuel Receipts and Cost, Schedules 2 (March 2012).

U.S. Energy Information Administration. 2011. Technology drives natural gas production growth from shale gas formations. *Today in Energy.* July 12, 2011. Available from http://www.eia.gov/todayinenergy/detail.cfm?id=2170.

United Nations Environment Programme. 2012. *Mercury Control from Coal Combustion.* United Nations [cited August 28 2012]. Available from http://www.unep.org/hazardous-substances/Mercury/PrioritiesforAction/Coalcombustion/tabid/3530/Default.aspx

United States Geological Survey. The U.S. Geological Survey Gas Hydrates Project Gas Hydrates Primer. Retrieved 14 May 2013, from http://woodshole.er.usgs.gov/project-pages/hydrates/primer.html.

Weber, C.L. and C. Clavin. 2012. Life cycle carbon footprint of shale gas: Review of evidence and implications. *Environ. Sci. Technol.* 46 (11):5688–5695.

Yampolskii, Y. 2012. Polymeric gas separation membranes. *Macromolecules* 45 (8):3298–3311.

Yumura, T., Nanba, T., Torigoe, H. et al. 2011. Behavior of Ag_3 clusters inside a nanometer-sized space of ZSM-5 zeolite. *Inorg. Chem.* 50(14): 6533–6542.

3 Thermodynamics

The first Law says you can't win, the second Law says you can't even break even.

C.P. Snow

3.1 INTRODUCTION

Understanding thermodynamics is critical to being able to progress in any area of chemistry or energy. Furthermore, sustainability is irreversibly tied to efficiency, and efficiency depends entirely upon the physical and chemical laws governing the energy conversion process, that is, thermodynamics. How much power can we derive from an engine? What is the maximum efficiency for a photovoltaic device? Is ethanol a viable replacement for gasoline in terms of overall energy value? All of these are ultimately questions about thermodynamics. Equilibrium thermodynamics deals with processes that exchange *heat* or *work* (or both) in systems as they approach equilibrium, and heat and work are both manifestations of energy. The purpose of this chapter is to provide a basic overview of these concepts of thermodynamics in order to understand the efficiency of energy conversion devices.

When discussing equilibrium thermodynamics, we often begin with a very clear, idealized model system and focus on the changes that happen in this system in going from an initial state to a final state. Inputs (in the form of some kind of energy) are compared to outputs (in the form of heat and/or work) in order to determine spontaneity and efficiency. Ultimately we can base our understanding of real systems on what we glean from the model system. While that seems straightforward enough, some fundamental terms must be precisely defined.

System. The *system* is the well-defined piece of the physical world that is the focus in a thermodynamic study. It is partitioned off from the rest of the universe (the *surroundings*) by some sort of clear boundaries, whether real or imaginary. Systems can be *open* or *closed*. A classic example of a closed system is the chamber of a piston—the only exchange that can happen between the system and the surroundings is heat or work—no mass can be exchanged. A typical open system is a beaker: not only can energy be exchanged with the surroundings (e.g., in the form of heat) but also mass (as in a liquid evaporating from the beaker).

State. Not only does the system we are studying have to be precisely defined, so does its *state*. The state of a system is described by its explicit properties, where a property is a measured quantity such as temperature, pressure, or volume. If, for example, our system is an ideal gas, then the state can be defined by the exact pressure, volume, temperature, and molar amount of the particular gas. An *equation of state* is the mathematical equation that defines the relationship between these properties, for example, the ideal gas law,

$$PV = nRT \tag{3.1}$$

Obviously, the state of the system changes when these properties change. When reality creeps in (i.e., the system deviates from ideal behavior), modified equations of state incorporating empirically derived proportionality constants are used.

A *state function* is one whose change is independent of path. Another way of stating this is that a change in a state function depends solely on the initial and final state of the system—*how* the system got to that final state is immaterial. Thus, energy is a state function—it is intrinsic to the system regardless of how the system came to that state; its quantity depends only on the state of the system. What is *not* a state function? Work, for example: a boulder on a hillside has some finite potential energy associated with it, but we do not know how much work was expended in getting it up there to give it that energy. The amount of work expended depends on the path by which it got there.

Energy (E): Work, heat, and energy are intimately intertwined. During a change in state, heat may be exchanged with the surroundings and/or work may be done. The state function that relates the total change in the amount of heat and/or work done on the system is its *thermodynamic energy*. The first and second laws of thermodynamics unite these concepts of heat, work, and energy.

Work (w): This is where the importance of understanding thermodynamics, even if in a very basic nonquantitative way, becomes obvious. In the classic illustration, work is a quantity that can be manifested as the lifting of a weight in the surroundings. It is energy made visible and, as such, carries the units of joules. In the realm of energy conversion, work is clearly what we are interested in: converting our source of energy into some process that produces work. Since we want to produce useful work (which requires the expenditure of energy), we need to understand how work "flows." *Work only appears at the boundary of a system and only appears during a change in system's state.* The obvious thing about work is that it is observed as an effect in the surroundings. The less obvious thing about work is that it is any interaction between the system and surroundings that is not heat.

Heat (q): A familiar form of energy is thermal energy, that is, heat. A temperature differential between a system and its surroundings leads to the flow of a quantity of heat across a boundary. The amount of heat transferred is related to the amount of material and its particular nature as reflected in its *heat capacity*. The SI unit for quantity of heat is the joule, thus heat and work are interconvertible. Like work, the amount of heat that flows depends upon the pathway—it is not a state function. We may know the initial and final states of a system, but we cannot deduce from this either the quantity of heat or the quantity of work that were involved in the transition from initial to final state. If no heat is exchanged between the system and surroundings (because the system is perfectly insulated from the surroundings), the system is said to be *adiabatic* ($q = 0$).

3.2   FIRST LAW OF THERMODYNAMICS

The first law of thermodynamics states that the total quantity of mass and energy is the same before and after its conversion. It is perhaps more familiar as the total amount of matter plus energy (for the system plus surroundings) is constant. A joule is a joule is a joule, whether it is from burning a lump of coal to warm a room or eating a bacon double cheeseburger. If a system (coal/cheeseburger) loses energy to the surroundings (room/body), the surroundings pick up that equivalent amount of energy—no more, no less: the total quantity of energy is a constant, period. The mathematical expression of the first law is given in Equation 3.2,

$$\Delta E = q + w \tag{3.2}$$

where E denotes the internal energy of a system, q is heat, and w is work. This equation is the *sine qua non* of energy conversions. For any given system, the change in energy is given by the difference between the heat absorbed ($\int dq$, representing the sum of all infinitesimal increments of heat, dq, along the path) by the system. [NOTE: calculations involving work can get pretty confusing because of sign: if you are a chemist, work is positive when *work is done on the system* (work has been produced in the surroundings and flows to the system). Work is negative when *work is done on the surroundings* (work has been produced in the system and flows to the surroundings). Physicists use the signs in the opposite sense.]

3.3   SECOND LAW AND THERMODYNAMIC CYCLES: THE CARNOT EFFICIENCY

The problem with the first law is that we cannot necessarily *make use* of the total amount of energy in an energy conversion. The second law concerns itself with this quandary, that is, the reality that the total amount of energy is not completely available to produce useful work. The second law is at the root of any discussion about efficiency: what is the maximum theoretical amount of useful energy we can get from any device, process, fuel, etc.? It is a required complement to the first law or else we would have perpetual motion machines. For example, suppose we want to use thermal energy to power some machine that does work (a *heat engine*). If the quantity of heat flowing from the surroundings into the system ($-q$) is converted into work in a precisely equivalent amount (w), then the change in energy from the initial state to the final state (according to the first law), ΔE, is zero. An unlimited source of heat could be turned into unlimited work! There would be no degradation of energy, and as a result, 100% efficiency. Nice idea, but also not real. Some amount of "waste heat" is always generated, limiting the amount of work that can ever be produced: this is the upshot of the second law.

This concept of heat or energy that is wasted or lost is where *entropy* enters in. In introductory chemistry, we learn that entropy is a measure of the disorder of the system, a concept that is in no way invalid here—entropy can be pictured as degraded energy, waste heat, or decrease in order. No matter what the perspective, when the maximum entropy of a system is attained, the system has achieved thermodynamic equilibrium and no further spontaneous change is possible. In the current context, the macroscopic focus on the inefficiency at which energy is converted into work is the most relevant. Any time that energy is transformed to heat or work it degrades and generates entropy: 100% efficiency in energy conversions is an impossibility.

Thermodynamic cycles are the standard means of illustrating energy conversion processes and a mechanical heat engine is a simplified model that helps to illustrate the concept of theoretical efficiency and waste energy. A mechanical heat engine is a device that converts heat into work as in, for example, an automobile engine. There are four stages in any mechanical heat engine:

1. Compression
2. Heat intake
3. Expansion
4. Heat rejection

There are a multitude of engine designs that vary the ways in which these steps are carried out, typically by describing the four steps in the form of a system consisting of a frictionless piston and an ideal gas, as described later for the Carnot engine. Because the system involves an ideal gas, the ideal gas law is our equation of state. The surroundings are a heat reservoir that provides the heat to be converted into work by a cyclic, reversible (*vida infra*) process.

The Carnot engine is based on the Carnot cycle, devised by Sadi Carnot (1796–1832). In order to understand the thermodynamics of the Carnot cycle, we have to understand the concept of reversibility. A reversible process simply means that whatever is "done" can be "undone" with no loss along the way. Thus a reversible process is made up of infinitesimal steps, each one of which can be reversed to its previous state without any deviation in any way. In a reversible cycle, a system and surroundings begin at an initial state, proceed through the steps of the cycle, and end at a final state which is identical to the initial state.

A representation of the reversible Carnot cycle relating the thermodynamic quantities of heat, work, energy, and entropy is shown in Figure 3.1a and b. Note that representations a and b focus on the exact same cycle but simply present different relationships (pressure–volume vs. temperature–entropy).

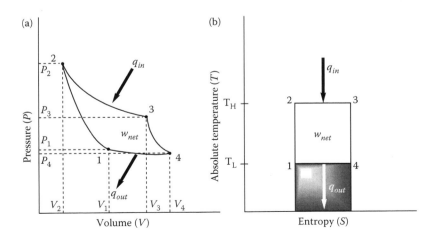

FIGURE 3.1   The Carnot cycle (a) as a function of pressure and volume and (b) as a function of absolute temperature and entropy.

- Step $1 \to 2$ represents adiabatic compression: the system is perfectly insulated so that there can be no exchange of heat with the surroundings ($dq = 0$). Keep in mind that this does not imply that there is no change in temperature: because the heat cannot escape into the surroundings, the temperature increases. Because the change in entropy for a reversible process is related to heat and temperature by

$$dS = dq_{rev}/T \tag{3.3}$$

and $dq_{rev} = 0$, there is no change in the quantity of entropy (Figure 3.1b). The volume decreases as the piston is compressed and as a result, the pressure increases (Figure 3.1a).

- Step $2 \to 3$ represents heat addition. The insulation is removed and the piston expands while heat is drawn from the surroundings into the system (q_{in}). The temperature is constant (see Figure 3.1b), making this an isothermal step. (The pressure decreases, the volume increases and as a result, the entropy is increased.)
- Step $3 \to 4$ is expansion yet again (volume increase, pressure decrease) yet again, but with insulation reapplied so that heat exchange is prevented, making this an adiabatic expansion. As a result, the work that is done to expand the piston is done at the expense of the internal energy from the gas, with the result that the temperature drops. Entropy again stays constant because no heat is drawn from the surroundings.
- Finally, step $4 \to 1$ is isothermal heat rejection. The insulation is removed and heat exchange between the system and surroundings is once again allowed. The piston is compressed as the temperature is held constant, so the excess heat flows into the surroundings. This is *wasted heat* (as indicated by q_{out}) and the amount is given by the shaded area.
- The net amount of work done by the Carnot cycle is indicated by the area bounded by the four steps of the cycle (w_{net}).

It can be shown that the net total work for the reversible cycle is given by

$$w_{net} = (T_L - T_H)R \ln (V_4/V_1) \tag{3.4}$$

where T_H and T_L represent the maximum and minimum absolute temperatures (K) and V_4 and V_1 are the initial and final volumes of the step $4 \to 1$ in the cycle. Since T_L-T_H is negative and $V_4 > V_1$, the net work (w_{net}) is invariably negative.

What is the significance of the Carnot cycle? Since the net work is negative in this ideal process, the Carnot engine represents the *maximum possible work* for any given expenditure of energy for a mechanical heat engine. Even in this perfect system some heat (q_{out}) must be dissipated to the surroundings with the result that the corresponding amount of work produced is less than the amount of energy expended to produce it. *This consequence of the second law cannot be overstated*:

If you are a newcomer to thermodynamics, note carefully the implications of this concept. *There are limits on the ability of even a flawless heat engine to convert heat*

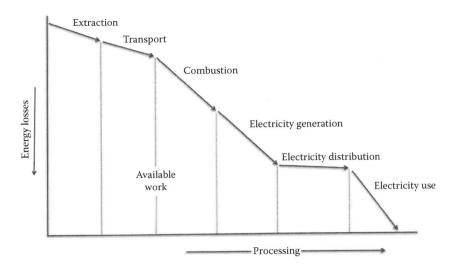

FIGURE 3.2   Degradation of energy resulting in losses in the availability of work. (From Winterton, N. 2011. *Chemistry for Sustainable Technologies. A Foundation.* Cambridge, UK. Reproduced by permission of The Royal Society of Chemistry.)

into useful work. No amount of technical innovation or political intervention will further improve the machine's performance beyond this ideal. (Tester et al. 2005, p. 89)

How, then, does this fit in with the first law? The quantity of energy *is* constant (first law)—but the quality is not. Energy is degraded in every energy conversion process and entropy is generated (second law). Figure 3.2 presents this concept in a very practical sense: the energy value associated with energy derived from the sun and ultimately transformed into fossil fuels, then electricity, degrades significantly at each step of energy conversion.

The maximum efficiency (η) of a heat engine (the *Carnot efficiency*) is represented mathematically by

$$\eta = w_{net}/q_{in} = T_H - T_L/T_H \text{ or } \eta = 1 - T_L/T_H \qquad (3.5)$$

where the temperature is expressed in degrees Kelvin. Note the significance of the temperature limits: a heat engine produces work because of the difference in the temperature limits; the larger the difference in temperature, the higher the efficiency. An apt analogy is the work provided by a water wheel. Just as water flows from a higher level to a lower level to produce work in a water wheel, so heat flows to produce work in a heat engine. In each case, the amount of work is related to the difference in energy (heat energy or potential energy). In both examples, the efficiency approaches 100% only at the extreme, that is, as the low temperature approaches absolute zero for the heat engine, and in an infinite separation of water levels for the water wheel. If there is no differential, the efficiency is zero and the system is at thermodynamic equilibrium.

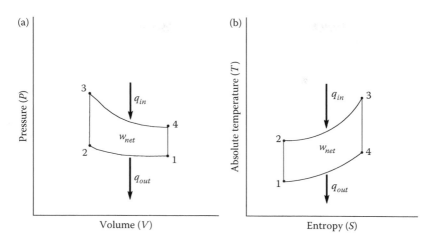

FIGURE 3.3   The Otto cycle (a) as a function of pressure and volume and (b) as a function of absolute temperature and entropy

Mechanical heat engines are ubiquitous in energy conversions. Numerous other thermodynamic cycles exist on which several engine designs are based (e.g., Brayton, Otto, Diesel, Stirling, Ericsson, Rankine). The Otto cycle (Figure 3.3a and b) is the model for many automobile engines and consists of a four-stroke cycle: adiabatic compression ($1 \rightarrow 2$), constant-volume heating ($2 \rightarrow 3$), adiabatic expansion ($3 \rightarrow 4$), and constant volume cooling ($4 \rightarrow 1$) (Fenn 1982). The efficiency of a typical gasoline-fueled internal combustion engine is about 30%; a diesel engine is more efficient at ≈45% (U.S. Department of Energy Energy Efficiency and Renewable Energy 2003).

Improving engine efficiency is an active area of research. A more recent incarnation is the "X2 rotary engine" whose designers claim can deliver an efficiency of 75%. This engine, based on the "high efficiency hybrid cycle" design, combines features of the Otto, Diesel, Rankine, and Atkinson cycles and is closely related to the rotary engine once featured in Mazda automobiles. The key to the new design is constant volume combustion plus overexpansion of the burning fuel/air mixture. This allows all of the fuel to combust, capturing almost all of the energy as work (Szondy 2012).

The Carnot efficiency for a natural gas-fired power plant is 63% based on the lower heating value (LHV) for natural gas (Higman and van der Burgt 2003) and 49% for a nuclear reactor operating at 300°C (and cooling to 20°C). Given the theoretical limits to energy conversion based on the physical laws of thermodynamics, what can we do? One approach is to operate energy generation devices at even higher temperatures, as in next-generation nuclear power plants (Chapter 9). Another tactic is to either minimize the generation of waste heat or make use of it to improve the efficiency. There are ways we can eliminate the mechanical middlemen and convert fuel directly into electricity, as in voltaic cells: without the intermediacy of thermal waste, these devices are not limited by the Carnot efficiency ... although they must still obey the second law!

3.4 EXERGY AND LIFE-CYCLE ASSESSMENT

We will look at specific applications of thermodynamics in regard to the efficiency of energy conversion processes throughout the text. Although the first law of thermodynamics requires that energy is conserved, the second law shows that there is a limit to the amount of useful work that can be produced, thanks to entropy. The limited energy corresponding to this maximum is known as *exergy* which, like energy, has the units of Joules. This concept is more familiar in the realm of chemistry as *Gibbs free energy*. Thus, like free energy, the amount of exergy depends entirely upon the reference environment of which it is a part. For example, a beaker of water at 100°C on a 0°C day has a greater exergy content than it does on a 35°C day—as we would expect, given the discussion of heat engine efficiencies earlier (Dincer and Rosen 2007). Exergy is the degraded energy whose amount is proportional to the amount of entropy produced: it is the amount of energy that is actually available to do work. Thus, in Equations 3.4 and 3.5, "w_{net}" really represents the exergy of the process.

Exergy as a broadly applied (and somewhat fuzzily defined) concept plays an important role in sustainability analyses, having been applied to ecosystems, industrial systems, and even in economic studies in order to assess their sustainability (Dewulf et al. 2008). It is a critical facet of life cycle assessment (LCA), a "cradle-to-grave" approach to assessing the impact of any process on the environment (where environment is environment *writ large*: the system plus surroundings). Exergy is used because it is the practical, accurate reflection of real, irreversible technology. Thus a prototypical LCA would take into account the inputs and outputs into and from a system in terms of exergy and materials. The extraction, processing, manufacture, distribution, use, and disposal or recycle of raw material inputs are all accounted for. All outputs are similarly considered, including waste emissions and any byproducts that are formed. While an in-depth discussion of LCA is not germane to this chapter, they are important and monumental undertakings that attempt to capture the big picture of sustainable processes and sustainable development, and are crucial to the development of any sustainable system.

OTHER RESOURCES

BOOKS

Kirwan, A. D. 2000. *Mother Nature's Two Laws: Ringmasters for Circus Earth. Lessons on Entropy, Energy, Critical Thinking, and the Practice of Science.* River Edge, NJ: World Scientific.
Fenn, J. B. 1982. *Engines, Energy, and Entropy.* New York: W.H. Freeman and Company.

REFERENCES

Dewulf, J., H. Van Langenhove, and B. Muys et al. 2008. Exergy: Its potential and limitations in environmental science & technology. *Environ. Sci. Technol.* 42:2221–2232.
Dincer, I. and M.A. Rosen. 2007. Chapter 1—Thermodynamic fundamentals. In *Exergy*, 1–22. Amsterdam: Elsevier.

Energy Efficiency and Renewable Energy. *Just the Basics—Diesel Engine.* 2003. Washington, DC, U.S. Department of Energy, http://www1.eere.energy.gov/vehiclesandfuels/technologies/fcvt_basics.html.

Higman, C. and M. van der Burgt. 2003. *Gasification.* Burlington, MA: Elsevier (Gulf Professional Publishing).

Szondy, D. 2012. Liquid piston unveils 40-bhp X2 rotary engine with 75 percent thermal efficiency. *Gizmag,* http://www.gizmag.com/liquidpistol-rotary/24623/

Tester, J.W., E.M. Drake, and M.J. Driscoll et al. 2005. *Sustainable Energy. Choosing Among Options.* Cambridge, MA: MIT Press.

Winterton, N. 2011. *Chemistry for Sustainable Technologies. A Foundation.* Cambridge, UK: RSC Publishing.

4 Polymers and Sustainable Energy

4.1 POLYMER BASICS

Polymers are an integral part of all aspects of sustainable energy solutions. Everything from wind turbine blades to biomass is made up of polymers, the high molecular-weight macromolecules that are a ubiquitous part of our lives. This chapter presents the basics of polymer chemistry with a few examples in the realm of CO_2 separation (Section 2.4.1.1). More depth is provided in the context of the composite polymers used in the fabrication of wind turbine blades. Additional applications of polymer chemistry in sustainable energy are introduced in later chapters in the context of fuel cells (Chapter 6), solar photovoltaics (Chapter 7), and biomass (Chapter 8).

Unlike typical small molecules studied in introductory chemistry courses, polymers are macromolecules with molecular weights hundreds of times larger, typically in the range of 10,000 and 1,000,000 amu. They may be synthetic (e.g., polystyrene) or natural (e.g., cellulose) or they may be characterized by their behavior: *thermoplastics* soften when heated and return to their original state by cooling, whereas *thermosetting polymers* do not become pliable with heating. The difference in this behavior is due to differences at the molecular level: in thermoset polymers, the individual polymer chains have been covalently cross-linked and covalent bonds must be broken for the material to flow. Vulcanization of rubber is an example of cross-linking. In thermoplastics, intermolecular forces are primarily responsible for their behavior; addition of heat merely overcomes these intermolecular interactions that reform upon cooling.

Regardless of whether it is a thermoplastic or a thermoset polymer, these macromolecules are made up of hundreds to thousands of small-molecule *monomers* that are linked together, with the common name of the polymer stemming from the name of the monomer. For example, the monomer that makes up polyvinyl chloride (PVC) is, aptly enough, vinyl chloride (Figure 4.1; n = a large number, where * indicates that the polymer chain extends indefinitely). PVC is an example of a homopolymer, that is, a polymer made from a single monomer. *Copolymers*, in contrast, are made from more than one monomer. A protein would be an example of a natural copolymer that is made up of a variety of amino acid monomers. Most polymers are made up of a regular *repeating unit* that is equivalent to the monomer. In the case of PVC, the repeat unit is the –CH_2CHCl– segment. The polyamide nylon 6,6 is a regular copolymer made up of two monomers—1,6-hexanediamine and adipoyl chloride—that combine to make the adipoyl hexanediamine amide repeating unit (see Figure 4.2). Reference is also often made to *oligomers*, which are simply short fragments of a polymer, often consisting of just a few repeating units.

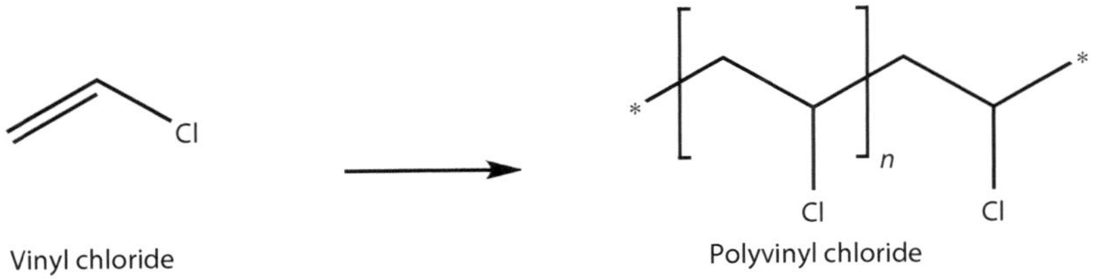

Vinyl chloride   Polyvinyl chloride

FIGURE 4.1 Polyvinyl chloride.

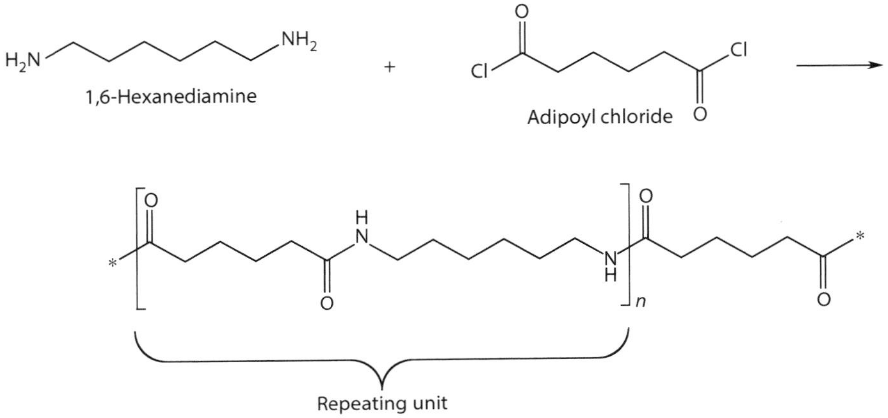

1,6-Hexanediamine

Adipoyl chloride

Repeating unit

FIGURE 4.2 Synthesis of nylon 6,6.

A key feature of synthetic polymers that makes their properties and behavior different from that of small molecules is the fact that a synthetic polymer is not a single, unique molecule: it is a conglomerate array of many molecules of varying lengths and (to some degree) structures. Some individual chains are shorter, some longer. Some may be branched, whereas others are linear. For each polymer, there is an associated statistical distribution of individual chains in the bulk sample. Unlike a small molecule, then, a typical polymer does not have a unique molecular weight or, even, a single determinate structure.

The bulk behavior of a polymer, therefore, encompasses the properties of all of the molecules that make up the sample. The statistical distribution and chemical heterogeneity, the length and structure of the individual units, the presence or absence of cross-linking, and the conformational flexibility all impact a polymer's properties such that polymers range from clear, brittle plastics to sticky adhesives or bouncy elastomers. Furthermore, given the sheer size of polymers, intermolecular forces play an outsized role. For example, a polymer may be elastic as a result of low intermolecular forces and chains with good conformational flexibility (styrene–butadiene rubber is a good example, Figure 4.3). If bulky substituents or stronger intermolecular forces are present, the polymer is likely to behave more like a "typical plastic." Very high intermolecular forces (e.g., hydrogen bonding) and crystallinity (as a result of the symmetry of the molecules) can lead to a polymer with

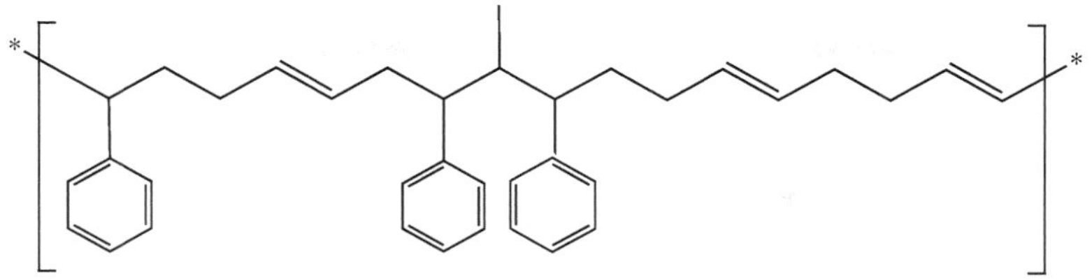

FIGURE 4.3   Styrene–butadiene rubber.

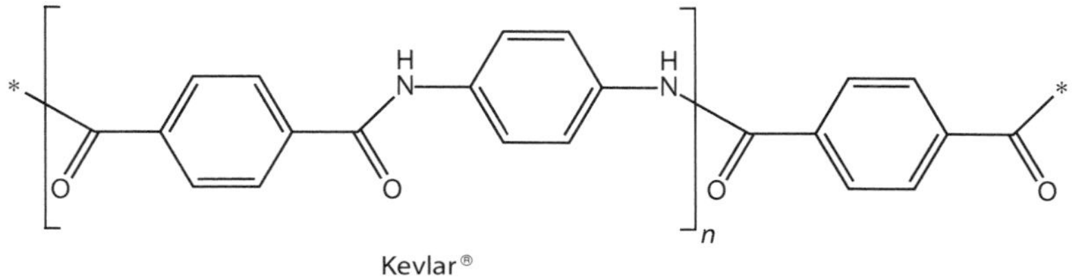

Kevlar®

FIGURE 4.4   A *para*-substituted polyaramide, Kevlar®.

excellent mechanical strength (as in Kevlar™, Figure 4.4, a material used in bullet-proof fabrics) (Billmeyer 1984).

The average molecular weight of a polymer can be calculated by taking into account the average *degree of polymerization* ($\overline{DP}$) and the molecular weight of the repeating unit. The degree of polymerization is equivalent to the number of repeat units, n. Thus, the average molecular weight of a strand of nylon 6,6 with $n = 10,000$ would be approximately 2.26×10^6 amu because the repeat unit $C_{12}H_{22}N_2O_2$ has a molecular weight of 226.

More commonly, the molecular weight of the bulk polymer is reported as a statistical average that takes into account the distribution of individual chains. This "molar mass distribution" is most commonly reported as either the *number-average* (M_n) or the *weight-average* (M_w) molecular weight. Imagine a hypothetical polymer consisting of 100 individual chains with the following molecular weight distribution:

10 chains @ 80 amu
10 chains @ 100 amu
10 chains @ 120 amu
60 chains @ 140 amu
10 chains @ 160 amu

The number-average molecular weight for this hypothetical polymer is the simple arithmetic mean, as given by Equation 4.1, where n_i is the number of chains with molecular weight M_i and Σn_i is the total number of individual chains.

$$M_n = \frac{\Sigma n_i M_i}{\Sigma n_i} \qquad (4.1)$$

In this example, the number-average molecular weight M_n is 130 (Equation 4.2).

$$M_n = \frac{(10 \times 80) + (10 \times 100) + (10 \times 120) + (60 \times 140) + (10 \times 160)}{100}$$

$$= \frac{13,000}{100} = 130 \tag{4.2}$$

As a simple counted quantity, M_n reflects the number of discrete particles in the sample and can, therefore, be determined by measuring colligative properties such as vapor pressure lowering, boiling point elevation, freezing point depression, or osmotic pressure for a very dilute solution of the polymer. More commonly, M_n is determined along with a complete molecular weight profile using size-exclusion chromatography (SEC) (*vide infra*).

The mathematical representation of weight-average molecular weight (M_w) is shown in Equation 4.3, where $w_i = n_i M_i$ = the weight of the chains with molecular weight M_i.

$$M_w = \frac{\Sigma w_i M_i}{\Sigma w_i} = \frac{\Sigma n_i M_i^2}{\Sigma n_i M_i} \tag{4.3}$$

In this case, the weight average of our hypothetical polymer sample would be 145 as shown in Equation 4.4.

$$M_w \frac{(10 \times 6400) + (10 \times 10,000) + (10 \times 14,400) + (60 \times 19,600) + (10 \times 25,600)}{(10 \times 80) + (10 \times 100) + (10 \times 120) + (60 \times 140) + (10 \times 160)}$$

$$= \frac{1,740,000}{12,000} = 145 \tag{4.4}$$

M_w is always greater than M_n since larger chains weigh more than smaller chains and molecular weight is squared in the M_w calculation. Thus, the higher-molecular-weight chains skew the average to a higher value (Figure 4.5). The weight-average molecular weight M_w can be determined using light-scattering techniques, ultracentrifugation or, as for M_n, by using the liquid chromatographic technique of SEC (*vide infra*).

Why are M_n and M_w important properties? It is the ratio of M_w to M_n (M_w/M_n; the *polydispersity* of the polymer) that gives an idea of the breadth of the molecular weight distribution and thus insight into the bulk properties of the polymer, including tensile strength, elasticity, hardness, and resistance to stress and cracking. For example, if a polymer contains many low-molecular-weight chains, those molecules can act like a plasticizer and soften the polymer. In contrast, higher-molecular-weight chains have a tendency to tangle more, increasing the viscosity of the polymer's liquid melt. If $M_w = M_n$, then the polymer is monodisperse, characteristically true in the case of natural polymers. Synthetic polymers range from narrowly to very broadly polydisperse, as shown below (Billmeyer 1984).

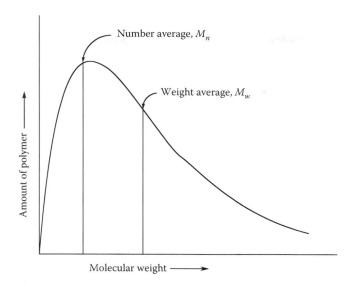

FIGURE 4.5 Distribution of molecular weights in a typical polymer. (Billmeyer, F.W. *Textbook of Polymer Science*, p. 17. New York: Wiley-Interscience. 1984. Copyright Wiley-VCH Verlag GmbH & Co. KGaA. Reproduced with permission.)

Type of Polymer	PDI
Monodisperse polymer	1.000
Actual monodisperse "living" polymer	1.01–1.05
Addition polymer (terminated by coupling)	1.5
Addition polymer (termination by chain transfer) or condensation polymer	2.0
High conversion vinyl polymers	2–5
Branched polymers	20–50

   The determination of polymer molecular weights and distribution of individual chain lengths are perfectly suited to the liquid chromatographic technique known as SEC. SEC (or the more specific term *gel permeation chromatography*, GPC, as it is referred to by material scientists) sorts particles (molecules) in solution according to their size. Like any liquid chromatographic technique, the method consists of a solid phase and a mobile phase, but unlike chromatographic techniques that rely on the adsorption of the analyte to the adsorbent, SEC separates the analytes in solution by mechanical means: the different-sized particles are sieved through a stationary phase "gel" (hence, *gel* permeation) consisting of cross-linked polymer beads. This gel is made up of millions of porous particles through which the dissolved analyte passes (see Figure 4.6). Larger polymers move through the gel more quickly as the smaller particles permeate the pores and are retained. Thus, polymers of different sizes exhibit different retention times, and a chromatogram that can be converted into a molecular mass distribution (MMD) results (Figure 4.7). The higher

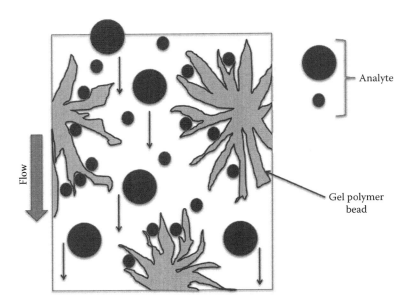

FIGURE 4.6   A cartoon of the gel permeation process.

the retention time of the molecular weight distribution, the lower the average molecular weight. The breadth and retention time(s) of the MMD are inextricably related to the physical properties of the final polymer product. It is important to note that, unlike the colligative measurements or light-scattering experiments that provide a direct measure of molecular weight, the use of GPC requires careful construction of a calibration curve using known molecular weight standards. If the polymer of

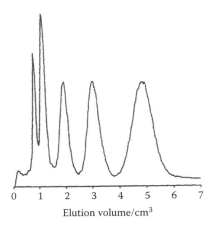

Elution volume/cm^3

FIGURE 4.7   Molar mass distribution from gel permeation chromatography. (From Nicholson, J.W. 2006. *The Chemistry of Polymers*. 3rd ed. Cambridge: Royal Society of Chemistry. Reproduced by permission of The Royal Society of Chemistry.)

interest and those used in the calibration curve are different, this method can only provide relative molecular weight values.

4.2 SYNTHESIS

Given that the properties of a polymer are related to the distribution of the individual chains that make up the bulk sample, a quick overview of the synthesis of polymers is necessary to help us understand how and why the dispersity of these materials can vary. There are several types of polymer synthesis but we will primarily focus on only the two most general: step or chain growth. A third area, *coordination polymerization*, takes advantage of transition metal organometallic catalyst systems. While it is an exceptionally powerful method for the polymerization of alkenes, it is arguably less relevant in the context of sustainable energy processes and so will not be covered in this text.

4.2.1 STEP-GROWTH POLYMERIZATION

Step-growth polymerizations are usually (but not always) condensation reactions, where a small molecule is expelled in the reaction between monomers. For example, the synthesis of polyethylene terephthalate (PET, a plastic often used in carbonated beverage bottles) is a typical esterification reaction in which water is expelled (Figure 4.8). The only difference between this polymerization and a small-molecule esterification is that the reacting species are polyfunctional, allowing for multiple reaction sites. The nature of the mechanism for step-growth polymerizations means that the polymer molecular weight will increase in a slow, step-like manner as the reaction proceeds. Hence, dimers form initially, then dimers can link to form tetramers, tetramer joins tetramer to form octomers, and so on. Of course, the real scenario is not so strictly regimented; a monomer can react with a dimer or a tetramer with a dimer, and so on. In any case, in order for a high-quality, high-molecular-weight polymer to be obtained, enough time must pass to allow the reaction to proceed to completion (or nearly so).

Polycarbonates (Figure 4.9a) are examples of polymers that can be synthesized by step-growth methods. These polycarbonates prepared with aromatic monomers such as bisphenol A (BPA, Figure 4.9b) are tough and transparent "engineering plastics" used as lenses, in CDs and DVDs, as construction materials, and as components in automobiles and aircraft (Pescarmona and Taherimehr 2012). The C_1 unit of polycarbonates makes them excellent prospects for the utilization of captured CO_2 (Section

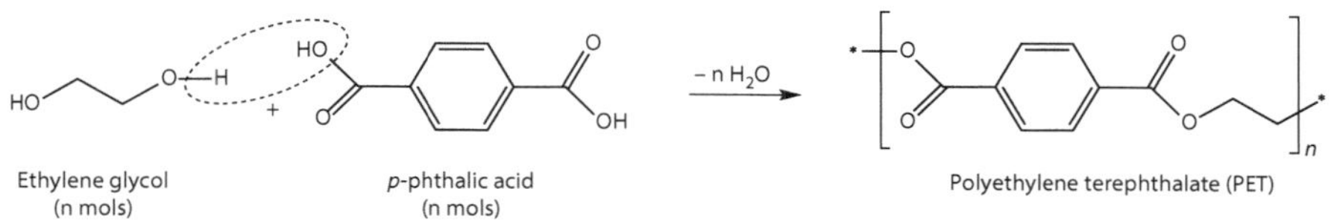

Ethylene glycol (n mols)   p-phthalic acid (n mols) $-n\,H_2O$ Polyethylene terephthalate (PET)

FIGURE 4.8 Synthesis of polyethene terephthalate via condensation.

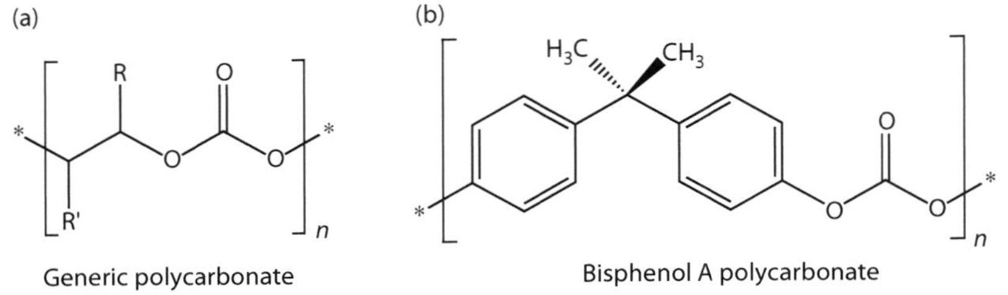

(a)   (b)

Generic polycarbonate   Bisphenol A polycarbonate

FIGURE 4.9 Examples of polycarbonate structures.

2.4.2.2) and indeed, an environmentally friendly (i.e., green; Section 4.6) industrial synthesis of polycarbonate has been developed via the scheme shown in Figure 4.10 (Fukuoka et al. 2003). Not only can this synthesis use captured CO_2 as one of the starting materials, it replaces the use of the highly toxic C_1 reagent, phosgene (ClC(O)Cl). The chemists developing this industrial route found that it was difficult to make the polymerization progress beyond a degree of polymerization of about 20 due to an extreme increase in the viscosity of the reaction mixture as the polymerization continued (a DP of 30–60 is required for the polycarbonate to possess the desired properties). This problem was remedied by developing a unique solid-state polymerization process utilizing a novel gravity-fed reactor, making this industrial process of polycarbonate manufacture especially benign (Fukuoka et al. 2003). It is worth noting here that aliphatic polycarbonates can be prepared from the reaction of CO_2 plus any of a number of epoxides. However, these polymers have lower rigidity and poorer thermal stability than aromatic polymers such as the BPA copolymer described above (Kember et al. 2011).

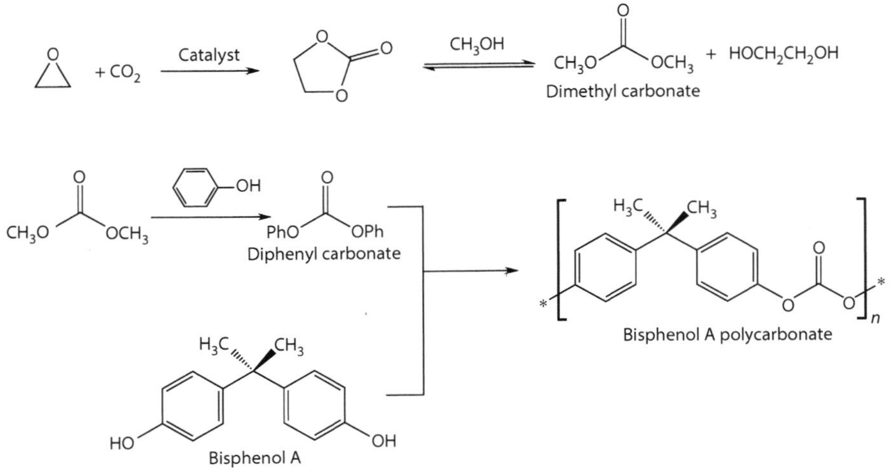

FIGURE 4.10 A green polycarbonate synthesis using captured CO_2.

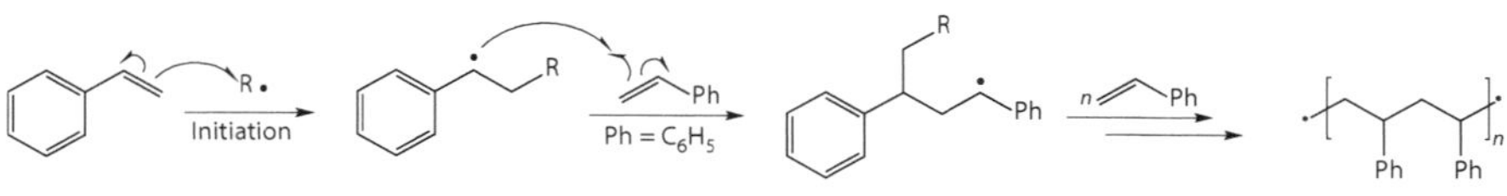

FIGURE 4.11   Polymerization of styrene by radical chain growth.

4.2.2   CHAIN-GROWTH POLYMERIZATION

Chain-growth polymerization is mechanistically distinct from stepwise polymerization. A classic example of chain polymerization is the radical-initiated polymerization of styrene to make polystyrene, well known as the familiar foam carry-out container (Figure 4.11). A wide variety of olefin monomers can be polymerized by the radical chain mechanism. As with all chain reactions, the reaction proceeds by three mechanistic steps: initiation, propagation, and termination. Unlike stepwise polymerizations, in chain-growth polymerizations, the monomer itself is generally unreactive: some sort of initiation is required to form the reactive intermediate. Note that chain polymerization can proceed via anionic (R–) or cationic (R+) intermediates as well as by radicals. For example, the synthesis of poly(p-phenylenevinylene) (PPV), a conjugated polymer with the potential for use in organic photovoltaics, is shown in Figure 4.12 (Cosemans et al. 2011). When the reaction is carried out in tetrahydrofuran, it proceeds by the anionic mechanism shown.

A significant difference between step-growth and chain-growth polymerization is that in chain polymerization, high-molecular-weight chains are formed virtually instantaneously as the monomer is rapidly added sequentially to the reactive terminus of each growing chain during the propagation. Thus while the *yield* increases with time, the molecular weight distribution is fairly stable over the course of the reaction. The polymerization is complete when all of the monomer is consumed or the reaction terminates. Termination can occur by the addition of a quenching agent or the active radicals can dimerize or react in a chain transfer reaction (see Figure 4.13). In contrast, in step-growth polymerization, the molecular weight increases steadily over the course of the reaction.

4.2.3   BLOCK COPOLYMERS AND CO$_2$ SEPARATION

It is often desirable to modify the properties of a polymer by incorporating other monomers into the polymer product to give a copolymer. This can take place during either the synthesis (copolymerization) or postsynthesis (by grafting additional molecular material to the already prepared polymer, giving a graft copolymer). Copolymers can be random (e.g., A–B–B–A–A–B–A–B–A–B–A–A–A), regular (e.g., A–B–B–A–B–B–A–), or block (e.g., A–A–A–A–B–B–B–B–). Block copolymers can be diblock, triblock, or even multiblock. The sulfonated polybenzophenone/poly(arylene ether) shown in Figure 4.14 is an example of a block copolymer synthesized for use in polymer electrolyte membrane fuel cells. The block nature of the copolymer leads to hydrophobic and hydrophilic domains that can lead, as we will see in Chapter 6, to good fuel cell performance (Miyahara et al. 2012).

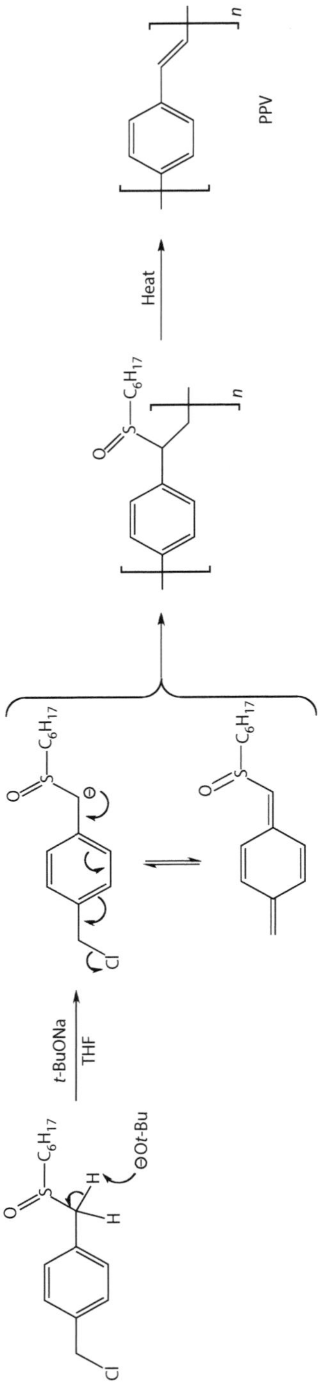

FIGURE 4.12 An example of an anionic chain-growth polymerization.

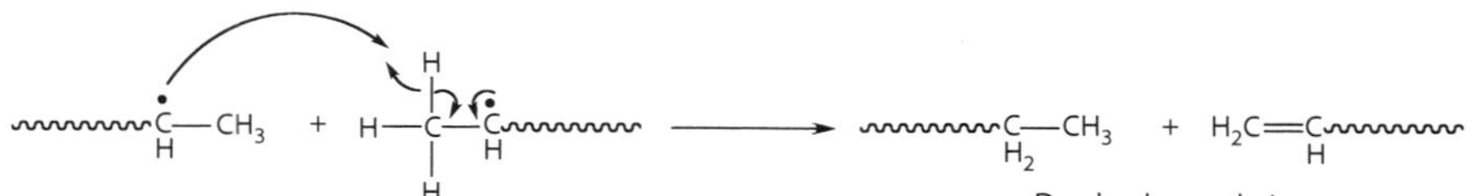

FIGURE 4.13   Termination via chain transfer.

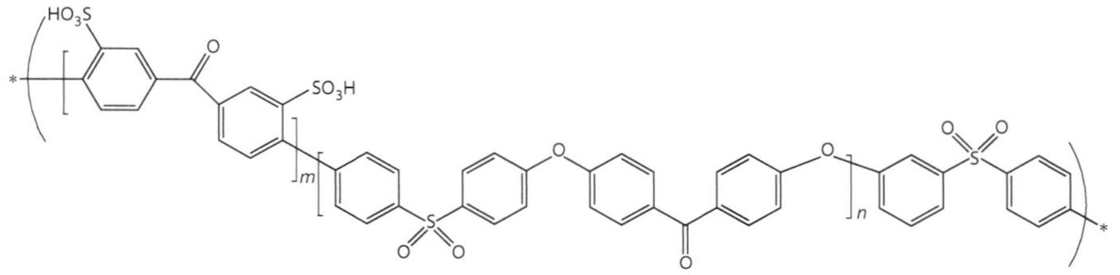

FIGURE 4.14   A block copolymer.

The feature of differing domains in block copolymers has led to the investigation of their use in gas separation membranes for CO_2 separation and capture. Recall from Section 2.4.1.1 that the effectiveness of a gas separation membrane is dependent upon both the gases' diffusion coefficient (based on permeability) and their solubility in the membrane material. The permeability of the membrane is very strongly impacted by the structure of the polymer: by altering the three-dimensional shape of the polymer chains, more tight or less tight packing between individual polymer chains can result. For example, the synthesis of a variety of polyimides used for CO_2/CH_4 separation yielded a large difference in the permeability and hence the selectivity for this particular application (Table 4.1).

The so-called PEBAX® membranes (a block copolymer consisting of polyamide and polyether blocks, Figure 4.15) are of particular interest in the CO_2 separation area because the polyamide segments provide rigidity (due to interchain hydrogen bonding), while the polyether segments' flexibility allows for greater gas permeability. Furthermore, CO_2 can interact strongly with the polyether domain leading to very large solubility coefficients for the gas. PEBAX derivatives have shown $\alpha(CO_2/N_2)$ of 48 and $\alpha(CO_2/H_2)$ of 30 at –20°C (Lin et al. 2006; Nguyen et al. 2010).

Addition of a wide variety of nanoscale fillers can also have a large impact on the performance of polymeric gas permeable membranes as a result of increased permeability. Carbon nanotubes (CNT; Figure 4.16), for example, have been shown to greatly improve the permeability of PEBAX membranes—incorporation of 2% multiwalled carbon nanotubes into the PEBAX polymer membrane (Figure 4.17) resulted in a sixfold increase in CO_2 permeability. Subsequent cross-linking with toluene 2,4-diisocyanate (Figure 4.18) reduced the permeability, but the selectivity of both CO_2/N_2 and CO_2/H_2 was increased with cross-linking, making these membranes promising materials for further study ($\alpha CO_2/N_2$ of 83.2 at 2% CNT and $\alpha CO_2/H_2$ of 1.4 at 2% CNT, respectively (Murali et al. 2010)).

TABLE 4.1

Permeability and Selectivity: Impact of Connector Groups

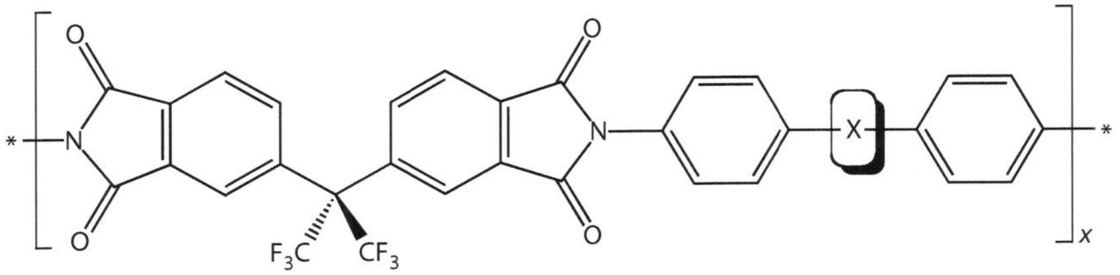

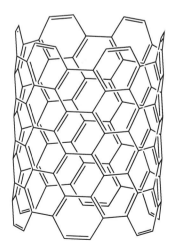

6FDA

Connector Group (X)	$P(CO_2)$, Barrer[a]	$\alpha(CO_2/CH_4)$
$-O-$	23	60.5
$-CH_2-$	19.3	44.9
$-C(CH_3)_2-$	30	42.9
$-C(CF_3)_2-$	63.9	39.9

Source: Reprinted with permission from Yampolskii, Y. 2012. Polymeric gas separation membranes. *Macromolecules* 45 (8):3298–3311. Copyright 2012, American Chemical Society.

[a] Barrer is the unit used to express the permeability coefficient of gases in polymeric membranes.

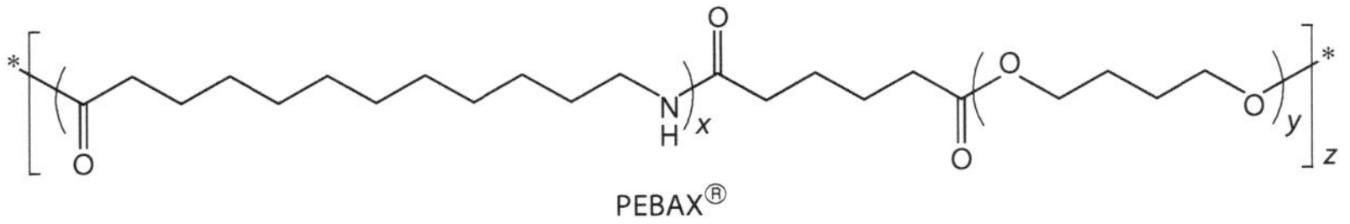

PEBAX®

FIGURE 4.15   PEBAX.

FIGURE 4.16   A single-walled carbon nanotube.

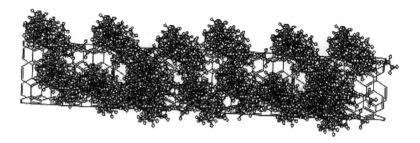

FIGURE 4.17 A hypothetical representation of the PEBAX polymer wound around a carbon nanotube. (Reprinted with permission from Murali, R.S., S. Sridhar, T. Sankarshana et al. 2010. *Ind. Eng. Chem. Res.* 49 (14):6530–6538. Copyright 2010. American Chemical Society.)

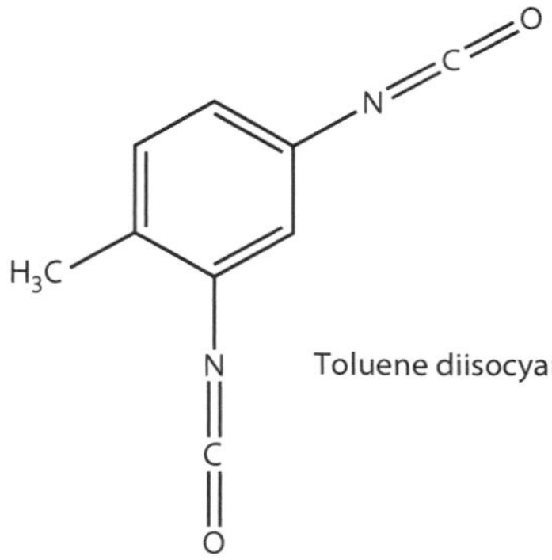

Toluene diisocyanate

H_3C

FIGURE 4.18 Toluene 2,4-diisocyanate, a cross-linking agent.

4.2.4 CONTROL IN POLYMER SYNTHESIS

Natural polymers are the ultimate example of control in polymer synthesis: proteins, DNA, and RNA are synthesized with essentially 100% precision in their molecular architecture, a polydispersity index (PDI) of 1.00 and a precise molecular weight. Synthetic polymer chemists, in contrast, must wrestle with many different aspects of control. For example, in a step-growth polymerization, does the initially formed dimer react sequentially with additional monomer or randomly (Figure 4.19)? If a regular A–B–A–B polymer is desired, then the two monomers—BPA and diphenyl carbonate—must alternate sequentially. With polyfunctional monomers (such as a trisphenol), the picture is even more complex. Furthermore, polycarbonates formed from the reaction of CO_2 with unsymmetrical epoxides illustrate the problem of *regiocontrol*. For example, in the preparation of poly(propylene)carbonate, polymerization can take place to form three different types of *regioregular* couplings (head-to-head, head-to-tail, or tail-to-tail, see Figure 4.20) or the coupling between the two monomers can be completely random (leading to a *regiorandom*) polymer. Each of these isomeric polymers

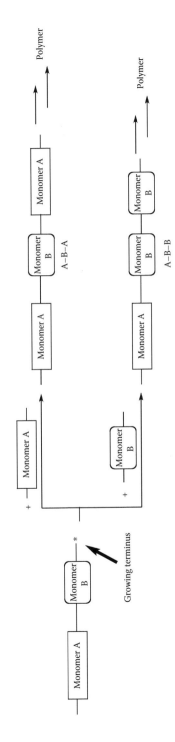

FIGURE 4.19 Possible outcomes in step-growth polymerization. (Reprinted with permission from Murali, R.S. et al. 2010. Gas permeation behavior of Pebax-1657 nanocomposite membrane incorporated with multiwalled carbon nanotubes. *Ind. Eng. Chem. Res.* 49 (14):6530–6538. Copyright 2010, American Chemical Society.)

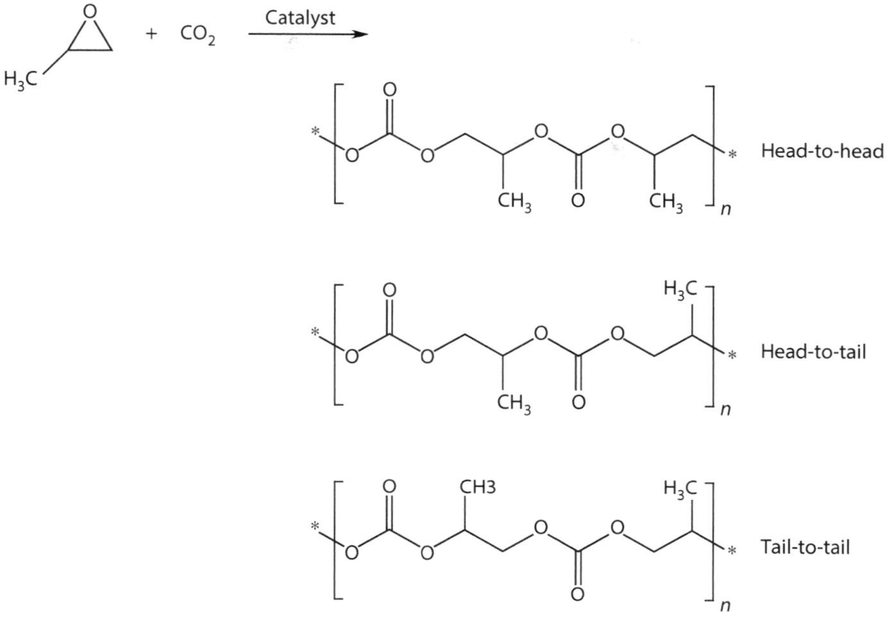

FIGURE 4.20 Regiochemistry of polycarbonate polymerization.

will exhibit differing properties. In the case of poly(propylene) carbonate, the regioregular head-to-tail polymer is desired as it possesses the highest crystallinity, thermal stability, toughness, and stiffness (Kember et al. 2011). As we will see in the photovoltaics chapter (Chapter 7), the regioregularity by which monomers link together in a conjugated polymer affects the overlap of the molecular orbitals and hence the electronic properties of the polymer (and, therefore, ultimately the efficiency of the photovoltaic device).

Chemists are making amazing progress in controlling the size and shape of polymers, incorporating specially designed substructures in copolymers to create cylindrical brush polymers, for example, and utilizing specific functionality to control polymer folding, much like protein folding (Stals et al. 2013). This very active area of research is known as "controlled polymerization" and subsumes the field of "living polymers." The International Union of Pure and Applied Chemistry (IUPAC) recommends that living polymerization be defined as "a chain polymerization from which irreversible chain transfer and irreversible chain termination (deactivation) are absent." Thus, if a chain polymerization is not quenched, a "living polymer" results, and addition of more monomer at a later time allows polymer growth to begin anew (Penczek 2002). This presents attractive options for controlling molecular architecture: addition of a different monomer to the living polymer allows for the formation of regular block copolymers, for example, or varying reaction conditions to fine tune the molecular weight. In addition, if initiation is rapid relative to propagation, a nearly monodisperse polymer can result.

Several methods of *controlled radical polymerization* (CRP) have resulted in polymers with very low polydispersity indices. These methods rely on careful manipulation of the reaction conditions as evidenced by their names: atom transfer radical polymerization, nitroxide-mediated polymerization, reversible addition-fragmentation transfer polymerization, and copper(0)-mediated single-electron transfer polymerization. Living/controlled polymerization processes are a very promising area of research for the design of well-defined polymers that have many applications in the realm of sustainable energy and beyond.

4.3 CHARACTERIZATION OF POLYMERS

As we will soon see, there are specific tests for all kinds of polymer properties, including thermal, electronic, and surface properties. Those properties are, ultimately, dependent upon the structure of the polymer, so knowing the structure is critical to predicting and designing its properties. We have already seen how SEC can provide information about molecular weight, but given the complexity of these macromolecules, how can we know their actual connectivity and functionality? As for other areas of structural characterization, spectroscopy is our primary tool for elucidating polymer structure. Spectroscopic techniques are very useful for revealing dynamic properties of the polymer and even configurational and conformational structure, but for our purposes we will limit the discussion to molecular connectivity and functional group analysis.

The vibrational spectroscopic techniques of infrared and Raman spectroscopy are frequently used in polymer characterization. Just as for small molecules, Fourier transform infrared spectroscopy is useful for functional group analysis of the polymer. Hence the IR spectrum of polyvinyl alcohol reveals the strong, broad O–H stretch at 3380 cm^{-1} as well as the attendant C–O stretching from 1000 to 1500 cm^{-1}, similar to any alcohol (Figure 4.21). Perhaps one of the most useful applications of FT-IR in polymer analysis is its coupling to GPC, allowing for the analysis of each separate fraction in a polymer mixture as it is eluted and detected.

Thin (0.001–0.005 mm) films cast from solution or hot pressed are often used for IR analysis (a thin film of polystyrene has long been provided as a convenient sample for the calibration of IR spectrometers, as the carbon–carbon bond stretching vibration at 1601 cm^{-1} is strong and well resolved). Newer techniques such as attenuated total reflectance (ATR) and diffuse reflectance infrared Fourier transform spectroscopy (DRIFTS) allow for the analysis of different forms of polymer samples, including powders and surface samples. Raman spectroscopy complements IR in that modes that are active in IR may be inactive in Raman spectroscopy (and vice versa).

The fact that polymers are typically insoluble solids initially hampered in-depth analysis by NMR spectroscopy. However, the advent of FT-NMR, higher magnet strength, and advanced techniques have led to better analysis of polymer structure and morphology. In some instances, valuable information can be obtained without resorting to complex techniques: in the synthesis of poly-3-hexylthiophenes (P3HT), coupling can take place in a regular or random fashion leading to a variety of isomers with unique properties. Just as for the polycarbonate example discussed in

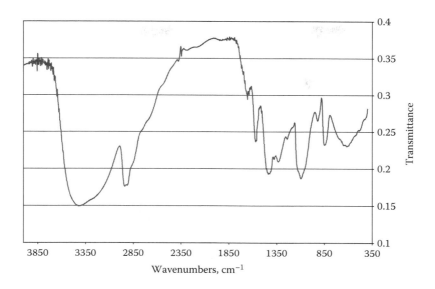

FIGURE 4.21   The infrared spectrum of polyvinyl alcohol.

Section 4.2.1, P3HT polymers can be *regioregular* (e.g., containing only head-to-tail linkages; Figure 4.22) or *regiorandom* (with a mix of head-to-head, head-to-tail, and tail-to-tail couplings). The regioselectivity of the coupling can be controlled by varying the synthetic method and can be easily estimated by examining and integrating the aromatic region in the ^{1}H NMR spectrum (see Figure 4.23 (Pappenfus et al. 2010)).

For insoluble polymers, the introduction of magic-angle spinning (MAS)—a technique whereby the solid samples spin at the "magic" angle of 54°44′—has led to relatively well-resolved NMR spectra that can provide information about internuclear interactions in the bulk state. For example, a detailed ^{1}H NMR study of benzoxazine dimers (as models for polybenzoxazines) utilized MAS at the fast spin rate of 35 Hz to determine the extent of inter- and intramolecular hydrogen bonding in the solid state and the melt for a series of alkyl-substituted dimers (Figure 4.24 (Schnell et al. 1998)). Although beyond the scope of this text, advanced NMR techniques such as double quantum filters, spin diffusion, and back-to-back recoupling sequences

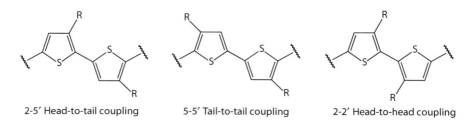

FIGURE 4.22   Regioselectivity of coupling for polyalkylthiophenes.

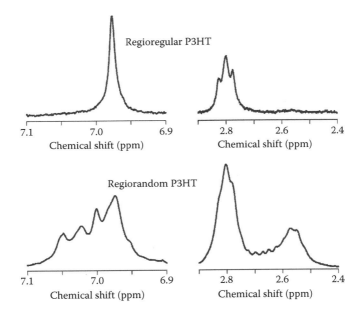

FIGURE 4.23 Difference in the aromatic region of the ^{1}H NMR spectra for P3HT isomers. (Reprinted with permission from Pappenfus, T.M. et al. 2010. Regiochemistry of poly(3-hexylthiophene): Synthesis and investigation of a conducting polymer. *J. Chem. Ed.* 87 (5):522–525. Copyright 2010, American Chemical Society.)

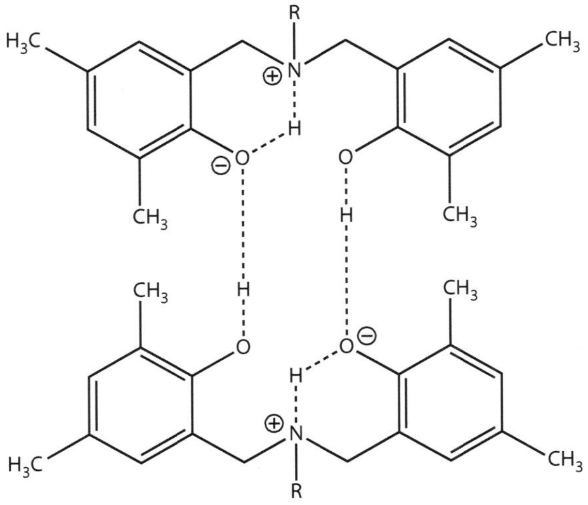

R = methyl, ethyl, *n*-propyl, and *n*-butyl

FIGURE 4.24 Hydrogen bonding in benzoxazine dimers.

(BABA) can aid in structure determination as well as in determining intermolecular distances and even information about the length of various domains (e.g., crystalline vs. amorphous) within the bulk sample (Cherry et al. 2005). NMR spectroscopy can even be used to determine the glass transition temperature of a polymer by measuring changes in the line shape (Carraher 2000).

Numerous other instrumental techniques have been used for polymer characterization, including UV-visible spectroscopy, x-ray diffraction, and electron spin resonance. Thermal analysis of polymers, particularly differential scanning calorimetry (DSC) and thermogravimetric analysis, is commonly used and provides data on the enthalpy and entropy changes, the glass transition temperature (T_g), and other properties. The reader is referred to specialized texts for further information (see, e.g., Campbell et al. 2000).

4.4 POLYMER PROPERTIES

In polymer science, the ultimate goal is to prepare a polymer with properties finely tuned to meet the requirements for the application in mind. Thermal, chemical, electronic, and mechanical properties are all critically important aspects of polymer behavior in terms of practical use. As we have already seen, the strength of the intermolecular forces between polymer chains plays a key role in the physicochemical properties of the polymer (e.g., thermoplastics vs. thermosetting polymers). Hydrocarbon polymers such as polyethylene and polypropylene are limited to weak van der Waals intermolecular forces, while Kevlar, an aromatic polyamide (recall Figure 4.4), is very highly oriented due to intermolecular hydrogen bonding between the nitrogen–hydrogen bond and carboxyl oxygen as well as the interchain π-stacking afforded by the benzene rings. As a result, aramids such as Kevlar demonstrate exceptional strength.

Branching, too, strongly influences the physical and mechanical properties of the polymer. The difference between pliable low-density polyethylene (LDPE, used in films and plastic bags) and stiffer high-density polyethylene (HDPE, most commonly found in bottles and containers) is the degree of branching (Figure 4.25). Even stereochemistry impacts the physical properties of a macromolecule (obvious examples of this concept can be found in biochemistry, as in enzyme activity). The stereochemical arrangement of substituent groups on a polymer chain is known as its tacticity. For example, three different forms of polypropylene are shown in Figure 4.26. *Isotactic* polypropylene is highly crystalline and melts at approximately 160°C, while *atactic* polypropylene is a soft, amorphous polymer.

Note that regular order—both connective and stereochemical—is essential for crystallinity, and strong intermolecular forces contribute to a polymer's tendency to exhibit crystallinity. But crystallinity may not be evenly distributed throughout the bulk material: some areas may be glassy and amorphous, while others show partial crystallinity. Polymers that are amorphous are translucent due to the absence of crystallites that would otherwise scatter the light.

Crystallinity also impacts thermal behavior. An important measure of polymer properties is the *glass transition temperature*, T_g. As with any molecule, upon heating

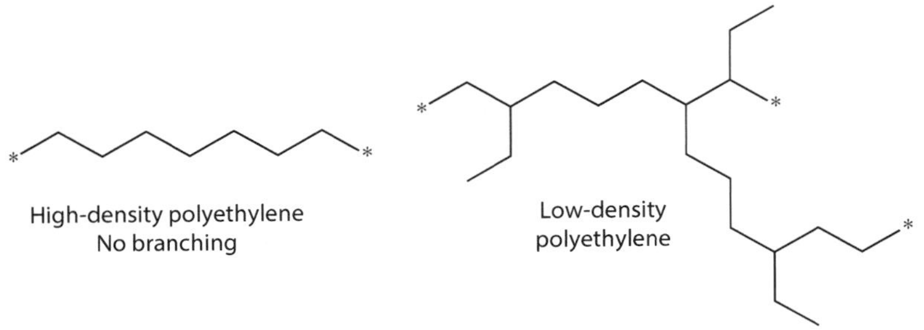

High-density polyethylene
No branching

Low-density
polyethylene

FIGURE 4.25 High-density versus low-density polyethylene.

the kinetic energy in the polymer increases allowing the individual molecules to become more mobile and approach the liquid state. However, most—but not all—polymers do not melt. Instead, the stiff, glass-like state of the solid polymer transforms sharply to a more mobile, almost rubbery state. The temperature at which this occurs is the glass transition temperature. The glass transition temperature for isotactic polypropylene is 100°C (373 K) but that of atactic polypropylene is –20°C (253 K). Unlike melting, this observable change is *not* a phase change. As the temperature is decreased below T_g, intermolecular motion ceases and the properties of the polymer become glass-like (hard, brittle, and transparent). The glass transition temperature can be determined using the method of DSC wherein a sample of the polymer is heated and the heat flow is measured as a function of time and temperature. The result is a DSC scan that can reveal T_g as well as several other thermal properties of the polymer.

Perhaps the most dramatic and tragic illustration of the significance of the glass transition is the Challenger disaster. On that cold January morning in 1986, an "o-ring failure in the right solid rocket booster" resulted in a massive explosion that claimed the lives of the seven astronauts aboard the shuttle. According to the NASA mission archives, "cold weather was determined to be a contributing factor"—in other words, the o-ring failure was a result of the operating temperature having fallen below the material's T_g (National Aeronautics and Space Administration 2007). In another

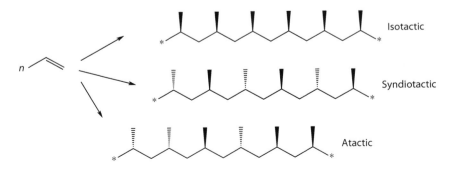

FIGURE 4.26 Stereoregularity of polypropylene isomers.

example, the T_g for polyalkylthiophenes (recall Figure 4.22) varies with the substitution on the thiophene ring, generally decreasing in a linear fashion with increasing carbon chain length (Ewbank et al. 2009). This can have an important impact on the properties of solar cells, which use this compound (Chapter 7). The glass transition temperature also strongly impacts the selectivity of membrane separation of CO_2 from hydrogen gas: a glassy polymer below its T_g will preferentially allow the smaller hydrogen molecules (relative to CO_2) to permeate the membrane, while a polymer above its T_g will allow more CO_2 to permeate (relative to H_2) because of the increased solubility of CO_2 in the membrane (Merkel et al. 2012).

In the case of immense wind turbine blades, the most important requirement for the polymer is strength, a feature of macromolecules that may be amplified by cross-linking. Covalent bonds are formed between individual polymer chains, either during the initial polymerization process or in a subsequent step, forming a strong three-dimensional network. Cross-linking introduced after the initial polymer is formed is known as *curing*. Uncross-linked polymers are typically thermoplastic. In contrast, cross-linked polymers do not melt or flow with heating due to the plethora of bonds between the polymer chains. In fact, "chains" is a misnomer here: a cross-linked polymer is more like a net or a three-dimensional lattice. While it may seem intuitive that more cross-linking means greater material strength, this is not necessarily the case. A highly cross-linked material has little conformational flexibility for relief of stress. As a result it can behave as a brittle material and fail catastrophically with minimal applied force.

Strength is one of many polymer properties that is important to quantify, but it is a complicated picture. How easily will this material break? *How* will it break—slowly, with elongation, or in a sharp snap? At what temperature will the material become brittle? Clearly, these are important questions to answer in designing a polymer for a wind turbine blade, for example! And, as usual, an in-depth discussion of the physical details of these determinations will not be covered here; interested readers can turn to any polymer chemistry text. Instead, we focus on tensile measurements (as opposed to compression or shear stress) and stress–strain curves as illustrations of the mechanical properties of a macromolecule.

Stress–strain curves provide a convenient snapshot of the mechanical behavior of a polymer when stretched. The *tensile strength* is the ability of the polymer to withstand a pulling stress: a force is applied to a sample of the material at a steady rate and specific temperature until it breaks. Measured in pascal, the tensile strength of a material is dependent upon the form of the material as seen in the following table in the case of graphite, where a "whisker" is a single crystalline fiber (Carraher 1996).

Graphite Form	Tensile Strength (MPa)
Bulk	1.00
Fiber	1.01–1.05
Whisker	1.5

Stress (force/unit cross-sectional area) is also measured in pascal, and since the force involved is one of elongation, the strain is dimensionless: the length of stretched

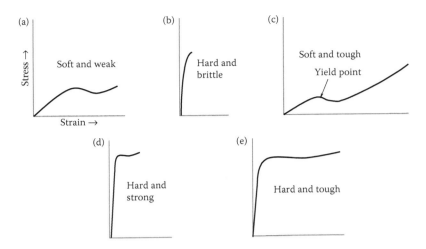

FIGURE 4.27   Representative tensile stress–strain curves for polymeric materials.

sample in mm is compared to the length of the original sample, also in mm. This ratio of stress to strain is known as Young's modulus (Equation 4.5).

$$\text{Young's modulus} = \frac{\text{stress (pascal)}}{\text{strain (mm/mm)}} \qquad (4.5)$$

The higher the value for Young's modulus, the more rigid the material and the more resistant it is to stretching. The stress–strain curve seen in Figure 4.27a is typical for a soft, weak polymer. The point at which the material yields to the stress is the *yield point*. Soft materials have low modulus and low yield point. The stress–strain curve for a hard, brittle polymer is shown in Figure 4.27b: note that the material breaks quickly, with little elongation and a high modulus. Stress–strain curves for hard and tough polymers are also shown in Figure 4.27; their categorization depends upon the amount of stress applied before the yield point.

For high-performance fibers such as the carbon fibers used in polymer matrix composites, an additional descriptor of strength is used: *tenacity*. This is a relative strength given in grams of breaking force per denier unit, where "denier" is the system used to measure the weight of the continuous-filament fiber. A low number indicates a finer fiber (Committee on High-Performance Structural Fibers for Advanced Polymer Matrix Composites 2005).

4.5   POLYMER CHEMISTRY AND WIND ENERGY

4.5.1   INTRODUCTION

As noted in the introduction to this chapter, polymer chemistry infuses the technology behind renewable/sustainable energy. At this point, we will focus on the fabrication of the massive polymer composite wind turbine blades that have allowed wind power to grow rapidly as a sustainable energy source. Additional applications will be covered as they are presented in later chapters.

When it comes to harnessing wind to create work, we have come a long way and polymer chemistry has played a huge role in that progress. Wind power has obviously been around since sailing ships and windmills were common in Europe in the 1700s. But, in the United States, creation of an energy grid and cheap fossil fuels led to the disappearance of the functioning farmstead windmill over the course of the nineteenth century. Only recently have technology and the view of wind power as a clean energy alternative led to the resurgence of wind energy, primarily in the form of modern wind turbines and expansive wind farms, both onshore and offshore. According to the American Wind Energy Association, in 2012 wind energy made up more than 40% of new electrical generating capacity in the United States (Rich 2013).

Historically, windmill blades were made of wood, wood and cloth, or metal (steel or aluminum). The weight and strength limitations of these materials limited the size of the blade and, hence, the maximum power output. Advancements in polymer chemistry have paved the way for the development of the large horizontal axis wind turbine (HAWT) for the generation of power over the past two decades (Figure 4.28; Figure 4.29 shows the nomenclature associated with the modern HAWT). Recent work has shown that the greater the capacity of the turbine, the greener the electricity: global warming potential per kWh for wind-generated electricity was reduced by

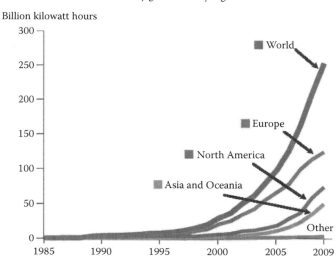

Wind electricity generation by region: 1985–2009

Billion kilowatt hours

Other: This category includes South and Central America, Eurasia, the Middle East, and Africa.

FIGURE 4.28 Growth in wind electricity generation. (From U.S. Energy Information Administration. 2013. *International Energy Statistics*. U.S. EIA 2011 [cited 2 July 2013]. Available from http://www.eia.gov/cfapps/ipdbproject/IEDIndex3.cfm?tid=6&pid=29&aid=12.)

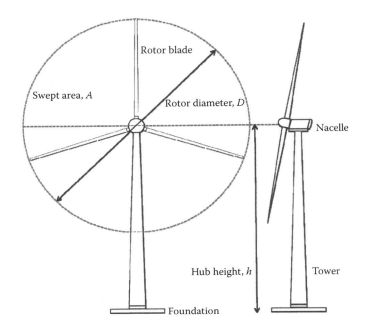

FIGURE 4.29 Nomenclature of wind turbines. (Reprinted with permission from Caduff, M. et al. 2012. Wind power electricity: The bigger the turbine, the greener the electricity? *Environ. Sci. Technol.* 46:4725–4733. Copyright 2012, American Chemical Society.)

14% with every cumulative production doubling (Caduff et al. 2012). Today's turbine blades are as long as 75 m (for comparison, the wingspan of an Airbus A380 is 80 m) and can weigh around 20 metric tons. The evolution of the modern wind turbine has been rapid, with small 50 kW machines being dwarfed by the 2–3 MW standard of today (Figure 4.30). The Enercon E126, rated at 7.58 MW, is one of the world's largest and has a hub height of 135 m with a rotor diameter of 127 m. It is predicted

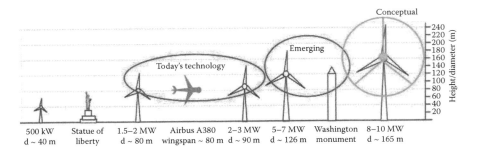

FIGURE 4.30 The evolution of wind power. (Reprinted with permission from Merugula, L., V. Khanna, and B.R. Bakshi. 2012. Reinforced wind turbine blades—An environmental life cycle evaluation. *Environ. Sci. Technol.* 46:9785–9792. Copyright 2012, American Chemical Society.)

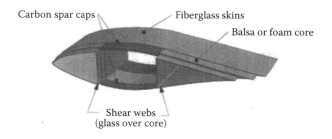

Carbon spar caps · Fiberglass skins

Balsa or foam core

Shear webs
(glass over core)

FIGURE 4.31 Structure of a wind turbine blade. (From Griffin, D.A. and T. Ashwill. 2009. Blade system design study part II: Final project report (GEC). Albuquerque, NM, US Department of Energy, 152.)

that by 2025, rotor diameters will be 160–165 m with blades of 100–125 m in length (New Energy Externalities Developments for Sustainability 2008).

This amazing growth in the length and strength of wind turbine components is a result of advances in materials science, including the development of polymer matrix composites. Polymer composites are strong, cross-linked thermoset polymer resins laminated with reinforcing fiber (usually glass or carbon, *vide infra*). The combination of the compressive and tensile strength of the fibers with the moldable properties of the polymer makes the combined matrix much stronger than the individual components, allowing for the dramatic growth of wind power as turbines improve in both power rating and efficiency. Today's ever-larger wind turbine blades must be tough and stiff but not brittle, with the blade overall exhibiting reasonable flexibility (the blades must not bend to the degree that they will hit the tower while rotating!). A typical turbine blade is an integrated whole made up of an overlaying shell, a foam core, and the structural spar that runs the length of the blade, with some form of protective skin that coats the blade (see Figure 4.31).

In addition to increasing the overall dimensions of the rotor diameter by manufacturing ever-larger blades, tower height is another focus for improvement in wind turbine efficiency (Giannis et al. 2008). Thus, wind turbines keep getting bigger (in terms of *swept width*) and taller. Increasing tower height has two advantages: there is less turbulence from interactions at ground level, and longer and better designed blades that have a larger sweep area can be used. In any case, these improvements are a result of the continuing evolution of the basic materials that make up these structural components: the polymer resins and the reinforcing fibers.

4.5.2 RESINS

Most of the polymers used in the manufacture of turbine blades are thermosetting polyester, vinyl ester, or epoxy resins. Orthophthalic or isophthalic unsaturated polyester resins (see Figure 4.32) incorporate alkene functionality that provides additional reactive sites for cross-linking. Addition polymerization is carried out in styrene to allow cross-linking (curing) to take place between the styrene monomer and the alkene units of the unsaturated polyester (Figure 4.33).

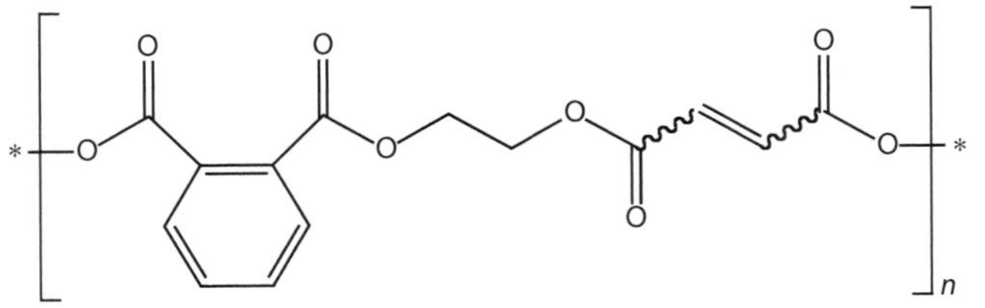

Orthophthalic polyester
(isophthalic = meta isomer)

FIGURE 4.32   Phthalic polyester resin structure.

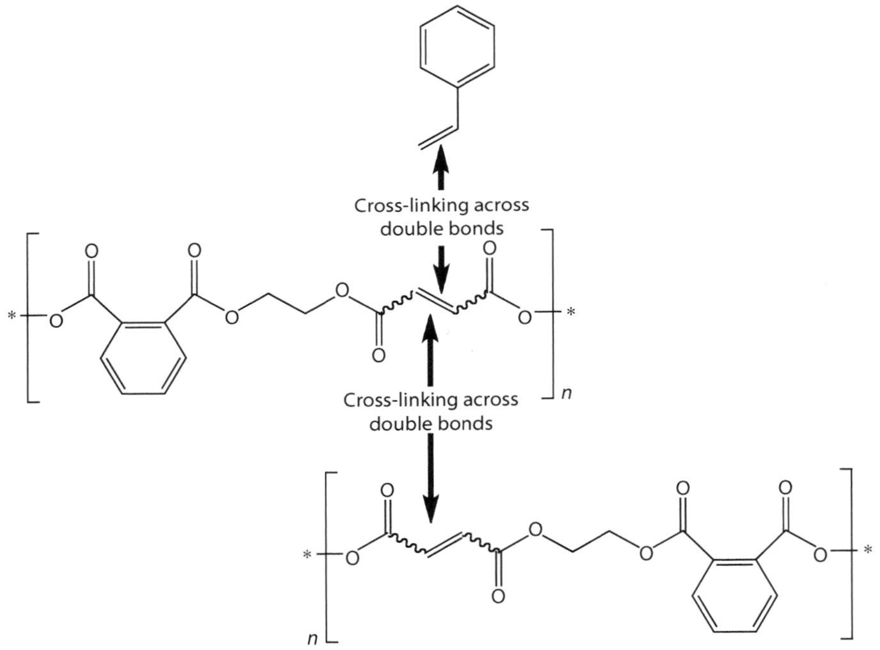

FIGURE 4.33   Sites for cross-linking between styrene and an unsaturated polymer.

Polymerization of vinyl esters (where the reactive vinyl substituent is at the ends of the main polymer chain) leads to a resin that is more resistant to hydrolysis due to the smaller proportion of ester linkages (Figure 4.34). Both unsaturated polyester and vinyl ester cross-linked resins suffer from shrinkage during the curing process, a fate to which epoxy resins are less prone. In addition, the absence of ester functionality makes these types of polymers especially resistant to hydrolysis, an important consideration in exterior applications such as turbine blades. Figure 4.35 shows an example of a bisphenol A epoxy prepolymer resin, prepared from the step-growth polymerization of bisphenol A to epichlorohydrin. Addition of a polyfunctional

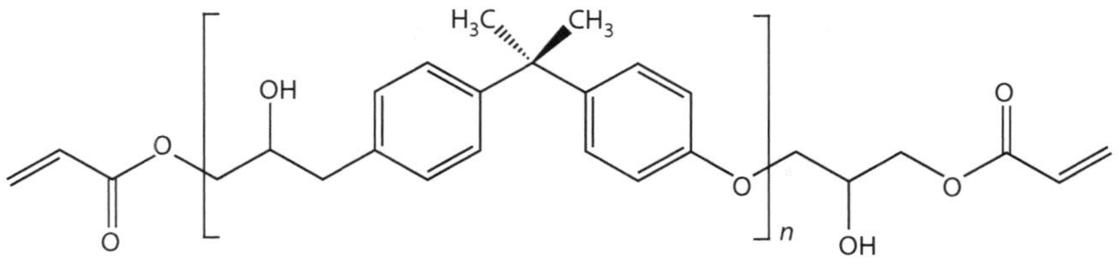

FIGURE 4.34 Example of a vinyl ester polymer.

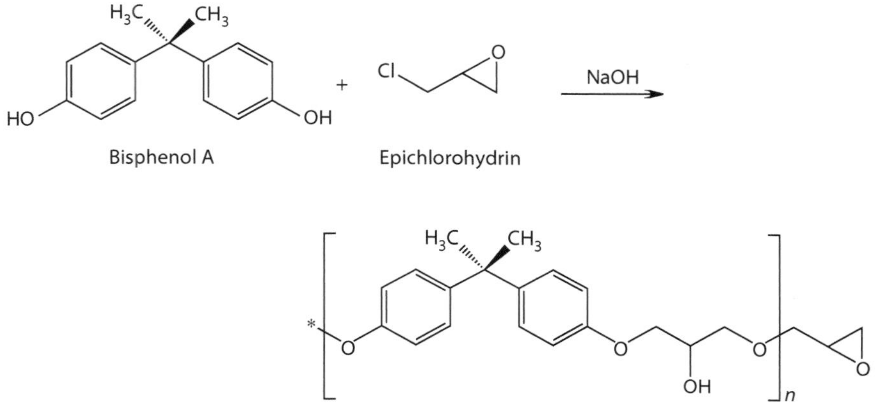

Bisphenol A   Epichlorohydrin

FIGURE 4.35 Epoxy resin made from epichlorohydrin and bisphenol A.

amine hardener such as diethylenetriamine ($H_2NCH_2CH_2NHCH_2CH_2NH_2$) cures the resin, leading to a throughly cross-linked, strong, and tough polymer network.

In an effort to generate a somewhat renewable version of an epoxy resin, researchers prepared a polymer composite made from an epoxidized soybean oil and 1,1,1-tris(p-hydroxyphenyl)ethane triglycidyl ether (THGE-PE) comatrix cross-linked with a variety of amine hardeners and strengthened by incorporation of flax fiber (Figure 4.36) (Liu et al. 2006). Use of the amine curing agent triethylenetetramine (TETA) gave the best results, providing a more environmentally benign polymer with sufficient strength to be used in, for example, the automotive or construction industries.

4.5.3 REINFORCING FIBERS

The polymer matrix is reinforced with fibers to make a synergistic whole that is stronger than each of the individual components. Glass and carbon are the two main types of fibers that have been used in polymer matrix composites, but natural fibers such as flax (as noted above) and jute, consisting mostly of cellulose (Figure 4.37), have been explored. Glass and carbon fiber each has its advantages and disadvantages and there is a wide variety of types within each category. Which fiber is used greatly impacts the weight, strength, and cost of the turbine blade. In general, carbon fibers have both a higher strength and modulus than glass, but glass fiber-reinforced

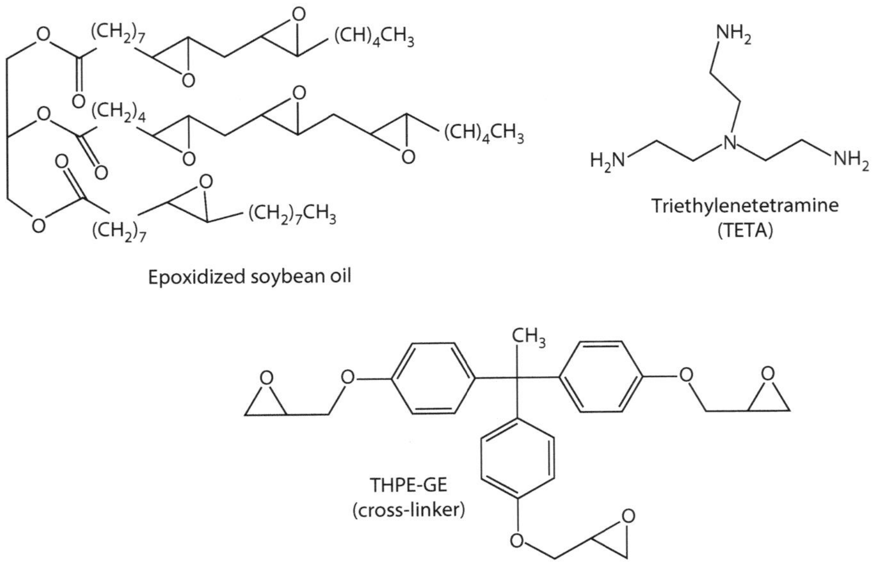

FIGURE 4.36 Components of a "green" polymer composite.

polymer matrices have the significant advantage of lower cost and ease of processing (the incorporation of glass fibers is more forgiving during the manufacturing process). Glass, however, is considerably heavier than carbon and the glass/epoxy composites are reaching their limit with respect to blade size and increased efficiency. The use of carbon fibers, while substantially more costly, can lead to a stiffer blade, allowing placement of the rotor closer to the tower. To minimize the cost of using carbon fiber and maximize the benefit of lightweight strength, these fibers are often incorporated at strategic structural areas in the blade (typically as a cap for the spar). As turbine blades become larger, the use of carbon fibers (or glass–carbon hybrids) is becoming increasingly common despite their increased costs, and it has been predicted that turbine blades will contain up to 50% carbon fiber by 2025 (New Energy Externalities Developments for Sustainability 2008).

The use of reinforcing fibers in polymer matrix composites two distinct areas of chemistry: inorganic (glass) and organic (carbon). Glass, of course, is an ancient ceramic material of many different compositions. Common soda lime glass is made by heating sand (the source of SiO_2) to the molten state along with various other inorganic additives (a flux, to lower the melting temperature, and a stabilizer, to increase the glass' resistance to attack by moisture). By allowing the molten mixture to cool back down to a transparent, rigid state, glass is formed. The cooling process is controlled so that ordered crystallization fails to take place and an amorphous solid results.

To make glass fibers, the molten glass is forced through a tiny hole to form fibers in diameters of 2–15 microns. As the molten glass is pulled, the three-dimensional structure becomes oriented, increasing the strength and stiffness of the fiber

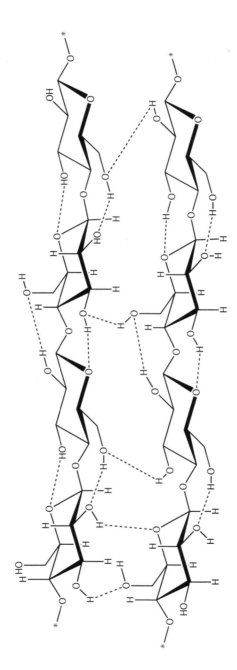

FIGURE 4.37 Cellulose (dashed lines indicate hydogen bonding).

TABLE 4.2
Glass Fiber Properties

Composition (wt.%)	E-Glass	S-Glass
SiO_2	52–56	64–66
Al_2O_3	12–16	24–25
CaO	16–25	0–0.2
MgO	0–5	9.5–10
$Na_2O + K_2O$	0–2	0–0.2
B_2O_3	5–10	—
Fe_2O_3	0–0.8	0–0.1
Tensile strength (MPa, 23°C)	3445	4890
Young's modulus (GPa, 23°C)	72.3	86.9

Source: Data from Hartman, D., M.E. Greenwood, and D.M. Miller. 1996. High strength glass fibers, *AGY Technical Paper.*

(Carraher 2000). The compositions of two types of glass commonly used in polymer matrix composites, E-glass and S-glass, are shown in Table 4.2. S-glass is a stiffer and stronger glass developed for use in more demanding conditions; as a result, S-glass is seeing greater use in wind turbine blade manufacture. However, it is more costly and difficult to process than E-glass.

Carbon fibers, while more recently developed, are (as noted above) increasingly replacing glass in turbine blade applications. Carbon fibers are most commonly produced from atactic polyacrylonitrile fibers (PAN, Figure 4.38) that contain at least 85% acrylonitrile with 6–9% of an additional acid comonomer (Committee on High-Performance Structural Fibers for Advanced Polymer Matrix Composites 2005; Peebles 1995). After polymerization to make the PAN precursor, the material is carefully treated in a sequence of steps that convert the polymer to filaments, then bundled filaments (a "tow") that are stretched to further orient the molecules along the axis and enhance the strength and modulus of the final product. Next, the stabilization step takes place by heating the acrylonitrile copolymer in air to 200–400°C. While the exact mechanism is still unknown, during this step inter- and intramolecular nitrile cross-linking occurs to stabilize the PAN fiber. Extensive spectroscopic studies have led to proposed models for the stabilized fibers; two examples are shown in Figure 4.39 (Johnson et al. 1972; Usami et al. 1990). In addition to stabilization, mechanical changes take place as the fiber is further oriented and strengthened at temperatures above its glass transition temperature (Peebles 1995). The stabilized

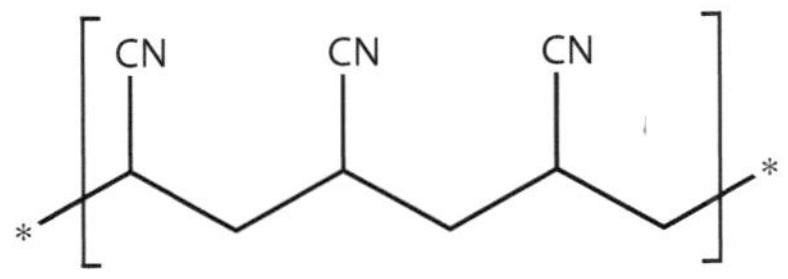

FIGURE 4.38   Polyacrylonitrile.

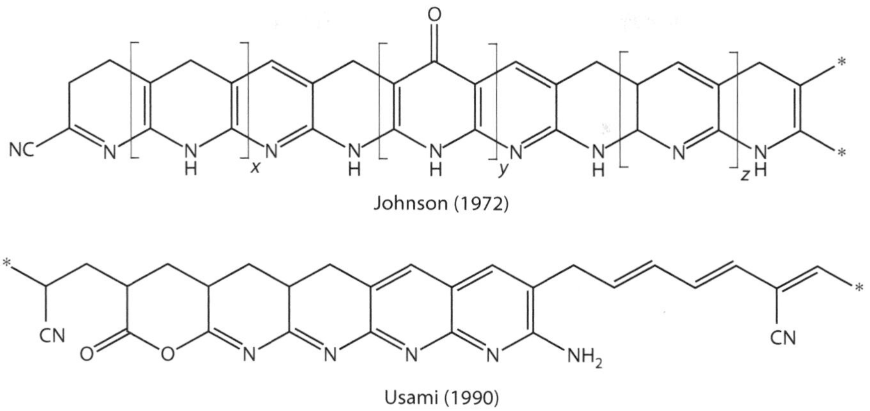

Johnson (1972)

Usami (1990)

FIGURE 4.39 Proposed structures of stabilized polyacrylonitrile.

fibers are then *carbonized* in an oxygen-free atmosphere at temperatures of up to 2600°C (Committee on High-Performance Structural Fibers for Advanced Polymer Matrix Composites 2005). Under these conditions, the fibers cannot combust but instead undergo a decomposition mechanism in which most of the heteroatoms (non-carbon atoms) are lost as small molecules such as carbon monoxide, carbon dioxide, ammonia, HCN, nitrogen, and water. As a result, primarily elemental carbon remains, with some nitrogen retained for flexibility (Peebles 1995).

The mechanical properties of carbon fibers are significantly different from those of glass and depend primarily on the degree of carbonization, the orientation of the carbon planes, and the degree to which the material is crystallized. The tensile strength for a carbon fiber has been reported to be as high as 290 GPa, whereas that for S-glass is 4.9 GPa (Table 4.1 and Committee on High-Performance Structural Fibers for Advanced Polymer Matrix Composites 2005).

4.5.4 CARBON NANOTUBES AND POLYMER MATRIX COMPOSITES

Recent studies have focused on polymer *nanocomposites* where at least one of the reinforcing materials in the matrix has dimensions in the nanoscale range (e.g., carbon nanofibers or nanotubes). Carbon nanotubes can be single-walled (SWCNT) (Figure 4.16), double-walled (DWCNT) (Figure 4.40), or multiwalled (MWCNT). They are attractive additives for polymer matrices because of their shape (they possess an extremely high ratio of height to width) and exceptional mechanical strength: SWCNTs alone have been shown to demonstrate Young's modulus on the order of 1 TPa (Schnorr and Swager 2011) and a tensile strength that is stronger than high-strength steel (Chou et al. 2005). Furthermore, their incorporation into polymer matrix composites imparts enhanced thermal stability as well.

The applications of CNT-reinforced polymer matrix composites extend throughout sustainable energy research, from turbine blades to fuel cells. However, this area of research is still in its infancy as there are considerable hurdles to overcome. One

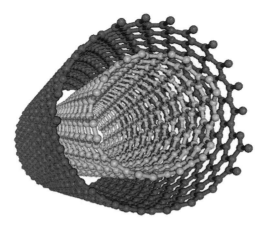

FIGURE 4.40   Image of a double-walled carbon nanotube. (Reprinted with permission from Sugai, T. et al. 2003. New synthesis of high-quality double-walled carbon nanotubes by high-temperature pulsed arc discharge. *Nano Lett.* 3 (6):769–773. Copyright 2003, American Chemical Society.)

challenge associated with the use of CNTs is that, as with glass and carbon fibers, the inherent strength of the CNT-reinforced composite depends on the orientation of the CNTs, a challenging processing problem. Yet perhaps the greatest challenge is the tendency for the CNTs to bundle together by virtue of their strong intermolecular attractions. As a result, it is difficult to disperse the CNTs throughout the polymer matrix and without excellent dispersion to distribute the load, the strength advantage of using CNTs is lost. This dispersion can be enhanced, however, by functionalizing the carbon nanotube. Zhu et al. modified SWNT by carboxylation with H_2SO_4/70% HNO_3 followed by fluorination, then mixed the functionalized SWNT with an epoxy resin of bisphenol A cured with an aromatic diamine. A 1 wt.% load of the nanotubes led to a remarkable 18% increase in tensile strength and a 24% increase in modulus over the untreated resin (see the following table) (Zhu et al. 2003).

Epoxy Formulation	Young's Modulus (MPa)	Tensile Strength (MPa)
Neat epoxy resin	2026	83.2
Epoxy resin + unfunctionalized SWNT	2123	79.9
Epoxy resin + functionalized SWNT	2632	95.0

The improved dispersion of the SWNT was attributed to covalent bonding of the functionalized nanotubes to the epoxy matrix through the carboxyl group. Similar improvements have been seen with the incorporation of amino-functionalized DWCNTs to a bisphenol-A epoxy matrix, at even lower load levels (0.1 wt.%) (Gojny et al. 2004).

The future of wind energy will be one of continued optimization in design and materials with the expectation that individual turbine capacity will increase to 10–12 MW. New designs continue to be advanced, including those based on jet

FIGURE 4.41 A vertical axis wind turbine. (With permission from Windspire Energy, LLC. www.windspireenergy.com)

engines (CleanTechnica 2010) and a bladeless design (Williams 2012). The vertical axis wind turbine (Figure 4.41) is likely to overtake the more conventional HAWT in the coming years due to the fact that it can produce up to 10 times more energy on the same amount of land (Islam et al. 2013). However, simultaneous development of energy storage technologies (Section 6.10) is imperative due to the sporadic nature of wind energy. The environmentally benign synthesis of the polymer composites is another imperative if wind energy is to be truly sustainable; this concept of "green chemistry" is briefly described in the next section.

4.6 GREEN CHEMISTRY

The astonishing growth in wind energy would not be possible without the development of polymer matrix composites, and future improvements in efficiency will require continued advancement in this area. As we will see in the later chapters, polymers are also integral to the development and improving efficacy of fuel cells and organic photovoltaic devices, and of course biomass (as in cellulose) is natural polymer chemistry. Design and synthesis of polymers with perfected properties of ion conductivity, absorption of the solar spectrum, strength, and processing ability will be crucial as sustainable energy solutions evolve. Perhaps even more important will be the development of new *green chemistry* methods of synthesizing polymers. While our focus is on energy, the importance of green chemistry cannot be overlooked in a sustainable future. Green chemistry means designing methods of preparing, purifying, and using chemicals that are energy efficient, water-conservative, and generally do not harm the environment. The 12 key principles of green chemistry are summarized as follows (Anastas and Warner 1998):

1. *Prevention.* It is better to prevent waste than to treat waste after it has been created.
2. *Atom economy.* Synthetic methods should be designed to maximize the incorporation of all materials used in the process into the final product.
3. *Less hazardous chemical syntheses.* Wherever practicable, synthetic methods should be designed to use and generate substances that possess little or no toxicity to human health and the environment.
4. *Designing safer chemicals.* Chemical products should be designed to affect their desired function while minimizing their toxicity.
5. *Safer solvents and auxiliaries.* The use of auxiliary substances (e.g., solvents and separation agents) should be made unnecessary wherever possible and innocuous when used.
6. *Design for energy efficiency.* Energy requirements of chemical processes should be recognized for their environmental and economic impacts and should be minimized. If possible, synthetic methods should be conducted at ambient temperature and pressure.
7. *Use of renewable feedstocks.* A raw material or feedstock should be renewable rather than depleting whenever technically and economically practicable.
8. *Reduce derivatives.* Unnecessary derivatization (use of blocking groups, protection/deprotection, temporary modification of physical/chemical processes) should be minimized or avoided if possible, because such steps require additional reagents and can generate waste.
9. *Catalysis.* Catalytic reagents (as selective as possible) are superior to stoichiometric reagents.
10. *Design for degradation.* Chemical products should be designed so that at the end of their function, they break down into innocuous degradation products and do not persist in the environment.
11. *Real-time analysis for pollution prevention.* Analytical methodologies need to be further developed to allow for real-time, in-process monitoring and control prior to the formation of hazardous substances.
12. *Inherently safer chemistry for accident prevention.* Substances and the form of a substance used in a chemical process should be chosen to minimize the potential for chemical accidents, including releases, explosions, and fires.

The synthesis of polycarbonate discussed in Section 4.2.1 is an excellent example of transitioning an environmentally harmful industrial synthesis to one that meets many of the principles noted above, particularly in the minimization of solvent use and the use of a feedstock, CO_2, that is clearly preferable to phosgene. The prevalence of polymers in virtually all aspects of sustainable energy generation requires synthesis of polymers in a green and sustainable way, lest our efforts at sustainability are undermined.

OTHER RESOURCES

Books

Anastas, P.T. and J.C. Warner. 1998. *Green Chemistry: Theory and Practice*. New York: Oxford University Press.

Campbell, D., R.A. Pethrick, and J.R. White. 2000. *Polymer Characterization: Physical Techniques*, 2nd ed. Cheltenham, UK: Stanley Thornes.

Carraher, C.E., Jr. 2000. *Polymer Chemistry*, 5th ed. New York, Basel: Marcel Dekker, Inc.

Gipe, P. 2004. *Wind Power*. White River Junction, VT: Chelsea Green Publishing Co.

Gupta, A. 2010. *Polymer Chemistry*. Meerut, India: Pragati Prakashan Publications.

Nicholson, J.W. 2006. *The Chemistry of Polymers*. Cambridge: Royal Society of Chemistry.

Wagner, H.-J. and J. Mathur. 2009. *Introduction to Wind Energy Systems. Basics, Technology and Operation*. Berlin/Heidelberg: Springer.

Online Resources

Global Wind Energy Council: http://www.gwec.net/

International Energy Agency/Wind power: http://www.iea.org/topics/windpower/

U.S. Department of Energy/Energy Efficiency and Renewable Energy/Wind: http://www.eere.energy.gov/topics/wind.html

REFERENCES

Anastas, P.T. and J.C. Warner. 1998. *Green Chemistry: Theory and Practice*. New York: Oxford University Press.

Billmeyer, F.W. 1984. *Textbook of Polymer Science*. 3rd ed. New York: Wiley-Interscience.

Caduff, M., M.A.J. Huijbregts, H.-J. Althaus et al. 2012. Wind power electricity: The bigger the turbine, the greener the electricity? *Environ. Sci. Technol.* 46:4725–4733.

Campbell, D., R.A. Pethrick et al. 2000. *Polymer Characterization: Physical Techniques*, 2nd ed. Cheltenham, UK: Stanley Thornes.

Carraher, C.E. 1996. *Seymour/Carraher's Polymer Chemistry: An Introduction*. 4th ed. New York: Marcel Dekker, Inc.

Carraher, C.E., Jr. 2000. *Seymour/Carraher's Polymer Chemistry: An Introduction*. 5th ed. New York, Basel: Marcel Dekker, Inc.

Cherry, B.R., C.H. Fujimoto, C.J. Cornelius et al. 2005. Investigation of domain size in polymer membranes using double-quantum-filtered spin diffusion magic angle spinning NMR. *Macromolecules* 38:1201–1206.

Chou, T.-W., E.T. Thostenson, and C. Li. 2005. Nanocomposites in context. *Compos. Sci. Technol.* 65:491–516.

CleanTechnica. *Wind Turbines Based on Jet Engines 3-4 Times More Efficient and Coming to Market?* 2010 [cited 12 November 2012]. Available from http://cleantechnica.com/2010/01/26/

Committee on High-Performance Structural Fibers for Advanced Polymer Matrix Composites. 2005. *High-Performance Structural Fibers for Advanced Polymer Matrix Composites*. Washington DC: The National Academies Press/500 Fifth St. N.W. 20001.

Cosemans, I., L. Hontis, D. Van Den Berghe et al. 2011. Discovery of an anionic polymerization mechanism for high molecular weight PPV derivatives via the sulfinyl precursor route. *Macromolecules* 44 (19):7610–7616.

Ewbank, P.C., D. Laird, and R.D. McCullough. 2009. Regioregular polythiophene solar cells: Material properties and performance. In *Organic Photovoltaics* 1–55. Weinheim, FRG: Wiley-VCH Verlag GmbH & Co. KGaA.

Fukuoka, S., M. Kawamura, K. Komiya et al. 2003. A novel non-phosgene polycarbonate production process using by-product CO_2 as starting material. *Green Chem.* 5:497–507.

Giannis, S., P.L. Hansen, R.H. Martin et al. 2008. Mode I quasi-static and fatigue delamination characterisation of polymer composites for wind turbine blade applications. *Energy Mater.* 3 (4):8.

Gojny, F.H., J.H.G. Wichmann, U. Köpke et al. 2004. Carbon nanotube-reinforced epoxy-composites: Enhanced stiffness and fracture toughness at low nanotube content. *Compos. Sci. Technol.* 64 (15):2363–2371.

Griffin, D.A. and T. Ashwill. 2009. Blade system design study part II: Final project report (GEC). Albuquerque, NM, US Department of Energy, 152.

Hartman, D., M.E. Greenwood, and D.M. Miller. 1996. High strength glass fibers, *AGY Technical Paper*.

Islam, M.R., S. Mekhilef, and R. Saidur. 2013. Progress and recent trends of wind energy technology. *Renew. Sustain. Energy Rev.* 21 (0):456–468.

Johnson, J.W., W. Potter, P.S. Rose et al. 1972. Stabilization of polyacrylonitrile by oxidative transformation. *Br. Polym. J.* 4:527–540.

Kember, M.R., A. Buchard, and C.K. Williams. 2011. Catalysts for CO_2/epoxide copolymerisation. *Chem. Commun.* 47 (1):141–163.

Lin, H., E. Van Wagner, B.D. Freeman et al. 2006. Plasticization-enhanced hydrogen purification using polymeric membranes. *Science* 311 (5761):639–642.

Liu, Z., S.Z. Erhan, D.E. Akin et al. 2006. Green composites from renewable resources: Preparation of epoxidized soybean oil and flax fiber composites. *J. Agric. Food Chem.* 54 (na):2134–2137.

Merkel, T.C., M. Zhou, and R.W. Baker. 2012. Carbon dioxide capture with membranes at an IGCC power plant. *J. Memb. Sci.* 389 (0):441–450.

Merugula, L., V. Khanna, and B.R. Bakshi. 2012. Reinforced wind turbine blades—An environmental life cycle evaluation. *Environ. Sci. Technol.* 46:9785–9792.

Miyahara, T., T. Hayano, S. Matsuno et al. 2012. Sulfonated polybenzophenone/poly(arylene ether) block copolymer membranes for fuel cell applications. *ACS Appl. Mater. Interfaces* 4 (6):2881–2884.

Murali, R.S., S. Sridhar, T. Sankarshana et al. 2010. Gas permeation behavior of Pebax-1657 nanocomposite membrane incorporated with multiwalled carbon nanotubes. *Ind. Eng. Chem. Res.* 49 (14):6530–6538.

National Aeronautics and Space Administration. 2012. *Space Shuttle/Mission Archives STS-51l* 2007 [cited 31 August 2012]. Available from http://www.nasa.gov/mission_pages/shuttle/shuttlemissions/archives/sts-51 L.html

New Energy Externalities Developments for Sustainability. 2012. *Needs Project 502687. 2008. RS 1a: Life Cycle Approaches to Assess Emerging Energy Technologies. Final Report on Offshore Wind Technology.* 2008 [cited 16 November 2012]. Available from http://www.needs-project.org/index.php?option=com_content&task=view&id=42&Itemid=66

Nguyen, Q.T., J. Sublet, D. Langevin et al. 2010. CO_2 permeation with Pebax®-based membranes for global warming reduction. Edited by Y. Yampolskii and B.D. Freeman, *Membrane Gas Separation*. Chichester, UK: Wiley.

Pappenfus, T.M., D.L. Hermanson, S.G. Kohl et al. 2010. Regiochemistry of poly(3-hexylthiophene): Synthesis and investigation of a conducting polymer. *J. Chem. Ed.* 87 (5):522–525.

Peebles, L.H. 1995. *Carbon Fibers. Formation, Structure, and Properties.* Boca Raton, FL: CRC Press.

Penczek, S. 2002. Terminology of kinetics, thermodynamics, and mechanisms of polymerization. *J. Polym. Sci., Part A: Polym. Chem.* 40:1665–1676.

Pescarmona, P.P. and M. Taherimehr. 2012. Challenges in the catalytic synthesis of cyclic and polymeric carbonates from epoxides and CO_2. *Catal. Sci. Tech.* 2 (11):2169–2187.

Rich, S. 2013. 2012 Strongest Year Ever for U.S. Wind Energy Industry. *EcoGeek* February 14, 2013.

Schnell, I., S.P. Brown, H.Y. Low et al. 1998. An investigation of hydrogen bonding in benzoxazine dimers by fast magic-angle spinning and double-quantum ^1H NMR spectroscopy. *J. Am. Chem. Soc.* 120:11784–11795.

Schnorr, J.M., and T.M. Swager. 2011. Emerging applications of carbon nanotubes. *Chem. Mater.* 23:646–657.

Stals, P.J.M., Y. Li, J. Burdynska et al. 2013. How far can we push polymer architectures? *J. Am. Chem. Soc.* 135(31):11421–11424.

Sugai, T., H. Yoshida, T. Shimada et al. 2003. New synthesis of high-quality double-walled carbon nanotubes by high-temperature pulsed arc discharge. *Nano Lett.* 3 (6):769–773.

U.S. Energy Information Administration. 2013. International Energy Statistics. U.S. EIA 2011 [cited 2 July 2013]. Available from http://www.eia.gov/cfapps/ipdbproject/IEDIndex3.cfm?tid=6&pid=29&aid=12.

Usami, T., T. Itoh, H. Ohtani et al. 1990. Structural study of polyacrylonitrile fibers during oxidative thermal degradation by pyrolysis-gas chromatography, solid-state carbon-13 NMR, and Fourier-transform infrared spectroscopy. *Macromolecules* 23 (9):2460–2465.

Williams, A. 2012. Saphonian bladeless turbine boasts impressive efficiency, low cost. *Gizmag*, http://www.gizmag.com/saphonian-bladeless-wind-turbine/24890/

Yampolskii, Y. 2012. Polymeric gas separation membranes. *Macromolecules* 45 (8):3298–3311.

Zhu, J., J. Kim, H. Peng et al. 2003. Improving the dispersion and integration of single-walled carbon nanotubes in epoxy composites through functionalization. *Nano Lett.* 3 (8):1107–1113.

5 Catalysis and Hydrogen Production

Another area of chemistry that is deeply interwoven throughout all areas of sustainable energy is catalysis. As noted with respect to green chemistry in Section 4.6, catalysis enables more environmentally benign production methods and cuts the cost of manufacture by dramatically reducing the amount of energy to carry out chemical transformations. In addition, progress in catalysis has allowed the synthesis of ever-more complex materials with greater selectivity. In the energy arena, catalysts are fundamental to the refining of petroleum (as we have already seen), the operation of fuel cells, hydrogen production, and the synthesis of most polymers. This chapter will provide a general overview of how catalysts work and then specifically focus on the production of hydrogen gas, where catalysts play a crucial role. The efficient generation of hydrogen gas is necessary for its use in fuel cells, as we will see in Chapter 6.

5.1 CATALYSIS

A catalyst is a compound that increases the rate of reaction by lowering the energy of activation: the catalyst enters into the reaction, interacts with the reacting species in some way that lowers the energy at the transition state of the rate-determining step, allows the reaction to proceed, and is then regenerated. Whether it is an inorganic catalyst in an industrial reactor or an enzyme in a biochemical system, the principle of rate enhancement by catalysis is the same: the mechanism by which the transformation is carried out is changed, with a resultant lowering of the transition state energy and increase in rate. In the prototypical energy diagram of Figure 5.1, the formation of low-energy catalyst–substrate and catalyst–product complexes via a completely different mechanistic pathway allows for substantial rate enhancement, as illustrated by the difference in the free energy of activation between the two pathways ($\Delta\Delta G^{\ddagger}$).

Catalysts can be generally categorized as homogeneous or heterogeneous, where homogeneous catalysts are soluble in the reaction medium and heterogeneous catalysts are not. Heterogeneous catalysts are usually *supported*, meaning that the active catalyst is somehow embedded upon an inexpensive, inert solid particulate support such as aluminum oxide or a porous carbon material. In this way, the surface area of the catalyst—and, therefore, active sites required for the catalyst activity—can be increased dramatically without a parallel increase in cost. The number of active sites on the catalyst correlates with the catalyst activity, where the *turnover frequency* (TOF) gives an indication of a catalyst's effectiveness. TOF defines how frequently a specific catalytic active site is able to cycle through the reaction sequence and is reported as a specific reaction rate in units of molecules per site second. As a result, it must be measured for a specific reaction under designated conditions. The higher the

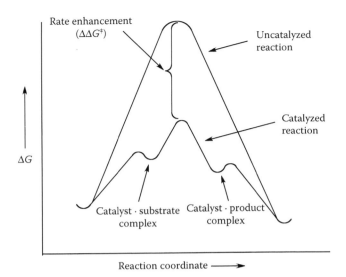

FIGURE 5.1   Energy diagram showing rate enhancement due to catalysis.

TOF, the more active the catalyst. Commercial catalysts typically have TOF values around 10^{-3}–10/s, while enzymes are amazingly more active, with TOFs in the range of 10^2–10^4/s (Bartholomew and Farrauto 2006).

The reactivity of a heterogeneous catalyst is dependent on a number of physical steps: the reacting species must diffuse into the porous network of the catalyst, be adsorbed onto the surface of the catalyst, react, and then diffuse back off the surface and through the porous network into the reaction medium. Understanding these processes at the molecular level is important for optimizing the performance of a given heterogeneous catalyst.

In addition to categorization by the physical state of the catalyst (homogeneous vs. heterogeneous), catalysts can be sorted by chemical designation, for example, transition metal (coordination) catalysts or acid–base catalysts, with acid–base catalysts including either Brønsted or Lewis types. Enzymes are the classic examples of catalysts that operate by proton donation or acceptance, and Lewis acids—in the form of zeolites—are the important catalysts in the petroleum industry (Section 2.3.1). Transition metals are tremendously important in catalysis; it is imperative to try and understand how they work. We will do this by looking at a simple and well-understood example: the hydrogenation of an alkene by Wilkinson's catalyst (**1**, Figure 5.2). Wilkinson's catalyst, $(PPh_3)_3RhCl$, is a representative homogeneous dihydride catalyst that activates H_2 to add across a double bond, thus illustrating the real power of a catalyst: converting an inert molecule into a reactive molecule. Hydrogen does not react with an alkene to any appreciable extent without the intermediacy of a catalyst.

How does it work? Wilkinson's catalyst is considered *coordinatively unsaturated* (and therefore more reactive) by virtue of the fact that it has fewer than the ideal number of electrons (18) in its coordination sphere. For most transition metals,

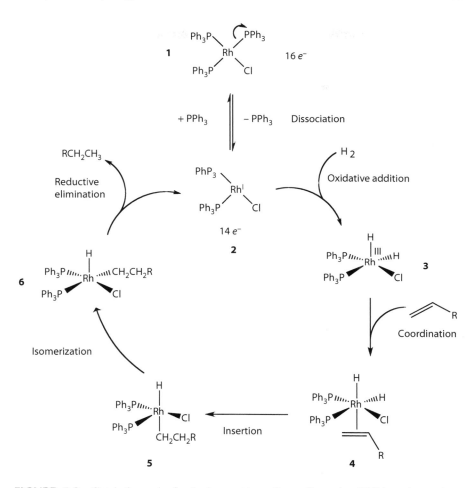

FIGURE 5.2 Catalytic cycle for hydrogenation of an alkene by Wilkinson's catalyst. (Reprinted with permission from Tranchemontagne, D.J. et al. 2012. Hydrogen storage in new metal–organic frameworks. *J. Phys. Chem. C* 116(24):13143–13151. Copyright 2012, American Chemical Society.)

a *saturated* complex is one that possesses 18 electrons, a number arrived at by a formalism based on the electron count in the metal's *d* orbitals and the electrons donated by the attached ligands (the phosphines and the chloride ligand in the case of Wilkinson's catalyst). Wilkinson's catalyst has 16 electrons initially: two from each phosphorus atom, two from the chloride, and eight from rhodium(I):

Each phosphorus ligand contributes two electrons	$3 \times 2 = 6$ electrons
The chloride ligand contributes two electrons	$1 \times 2 = 2$ electrons
Rhodium(I)	$= 8$ electrons
(Rh is a d^9 element $\therefore$ Rh(I) contributes eight electrons)	
Total	$= 16$ electrons

In a step fundamental to most transition metal-catalyzed reactions, this metal complex undergoes ligand dissociation to make the kinetically active unsaturated 14-electron complex **2**, shown in the catalytic cycle in Figure 5.2. This hungry complex readily reacts with the hydrogen molecule in a step known as *oxidative addition*, meaning that the metal center is oxidized (Rh(I)→Rh(III)) upon the cleavage of the H–H bond and addition of the two hydrogen atoms to the metal center (**3**). With the H_2 now activated by virtue of the cleaved H–H bond, the alkene can coordinate through the pi molecular orbitals, a bonding pattern common to transition metal complexes (shown as **4**). This sets up the key insertion of one of the hydrogen atoms *cis* to the alkene to make the new carbon–hydrogen bond (**5**). The isomerization of the inserted complex (**6**) followed by release of the alkane through a *reductive elimination* step regenerates the kinetically active species **2**, eliminating the alkane and reducing the rhodium atom back to Rh(I) to complete the catalytic cycle.

While transition metals are relatively expensive and many are in short supply (Section 1.2.3), one of the features that make them valuable is that they are highly reactive under mild conditions. This, in combination with their ability to carry out reactions with remarkable selectivity, makes transition metal catalysts ideal for many applications. On the other hand, many transition metal catalysts are prone to poisoning by impurities such as carbon monoxide and sulfur-containing species that can act as irreversible ligands and shut down the key dissociation step in the cycle. Thus, depending on the type of catalyst used, special precautions must be taken to purge the system of these potential inhibitors.

5.2 HYDROGEN PRODUCTION

Wilkinson's catalyst illustrates the power of transition metals in activating otherwise unreactive molecules to unusual reactivity patterns. Catalytic transformations are fundamental to the production of hydrogen and its possible use as the "energy carrier of the future" (Das and Veziroglu 2001). The so-called *hydrogen economy* is often touted as the solution to our energy and environmental woes, since hydrogen burns without increasing the global CO_2 concentration (the combustion product being water) and it has an extremely high-energy value of 142 MJ/kg (HHV; see Table 1.4). However, as a lightweight gas, hydrogen may have the highest energy density per unit mass, but it has a very low energy density by volume. Furthermore, large volumes of elemental hydrogen are not found directly on our planet: it is an *energy carrier*, not a primary fuel. It takes energy to produce hydrogen from some other primary source, and these other sources are usually fossil fuels. A third critical issue associated with the hydrogen economy is hydrogen storage. Nevertheless, "the shifting of fuels used all over the world from solid to liquid to gas, and the 'decarbonization' trend that has accompanied it, implies that the transition to hydrogen energy seems inevitable" (Haryanto et al. 2005, p. 9). For the hydrogen economy to live up to the hype, two goals must be achieved: (i) large-scale, pollution-free production of hydrogen gas and (ii) improved fuel cell technology to efficiently convert hydrogen into final (i.e., useful) energy. The generation and storage of hydrogen is the focus for the rest of this chapter; fuel cells are the topic of Chapter 6.

There are many pathways to the generation of hydrogen gas. On a small scale, the reaction of metal hydrides or elemental metals with water (under acidic or basic conditions) conveniently generates hydrogen gas as shown in Equations 5.1 through 5.3.

$$CaH_2 + 2H_2O \rightarrow Ca(OH)_2 + 2H_2(g) \tag{5.1}$$

$$Zn^\circ + 2HCl(aq) \rightarrow ZnCl_2 + H_2(g) \tag{5.2}$$

$$2Al^\circ + 6H_2O \, (HO^-) \rightarrow 2Al(OH)_3 + 3H_2(g) \tag{5.3}$$

However, metal hydrides are not a viable source of H_2 gas on an industrial scale (although much current research studies them for their potential as a hydrogen *storage* medium, *vide infra*). Industrially, the vast majority of hydrogen gas is, unfortunately, produced from fossil fuels (natural gas, coal, and petroleum) by gasification or steam reforming, as seen in Figure 5.3. Liquid feedstocks such as alcohols can also be converted into hydrogen by steam reforming, and biomass gasification (Chapter 8) provides a renewable source of hydrogen. Ammonia has a high proportion of hydrogen (18 wt.%), but its industrial synthesis is from hydrogen to begin with (in the Haber–Bosch process, Figure 5.4). Furthermore, it is a noxious gas and quite resistant to dehydrogenation. Enormous research effort continues to be devoted to the generation of hydrogen via the electrolysis ("splitting") of water, a carbon-free source, and next generation (Generation IV) nuclear reactors (Chapter 9) include the production of hydrogen as a fundamental aspect of their design. But all hydrogen production processes operate with a net energy loss. The efficiency of steam reforming of methane hovers at about 70%, while electrolysis processes are even less efficient at converting electricity into hydrogen (about 61%; U.S. Energy Information Administration 2008). Nevertheless, research toward achieving sustainable hydrogen production is very active.

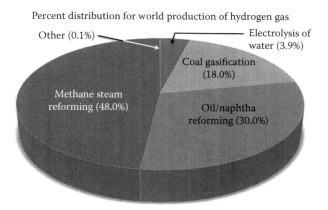

FIGURE 5.3 Percent distribution of various feedstocks for the production of hydrogen 1999. (Reprinted with permission from Ewan, B.C.R. and R.W.K. Allen. 2005. A figure of merit assessment of the routes to hydrogen. *Int. J. Hydrog. Energy* 30:809–819.)

$$3H_2 + N_2 \xrightleftharpoons[450°C/250\ atm]{\substack{Fe\ oxide\text{-}promoted \\ catalyst}} 2NH_3$$

FIGURE 5.4   The Haber process for the production of ammonia.

5.2.1 STEAM REFORMING

The current primary source of hydrogen is from the steam reforming of natural gas, also known as "SMR" for *steam reforming of methane*, the major component of natural gas. The overall reaction is shown in Equation 5.4.

$$C_nH_{2n+2} + nH_2O \rightarrow nCO + (2n + 1)H_2 \qquad (5.4)$$

The naphtha/light oil fraction of petroleum refining is also converted into hydrogen by steam reforming, but our focus will be on methane. The SMR process is part of an entire industry devoted to the conversion of natural gas and other hydrocarbons into a mixture of hydrogen, carbon dioxide, and carbon monoxide known as *synthesis gas*. Syngas (as it is commonly known) is a valuable industrial commodity across the globe since it can be used as a low-quality fuel, purified to make hydrogen gas, or converted into higher-molecular-weight, value-added chemical compounds. Among the more well-known processes for the conversion into higher-molecular-weight chemicals is the Fischer–Tropsch process (abbreviated as FT; see Section 5.2.2). Figure 5.5 provides an overview of the cycle of processes that convert fossil fuels into syngas and syngas into other, more valuable hydrocarbons. The chemistry underlying all these processes is fundamentally related, but we will concentrate on the production of hydrogen.

The production of hydrogen by steam reforming is a balancing act of reaction conditions, reactant ratio (H_2O/CH_4), reactor design, and catalyst choice. The central reactions for SMR are shown in Equations 5.5 through 5.7:

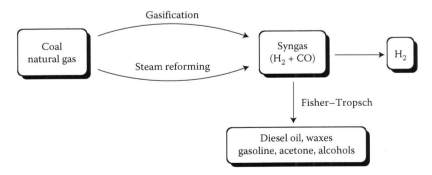

FIGURE 5.5   An overview of synthesis gas production and use. (Reprinted from Rostrup-Nielsen, J.R. 2002. Syngas in perspective. *Catal. Today* 71:243–247. With permission from Elsevier.)

$$CH_4 + H_2O \rightleftarrows CO + 3H_2 \quad (\Delta H^\circ = +205 \text{ kJ/mol}) \tag{5.5}$$

$$CO + H_2O \rightleftarrows CO_2 + H_2 \quad (\Delta H^\circ = -42 \text{ kJ/mol; water gas shift reaction}) \tag{5.6}$$

$$CO + 3H_2 \rightarrow CH_4 + H_2O \quad (\Delta H^\circ = -217 \text{ kJ/mol}) \tag{5.7}$$

A typical SMR process is made up of several stages in linked reactors with each stage being enabled by catalysis. The water/natural gas reactant stream is preheated to about 400°C and fed into the first reactor for sulfur removal. Desulfurization of the hydrocarbon feedstock is carried out with CoMo (cobalt–molybdenum) catalysts that convert sulfur compounds into H_2S that is then removed by ZnO catalysts (Bartholomew and Farrauto 2006). The low-sulfur reactant stream is then heated to begin the steam reforming. The reforming reaction is highly endothermic and the product volume is higher than the reactant volume. As a result, the reforming stages are carried out at high temperature and low pressure to achieve a more favorable equilibrium. To favor the production of hydrogen, the *water gas shift* (WGS) reaction (Equation 5.6) is utilized to oxidize the carbon monoxide to carbon dioxide, reducing the $H_2O(g)$ with the concurrent release of additional hydrogen. Since the WGS reaction is exothermic, the product stream leaving the secondary reformer is cooled to below 350°C for the *high-temperature WGS* and is cooled even further to carry out the *low-temperature WGS*. At the higher temperature, the kinetics of the reaction is more favorable, but the secondary low-temperature WGS step propels the reaction forward because of the favorable thermodynamics: the equilibrium constant increases from 11.7 at 400°C to 211 at 100°C (Lloyd et al. 1989).

The catalysts used for the reforming reactions in SMR are varied. Several general considerations apply to the selection of the catalyst:

- Given the high operating temperatures, these catalysts must be thermally stable as well as mechanically stable. This means they should have a high density and melting point and should not be prone to sintering, in which case the catalyst particles would stick together with an accompanying decrease in activity.
- Compositionally, they need to include both a metal (for activation of the hydrocarbon) and an oxide promoter (for activation of the $H_2O(g)$).
- Furthermore, the support should be of low acidity to prevent carbon formation (*coking*) and subsequent deactivation of the catalyst.

Ultimately, it is a balancing act to match high thermal stability, low acidity, and good mechanical strength with the right catalytic properties. The end result for SMR is that the metal is typically nickel (stable at high temperatures and relatively inexpensive), with MgO, $MgAl_2O_4$, $CaAl_2O_4$, or α-Al_2O_3 as the support, and calcium (or magnesium) oxide or potassium aluminum silicate ($KAlSi_3O_8$) as the promoter. For the high-temperature WGS reaction, magnetite (Fe_3O_4) is the active metal and chromium oxide (Cr_2O_3) is present to minimize sintering. The iron catalyst in use today is inexpensive, robust, and insensitive to contaminants, and is essentially the same catalyst developed by the German chemical company BASF in 1915 (Bartholomew and Farrauto 2006)! The low-temperature WGS uses a CuO/ZnO/alumina catalyst that exhibits high activity and good selectivity for the shift reaction. The active metal is copper, while the zinc oxide is believed to add mechanical stability. Both the zinc and aluminum oxide scavenge poisons such as sulfur and chloride.

The development of the right catalyst begins with empirical studies. Once an optimal catalyst is discovered, mechanistic studies can take over so that the evolution of the "right" catalyst can be based on intelligent design. That said, mechanistic studies are not always straightforward. After all, how can the presence of a catalyst make a stable hydrocarbon such as methane rearrange its carbon–hydrogen bonds to give H_2 and CO gas? Keeping in mind that these are harsh conditions and that mechanistic studies continue, it is still possible to provide a general picture of the mechanism. First, the reacting species must be adsorbed onto the catalyst surface, just as hydrogen has to interact with the rhodium atom in Wilkinson's catalyst. The activation of methane requires adsorption onto the nickel surface, whereas the steam requires adsorption onto the oxide. Upon adsorption, each molecule can undergo dissociation to CH_x fragments (e.g., $\cdot CH_3$ in the case of methane) and H$\cdot$ and HO$\cdot$ radicals in the case of water. It is then believed that these adsorbed radical fragments combine in a variety of elementary steps to form the observed gaseous products upon desorption from the catalyst surface. The bottom line is that the chemistry is very complex, is carried out under fairly extreme conditions, and requires a catalyst to make it happen.

Steam reforming of ethanol (sometimes referred to as SRE or ESR) has been examined as an alternative to the use of fossil fuels in the production of hydrogen. The advantages of using ethanol as the hydrogen source for the reforming process are (a) it can be produced from renewable resources (so-called "bioethanol," Chapter 8), (b) it is a low-toxicity, low-hazard liquid feedstock, (c) it is free from catalyst poisons, and (d) it is considered by many to be environmentally benign. Like methane, the steam reforming of ethanol to hydrogen is highly endothermic, as shown in Equation 5.8.

$$CH_3CH_2OH(g) + 3H_2O(g) \rightleftarrows 2CO_2(g) + 6H_2(g)\,(\Delta H^\circ = +347\,kJ/mol)\ (5.8)$$

The reaction pathways that take place during the steam-reforming process of ethanol are quite complex, as seen in Figure 5.6 (Haryanto et al. 2005). Although several of the pathways are common to the steam reforming of methane, the presence of the alcohol functionality presents additional potential issues, such as elimination to form the alkene (followed by polymerization), or dehydrogenation to the carbonyl compound with subsequent aldol reactions.

A very wide variety of compounds has been found to be active as ESR catalysts, with noble metals such as rhodium, ruthenium, palladium, and platinum showing high activity and selectivity for hydrogen production. However, the more economical nickel- and cobalt-based catalysts such as Co/ZnO, Co/CeO_2, and $Ni/La_2O_3–Al_2O_3$ are among the plethora of metal/oxide catalyst systems that have

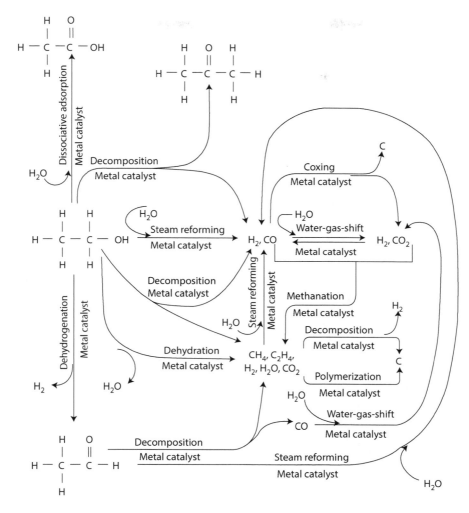

FIGURE 5.6 The many reactions of ethanol steam reforming. (Reprinted with permission from Haryanto, A. et al. 2005. Current status of hydrogen production techniques by steam reforming of ethanol: A review. *Energy Fuels* 19 (5):2098–2106. Copyright 2005, American Chemical Society.)

also shown to be good choices for ESR (Haryanto et al. 2005). Much research is ongoing with respect to optimizing the ESR process for the production of hydrogen, including examination of other catalysts, supports, feedstock ratios, and reaction conditions.

It should also be noted that these methods for the production of hydrogen require the separation of hydrogen from the other gases present in the product stream. As we have seen in Section 2.4.1, the separation of H_2 from CO_2 is a challenge but one whose solution is crucial to the development of the hydrogen economy.

5.2.2   ASIDE: THE FISCHER–TROPSCH PROCESS

While the FT synthesis is not about energy conversion per se, it is closely related to the production of syngas and is an important sustainable chemistry topic in its own right. The FT synthesis is an industrial process developed in the early twentieth century by Franz Fischer and Hans Tropsch for the conversion of coal into liquid products—a *coal liquefaction* (or *coal-to-liquid*, CTL) process. (As we will see in Chapter 8, gasification of biomass can be carried out to generate syngas; thus, the FT process is of significant interest for *biomass-to-liquid* conversions as well.) The major driving force for the development of the FT process has been the production of liquid fuels, aptly named *synfuels* due to their production from syngas (Equation 5.9):

$$C_xH_y(\text{coal}) \rightarrow CO + H_2(\text{syngas}) \xrightarrow{\text{Fischer–Tropsch}} C_nH_{2n+2}(\text{synfuels}) + H_2O \quad (5.9)$$

The FT process consists of steam gasification of coal (or biomass) to produce syngas followed by the catalytic conversion of the syngas mixture into (preferably) liquid fuels at reasonably low pressures and temperatures. It was the development of the cobalt catalysts that carry out the syngas-to-synfuel conversion by Fischer and his coworkers that was key to its success (Stranges 2007). Today's FT catalysts are primarily iron, cobalt, nickel, or ruthenium based and the choice of a catalyst is based on weighing the cost versus product distribution.

The product stream from FT synthesis can vary from gasoline-like products to diesel or higher-molecular-weight lubricants and even waxes, or the process can be tailored to produce small-molecule commodity chemicals such as alkenes and alcohols. In essence, the FT synthesis amounts to a polymerization reaction of carbon monoxide on a transition metal catalyst surface. The mechanism is not yet fully elucidated, but some elemental steps are postulated to be the following (Schulz 1999):

- Coordination of CO on the metal surface
- Cleavage of the CO bond
- Coordination of H_2 to the metal
- Reaction of H_2 with oxygen of the coordinated CO to generate H_2O followed by
- Dissociation of water leaving a surface carbene (CH_2)
- Dimerization, trimerization, and so on of surface-bound $-CH_2-$ groups to make the hydrocarbons

As fossil fuel supplies dwindle, further optimization of the FT process for synthesis of all kinds of carbon-containing chemicals will take on even greater importance.

5.2.3   GASIFICATION

Another route to hydrogen via synthesis gas is the gasification of coal. Gasification subsumes steam reforming and is a more general thermochemical conversion process that is primarily applied to solid feedstocks. Like it or not, coal remains one of the most abundant supplies of fuel on the planet. While coal is the primary feedstock for the generation of electricity, its combustion is also an environmentally unsavory process, generating CO_2 as a by-product and ejecting sulfur and nitrogen oxides, mercury, and particulate pollutants into the environment. Gasification of coal to produce syngas presents a cleaner alternative to coal combustion, although the cogeneration of CO_2 remains a hurdle, *vide infra*. (Gasification of biomass to produce syngas is a process that we will examine in detail in Chapter 8.)

Gasification is a complex process with many different types of chemistry taking place in the reactor, including *combustion, pyrolysis*, and *gasification* of coal (among others). The key to gasification is to partially oxidize the coal by limiting the amount of oxidant (pure O_2 or air) present, while using the energy released from the small portion of combustion taking place to drive the process. Typically, roughly one-third to one-fifth the stoichiometric amount of oxidant is introduced into the reactant stream. The primary gasification reactions are shown in Equations 5.10 through 5.12, but many other reactions can take place depending, as usual, on the process conditions, reactor design, and so on (Equations 5.13 through 5.15).

$$C + H_2O \rightleftarrows CO + H_2 \quad (\Delta H° = +131\,kJ/mol) \tag{5.10}$$

$$C + \tfrac{1}{2}O_2 \rightleftarrows CO \quad (\Delta H° = -111\,kJ/mol)\ (combustion) \tag{5.11}$$

$$CO + H_2O \rightleftarrows CO_2 + H_2 \quad (water\ gas\ shift\ reaction; \Delta H = -42\,kJ/mol) \tag{5.12}$$

$$C + CO_2 \rightleftarrows 2CO \quad (Boudouard\ reaction; \Delta H = +171\,kJ/mol) \tag{5.13}$$

$$C + 2H_2 \rightleftarrows CH_4 \quad (hydrogenation; \Delta H = -208\,kJ/mol) \tag{5.14}$$

$$CO + 3H_2 \rightleftarrows CH_4 + H_2O \quad (methanation; \Delta H = -271\,kJ/mol) \tag{5.15}$$

Gasification is typically carried out at very high temperatures and pressures—up to 2000°C and 8000 kPa, depending on the type of reactor. Impurities in the coal are converted into hydrogen sulfide (from sulfur impurities), ammonia (from nitrogen impurities), and HCl (from chlorine in the feedstock). Inert mineral impurities tend to melt into a glass-like "slag" that is collected elsewhere in the reactor, or as ash that is blown out of the gasifier and captured.

Catalysts may or may not be used in the actual gasification step to increase efficiencies and minimize side reactions, but the harsh process conditions limit the options. Nonetheless, alkali carbonates and metal oxides of zinc, indium, and iron have been used to successfully gasify low-quality coal (An et al. 2012; Ohtsuka and Asami 1997). As with steam reforming, coal gasification takes advantage of the WGS reaction and the attendant catalysts to maximize the yield of hydrogen gas.

Coal gasification remains a very active area of research, despite the fact that it is an old technology. It is the basis for the more efficient integrated gasification combined cycle (IGCC) power plants mentioned in Chapter 2. Ultimately, further refinement of catalysts and the development of improved technologies for the separation of hydrogen gas from other gases will be required for the development of the hydrogen economy.

5.2.4 WATER AND THE BIOLOGICAL PRODUCTION OF HYDROGEN

The cogeneration of CO and CO_2 is a significant problem in both reforming and gasification of carbon-containing feedstocks as methods for the production of hydrogen fuel. We are already well aware of the problems associated with carbon dioxide, but carbon monoxide is also a problem: it is a serious contaminant when it comes to hydrogen-powered fuel cells since the CO poisons the platinum catalysts in the cells. As an alternative to reforming or gasification, the hydrogen by-product of thermal cracking in the petroleum-refining process is CO-free, and hydrogen can be extracted from ammonia or methanol (*vide infra*). But the production of hydrogen from "splitting" water by electrolysis is arguably the most attractive carbon-free source of hydrogen.

The production of hydrogen fuel from water is not a new concept—nor is the depletion of fossil fuels, for that matter. In 1874, Jules Verne published his work *L'Île Mystérieuse*, in which Captain (and engineer) Cyrus Harding predicts

Yes, my friends, I believe that water will one day be employed as fuel, that hydrogen and oxygen which constitute it, used simply or together, will furnish an inexhaustible source of heat and light, of an intensity of which coal is not capable. Some day the coalrooms of steamers and the tenders of locomotives will, instead of coal, be stored with these two condensed gases, which will burn in the furnaces with enormous calorific power. There is, therefore, nothing to fear. ... I believe, then, that when the deposits of coal are exhausted we shall heat and warm ourselves with water. Water will be the coal of the future.

Verne 1920

Electrolysis of water to generate hydrogen is currently done at scales from a few kilowatts to up to 2000 kW per electrolyzer (Lipman 2011). But a critical consideration in the use of water as a source of hydrogen is that clean water is a valuable and increasingly scarce resource (World Health Organization 2012). To replace the gasoline component of the U.S. Energy economy for *one day*, about 0.34 tonnes of hydrogen is needed. If water is the source of this hydrogen, the amount needed is equivalent to 1650 full-size swimming pools *per hour* (Armaroli and Balzani 2011). Furthermore, splitting of water generates oxygen as well as hydrogen and the demand for oxygen as a commercial product is limited.

The overall reaction for the generation of hydrogen from water (Equation 5.16) is the reverse of the reaction whereby electricity is generated in the hydrogen fuel cell, which we will discuss in great detail in Chapter 6.

$$H_2O + \text{electricity} \rightarrow H_2 + \tfrac{1}{2}O_2 \qquad (5.16)$$

This clearly presents an interesting cyclical dilemma: we use electricity to split water in an electrolyzer to produce hydrogen. We use hydrogen to produce electricity in a fuel cell. While the two processes are not necessarily coupled (our main interest in fuel cells is for hydrogen-powered transportation), the paradox remains germane.

TABLE 5.1
Types of Electrolyzers

Name/Acronym	Description
Alkaline	Makes use of inexpensive Ni° catalysts, but is limited to low-pressure H_2 generation
Polymer electrolyte membrane (PEM)	Can generate high-pressure H_2 but requires expensive noble metal electrodes
Solid oxide	Used for steam electrolysis. Strontium-doped lanthanum manganite (LSM) for anode and nickel zirconia cermets for cathode. See Chapter 6, Section 7.3. Can be used at very high operating temperatures.
Solar (PEC)	"Photoelectrochemical" electrolyzers link solar energy to electrolysis. Electricity for electrolysis is generated by photovoltaics. Sunlight + water $\rightarrow$ hydrogen + oxygen
Microbial (MEC)	Uses algae/photosynthetic bacteria to convert the biodegradable material into H_2

There are many kinds of electrolyzers for the production of hydrogen; Table 5.1 summarizes some of the most important ones. We will focus on microbial electrolyzers in this chapter because (a) they hold promise as a renewable and more sustainable alternative to hydrogen generation, (b) details for the other types of electrolyzers will be presented in Chapter 6 in the guise of fuel cells, and (c) we will take advantage of the topic of microbial electrolysis as an entry point for the discussion of some fascinating biochemical processes for the production of hydrogen (*vide infra*).

5.2.4.1 Microbial Electrolysis of Water

The process of generating hydrogen through a bacteria-catalyzed electrochemical reaction is referred to as *electrohydrogenesis* (also known as *microbial electrolysis*). A schematic representation of a *microbial electrolysis cell* (MEC) is shown in Figure 5.7. The key feature of these microbial electrolyzers is the use of bacteria (*exoelectrogens*) growing on the anode to catalyze the oxidation of biodegradable materials and generate the electrons that drive the reduction of protons at the cathode, producing hydrogen. The exhaust gases (CO_2 and H_2) are collected from the cell headspace. While the MEC is obviously an attractive approach to replacing fossil fuels in the generation of hydrogen gas, they have several shortcomings: first, the current generated at the anode is insufficient to overcome the unfavorable thermodynamics associated with the redox process (ΔG°_{rxn} for the reduction of acetate, for example, is +105 kJ/mol; Logan et al. 2008); hence, additional energy from an external power supply (PS in Figure 5.7) is required. Second, while the anode is typically made of some form of elemental carbon, the cathode required is often platinum, which, as we have already seen, raises questions about sustainability and contamination by poisons such as CO and sulfur compounds. Competing generation of methane (by microorganisms known as *methanogens*) is another problem seen in MECs. Nevertheless, MECs have real strengths as well: the presence of the exoelectrogen means that the additional energy required to overcome the thermodynamic threshold is much less than that which would be theoretically required to produce hydrogen

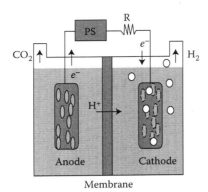

FIGURE 5.7 A two-chamber MEC depicting bacterial growth (ovals) on the flat anode (and a bacterial biocatalyst on the cathode). (Reprinted with permission from Logan, B.E. et al. 2008. Microbial electrolysis cells for high yield hydrogen gas production from organic matter. *Env. Sci. Tech.* 42 (23):8630–8640. Copyright 2008, American Chemical Society.)

from water (typically 0.4 V vs. >1.2 V; Logan et al. 2008). An additional advantage is that wastewater can be used as the water source (Rulkens 2008). Hydrogen production rates in prototype MECs of over 3 $m^3/m^3 \cdot day$ have been reported (Logan et al. 2008). Overall, the outlook for MECs is good, with promising results in the development of biocathodes, design of combined processes that increase the production efficiency, and in the use of other sustainable sources of energy (such as sunlight, *vide infra*) to provide the external power for the MEC.

5.2.4.2 Hydrogenases

Enzymes, bacteria, and algae—all manner of biological manifestations of carbon, nitrogen, and oxygen—are intertwined in a natural hydrogen web. A widespread class of enzymes known as *hydrogenases* catalyzes the reduction of protons to hydrogen gas (Equation 5.17).

$$2H^+ + 2e^- \rightleftarrows H_2 \qquad\qquad (5.17)$$

Couple hydrogenase with photosynthesis and you have a direct line from sunlight to hydrogen as we will see in the next section. Photosynthesis and hydrogenases bridge biochemistry and organometallic chemistry and provide a rich area for exploring renewable energy solutions.

Understanding the structure of hydrogenase enzymes and their role in proton reduction to dihydrogen has been a relatively recent development, with hundreds of studies—both computational and experimental—uncovering the basic molecular understanding of the process. Iron and nickel are the key redox partners in the active sites of these enzymes, of which there are three types: those with a [NiFe] active site (found in cyanobacteria), the more active dinuclear [FeFe] hydrogenases found in green algae (Magnuson et al. 2009), and mononuclear Fe hydrogenases.

Ball-and-stick depictions as well as structural representations of the active site of the [NiFe] and [FeFe] hydrogenases are shown in Figure 5.8. The active sites contain a bridging dithiolate ligand along with the common organometallic ligands CO and CN. For the [FeFe] hydrogenases, this iron–sulfur framework at the active site is known as the *H-cluster*.

While many questions are as yet unanswered, some central steps in the catalytic cycle for the [FeFe] hydrogenases have been proposed as outlined in Figure 5.9. Ligand dissociation at the metal center (**1**) followed by protonation at nitrogen (**2**) provides a vacant coordination site to allow for protonation at the metal (**3**), yielding a metal hydride species in equilibrium with the dihydride complex (and dihydrogen precursor) (**4**) (Gloaguen and Rauchfuss 2009; Tard and Pickett 2009). The close proximity of hydrogen on the protonated amine and on the metal facilitates the formation of the coordinated H_2 complex, a metal–hydrogen three-center, two-electron bond (**4** in Figure 5.9; Carroll et al. 2012). Irreversible oxidative addition of H_2 is not involved. The reverse step—heterolytic cleavage of the bound H_2—takes place either in an intra- or intermolecular manner and the Fe–S cluster facilitates the redox process by acting as an electron shuttle with the enzyme (Armstrong 2004; Gordon and Kubas 2010).

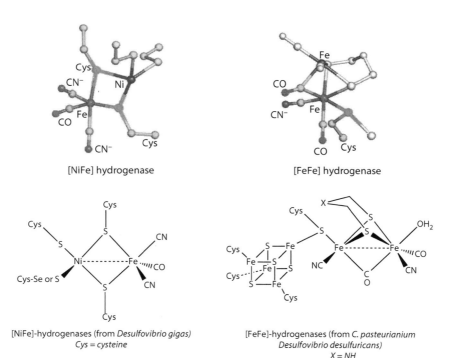

FIGURE 5.8 Representations of [NiFe] and [FeFe] hydrogenase active sites. (Reprinted with permission from Gordon, J.C. and G.J. Kubas. 2010. Perspectives on how nature employs the principles of organometallic chemistry in dihydrogen activation in hydrogenases. *Organometallics* 29 (21):4682–4701. Copyright 2010, American Chemical Society.)

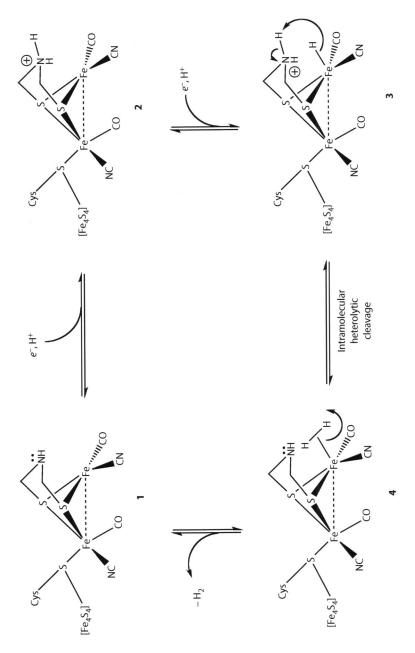

FIGURE 5.9 Proposed mechanism of the formation of hydrogen at the hydrogenase active site.

There are several features to note regarding this chemistry. First, the TOF for [FeFe] hydrogenases is an amazing $6000-9000$ s^{-1} (Ghirardi et al. 2009). Second, the presence of CO and $-CN$ ligands as part of this catalysis cycle was a surprising discovery as these are typically poisons in transition-metal-mediated processes (not to mention poisons to the host organism). The activity of the [FeFe] hydrogenase is carefully balanced by the ligands CO, CN, and H_2: the strongly bound CO activates the iron center to reversibly bind hydrogen, while the presence of CN is important for its impact on the electronic properties of the enzyme (Gordon and Kubas 2010). It is also remarkable that hydrogen gas competes effectively with water, nitrogen, and oxygen for activation by the enzyme.

5.2.4.3 Photochemical Electrolysis of Water

Perhaps the ideal source of hydrogen is to couple light energy from the sun with the biological generation of hydrogen. This is one aspect of the area of research known as "solar fuel": the attempt to efficiently harness the conversion of light into hydrogen or other potentially useful fuels. There is a plethora of approaches to photo-driven water splitting, many of which are synergistic with the advances being made in the realm of solar photovoltaics (Chapter 7). We will keep a tenuous connection to the biological production of hydrogen in this section by emphasizing the biomimetic approach of *artificial photosynthesis* (APS).

There are three main types of light-driven processes for the generation of hydrogen found in nature: (1) hydrogenase-catalyzed oxygenic photosynthesis, (2) nitrogenase-catalyzed oxygenic photosynthesis, and (3) nonoxygenic photosynthesis. Engineering microorganisms to produce a higher yield of H_2 (synthetic biology) and designing APS systems to improve light conversion efficiency (biomimetic chemistry) are the keys to success in this research into hydrogen production.

Photosynthesis harvests visible light (wavelength range of about $400-700$ nm) to effectively convert carbon dioxide and water into carbohydrates for energy use by the organism (Equation 5.18).

$$CO_2 + H_2O \xrightarrow{\text{Light}} (CH_2O)_n + O_2 \qquad (5.18)$$

To explain the complex biochemistry of photosynthesis in detail is well beyond the scope of this book, but a concise summary is required to understand the progress being made in the field of APS. Photosynthesis is carried out by two compartmentalized photosystems labeled *photosystem I* (PSI) and *photosystem II* (PSII). It is further divided into the "light reactions" (which use light energy) and the "dark reactions" (which take the energy formed from the light reactions to actually make the carbohydrates). It is in PSII that the light reaction takes place and water is "split," being oxidized to oxygen:

$$2H_2O \rightarrow O_2 + 4H^+ + 4e^- \qquad (5.19)$$

Ultimately, the four electrons so generated can be used to reduce protons to H_2.

The light energy that drives photosynthesis is harvested by an antennae-like array of chromophores consisting of chlorophyll pigments P680 and P700 (corresponding to the nanometer wavelength of light they absorb) and carotenoids

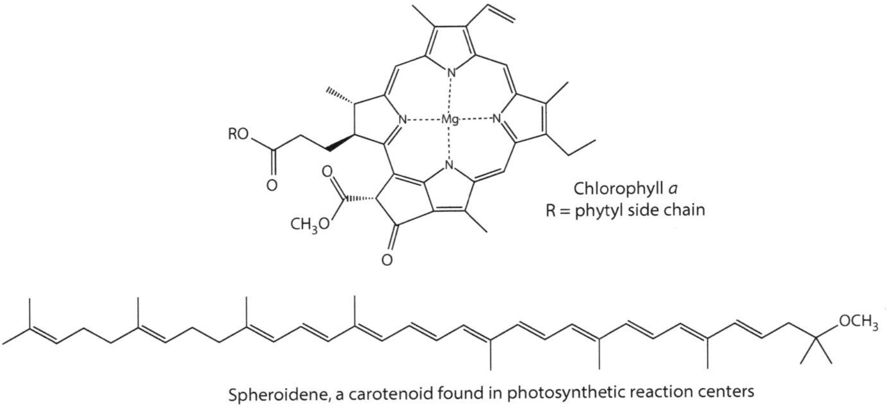

Chlorophyll *a*
R = phytyl side chain

Spheroidene, a carotenoid found in photosynthetic reaction centers

FIGURE 5.10   Examples of chlorophyll and carotenoid compounds.

(Figure 5.10). The absorption of photons of light by these chemicals promotes electrons to an excited state, thereby producing electron–hole pairs.* Through a series of steps along an electron transfer chain, the excited electron is moved from one complex to another down an energy gradient, thus producing a positively charged radical species ($R^{+\bullet}$) and a negatively charged radical partner ($R^{-\bullet}$), both still in the excited state. A simplified representation of this downhill electron transfer (the so-called *Z-scheme*) is shown in Figure 5.11. The resultant excited radical cation is the

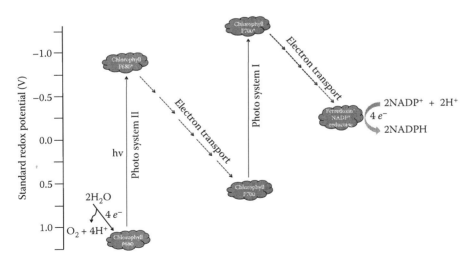

FIGURE 5.11   The photosynthetic Z-scheme.

---

* Electron–pair holes are also an important concept in photovoltaics. We will revisit them in greater detail in Chapter 7.

photooxidant that must eventually be reduced by some nearby electron donor, thus completing the redox cycle. In PS II, this electron donor is water, being oxidized to O_2 by the *oxygen-evolving complex* (OEC), a reaction center that contains a manganese–calcium oxide cluster (Mn_4Ca) (Nocera 2012; Tachibana et al. 2012).

It is useful to think of electron transfer in the physical sense to understand the role of intervening species: an electron travels from one compound to another through molecular orbitals. The barrier to electron transfer depends on the relative energy level of the reacting species as well as the distance between the partners and interference by insulating molecules such as the solvent. Crucial to the success of photosynthesis (and APS) is the separation of the reaction center from the light-harvesting apparatus so that a sufficient redox potential can be achieved to carry out the water-splitting reaction. On the basis of these principles of photosynthesis, the scientist involved in APS research with the goal of hydrogen production from water must design (a) a chromophore for harvesting light, (b) a pathway for electron transfer, and (c) a species to catalyze the reduction of protons to H_2 and oxidize water to O_2. Several recent examples follow:

1. Researchers coupled a ferrocene electron donor to a rhenium(I) photosystem and took advantage of a $(CO)_6Fe_2S_2$ cluster to act as a hydrogenase mimic (Figure 5.12). Upon irradiation, this triad accomplished the desired photoinduced electron transfer and successfully generated H_2, but noncatalytically (Wang et al. 2011).

2. In a related approach, an $[Fe_4S_4]$–cubane electron transport cluster was assembled within a tin sulfide cluster to create a functionalized chalcogenide aerogel. A chalcogenide is the 2^- ion of the Group VI/16 elements, for example, O, S, Se, Te. This porous *chalcogel* was then subjected to ion exchange to replace the tetraphenylphosphonium counterion with a photoactive Ru(bipyridyl)$_3^{2+}$ dye (Figure 5.13) to give a material that was capable of producing hydrogen when illuminated in the presence of a proton source (Yuhas et al. 2011).

3. Another promising example is seen in the use of a rhodamine dye analog as the photosystem, triethanolamine as the electron donor, and a cobalt glyoximate catalyst (Figure 5.14). This combination provided "unprecedented yields" of hydrogen from the photoreduction of water (McCormick et al. 2010).

4. "PDI" derivatives (perylene-3,4:9,10-bis(dicarboximide)) are excellent chromophores in that they are inexpensive, stable, and have a high molar absorptivity. Linking the PDI to an iridium complex (Figure 5.15) yields a potential sunlight-driven water oxidation catalyst: electron transfer from the iridium to the excited PDI chromophore took place in <10 picoseconds following excitation (Vagnini et al. 2012).

By elucidating a full molecular understanding of how nature uses photosynthesis and takes advantage of enzymes such as hydrogenase to generate hydrogen, progress in the area of APS is likely, but the transformation into a "hydrogen economy" still has enormous challenges to overcome, not the least of which is storing this energy-rich but low-density fuel. Research efforts toward this goal are the topic of the next section.

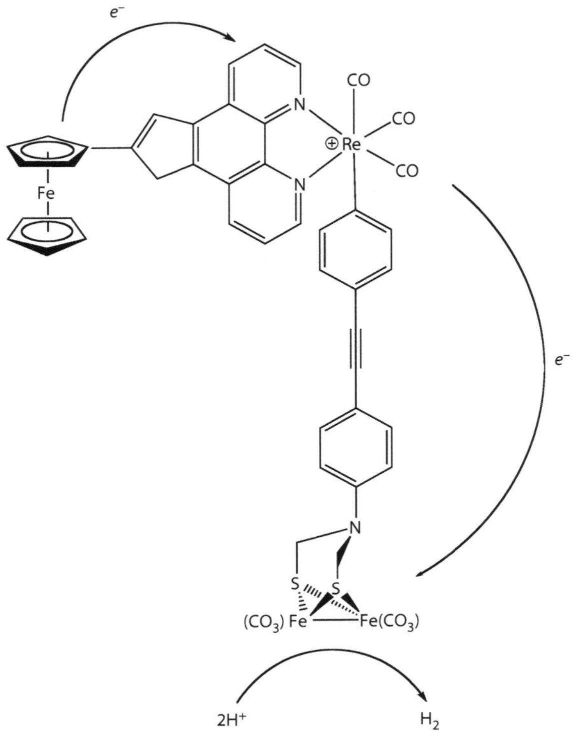

FIGURE 5.12 Iron–rhenium triad for APS. (Reprinted from Wang, F. et al. 2012. Artificial photosynthetic systems based on [FeFe]–hydrogenase mimics: The road to high efficiency for light-driven hydrogen evolution. *ACS Catal.* 2 (3):407–416. Copyright 2012, American Chemical Society.)

5.3 HYDROGEN STORAGE

Certainly, the most evident challenges associated with the use of hydrogen gas as a fuel are its low density, high volume, and flammability. To use H_2 as a replacement for fossil fuels in fuel cell-powered cars the hydrogen fuel must be safely delivered in an amount that will meet the driving-range requirements of the typical driver. Furthermore, providing this amount of hydrogen fuel to the vehicle must be accomplished without taking up a lot of space or adding excessive weight. Given hydrogen's properties, this is not trivial. The energy content of a kilogram of H_2 is roughly equivalent to 4 L of gasoline (based on its LHV), and for the same amount of energy, hydrogen gas takes up about three thousand times more volume (Armaroli and Balzani 2011; U.S. Department of Energy and Office of Energy Efficiency and Renewable Energy 2009). This means that it would take over 3400 L (around 900 U.S. gallons) to replace 4.5 L ($\approx$1 gal) of gasoline at standard temperature and pressure (Graetz 2009). Thus, not only do we need to figure out a different way to produce huge volumes of hydrogen fuel, but we also have to figure out a way to *store*

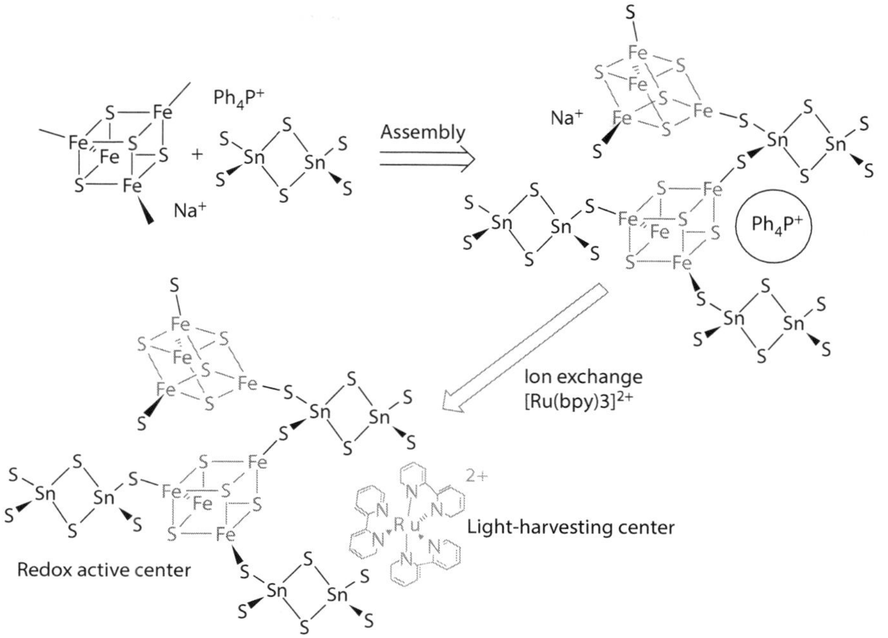

FIGURE 5.13 **(See color insert.)** Synthesis of electron transport chalcogel. (Reprinted with permission from Yuhas, B.D. et al. 2011. Biomimetic multifunctional porous chalcogels as solar fuel catalysts. *J. Am. Chem. Soc.* 133 (19):7252–7255. Copyright 2011, American Chemical Society.)

Rhodamine dye analogue
X = O, S, Se

Cobalt glyoximate
catalyst

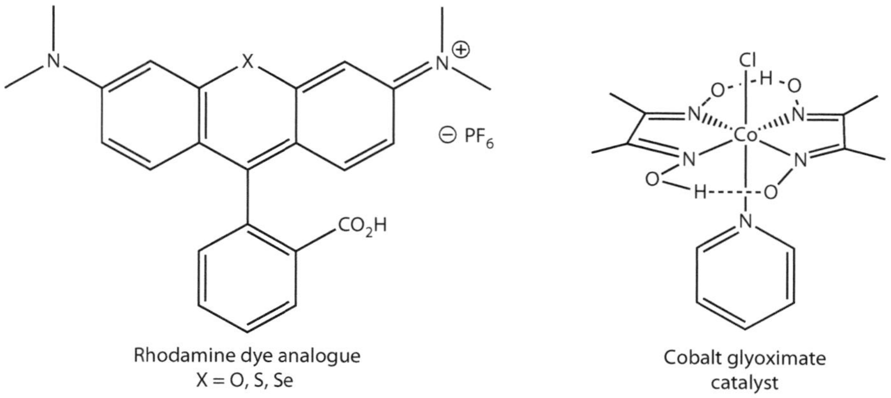

FIGURE 5.14 Cobalt glyoximate compound dye for photoreduction of water. (Reprinted with permission from McCormick, T.M. et al. 2010. Reductive side of water splitting in artificial photosynthesis: New homogeneous photosystems of great activity and mechanistic insight. *J. Am. Chem. Soc.* 132:15480–15483. Copyright 2010, American Chemical Society.)

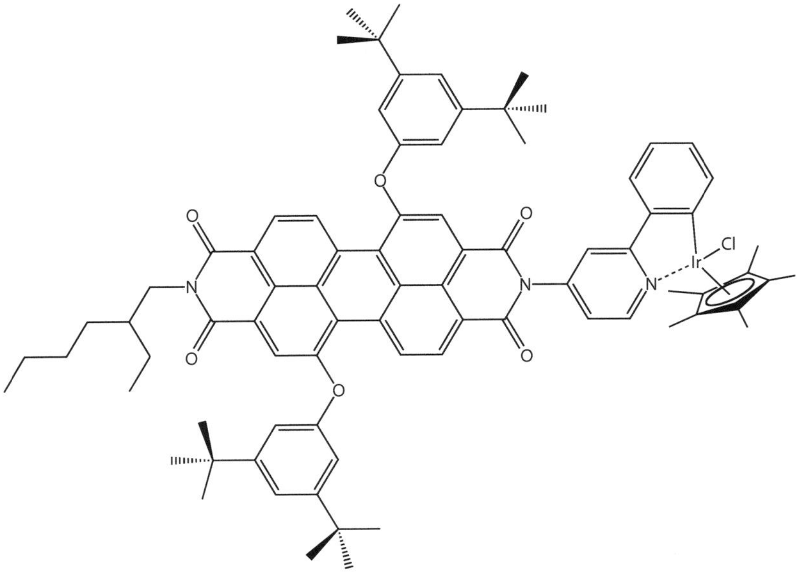

FIGURE 5.15   Perylene–iridium complex.

it safely and compactly so that it can be delivered on demand. Hydrogen storage is widely considered the biggest hurdle to the implementation of the hydrogen economy and is an area of explosive research that goes hand in hand with the development of the hydrogen fuel cell.

Hydrogen can be stored by physical means or by chemical means. To store hydrogen gas by physical compression to the liquid state is not commercially feasible for the transportation sector: the boiling point of liquid hydrogen is −253°C; so, compression is extremely expensive and boil-off is a huge issue. Furthermore, even liquid hydrogen has a lower energy density by volume than a liquid hydrocarbon such as gasoline. The compression of hydrogen gas up to 71 MPa (700 atm) is viable but costly and presents numerous engineering problems: the storage tank needs to be larger than a comparable gasoline tank; it has to be cylindrical to withstand the high pressure (making it difficult to design into the vehicle), and many materials fail to prevent hydrogen diffusion, posing a potentially dangerous leakage problem (Armaroli and Balzani 2011).

Chemical solutions to the hydrogen storage problem (*materials-based hydrogen storage* or CHS—*chemical hydrogen storage*) would ideally provide hydrogen fuel in a low-pressure, controlled manner that allows refueling: whatever chemical process is involved, the storage unit, once depleted, should be able to be refilled so that the vehicle can be powered over and over again with the hydrogen-generated electricity of the fuel cell. Three general chemical approaches to the challenge of hydrogen storage are presented here: the use of metal–organic frameworks (MOFs) to adsorb (and release when needed) hydrogen gas (Section 5.3.1), metal hydrides as hydrogen carriers (Section 5.3.2), and the use of hydrogen-rich liquids (Section 5.3.3) as surrogate hydrogen storage.

5.3.1 METAL–ORGANIC FRAMEWORKS

When you pull together the almost endless variety of organic ligands with the diversity of metal ions you enter the realm of *coordination network solids*, of which MOFs are a subgroup. (You also enter a nomenclature fray: the area(s) of *coordination polymers* and MOFs have simultaneously developed, with concurrent confusion in the realm of nomenclature (Batten et al. 2012). We will retain the use of MOF in referring to these interesting compounds.) The study of MOFs is a relatively new area of chemistry that has expanded rapidly since the late 1990s, combining solid-state chemistry with coordination chemistry to present a beautiful array of crystalline materials with well-defined and tunable pore sizes and channels. A material's capacity for H_2 storage strongly depends on its pore volume and surface area (see Section 2.4.1.3 for a review of surface area measurements and Figure 5.16 for a representation of MOF-177). The benchmark BET surface area of 5500 m^2/g for MOF-177 (an MOF consisting of a zinc oxide cluster, $Zn_4O_6^{6+}$, joined with 1,3,5-benzenetribenzoate linkers) was reported in 2007, and commercially available MOFs have BET surface areas in the 2000 m^2/g range (Furukawa et al. 2007).

MOFs are the ideal candidates for both hydrogen storage and proton conduction (which we will explore in the context of fuel cells in Chapter 6). The process whereby MOFs actually store hydrogen is one of physisorption, where the H_2 is absorbed into the pores of the MOF and trapped by weak van der Waals forces. As usual, the thermodynamics of the processes involved presents a precarious balancing act: the enthalpy of desorption should be relatively low so that the waste heat from the fuel cell can be sufficient to dislodge the hydrogen. However, if the enthalpy of desorption is too low, then high pressures will be required for repressurizing the framework. A range of 20–30 kJ/mol H_2 for the enthalpy of desorption is considered optimal based on efficiency considerations, but, at this point, few materials meet this criterion (Yang et al. 2010).

MOFs are built up by combining building units of metal ions with organic *linkers*. They can be exquisitely easy to prepare: simply dissolve the appropriate ratio of the metal precursor (e.g., $Zn(NO_3)_2 \cdot 4H_2O$) and linker in the suitable solvent(s) and heat. With any luck and a little bit of time, crystals precipitate out and you have your MOF. For example, Figure 5.17 shows two representations of MOF-5, a porous, three-dimensional framework with Zn_4O tetrahedra as the nodes and bridged by 1,4-benzenedicarboxylate anion struts (Ji et al. 2012). Zn(II) and Cu(II) are the most commonly used metal ions in MOFs and carboxylate ligands are similarly prevalent as linkers. MOFs can also be synthesized with mixed ligand systems or mixed metal ions, or can be modified postsynthesis (such as grafting in polymer synthesis). Since MOFs can be synthesized from this great diversity of metal ion and bridging ligand building blocks, it is straightforward to vary the pores and channels, although control of conditions (type of solvent, temperature, and rate of crystal growth) will also impact the morphology.

A variety of MOFs containing heterocyclic linkers was prepared as shown for MOF-326 in Figure 5.18 for potential use as hydrogen storage. The ligands the researchers used to fine tune the pore size, surface area, and polarity are shown in Figure 5.19 (Tranchemontagne et al. 2012). Table 5.2 presents some of the relevant

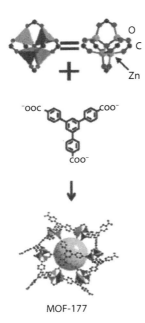

MOF-177

FIGURE 5.16   Synthesis of MOF-177. The central sphere represents the void space. (Adapted with permission from Xiang, Z. et al. 2011. CNT@Cu$_3$(BTC)$_2$ and metal–organic frameworks for separation of CO$_2$/CH$_4$ mixture. *J. Phys. Chem. C* 115 (40):19864–19871. Copyright 2011, American Chemical Society.)

data for these MOFs. The hydrogen uptake results for this series are presented in Figure 5.20 (Tranchemontagne et al. 2012). While the results are only marginally better than the benchmark MOF-177, this graph nicely illustrates the advantage of MOF storage of hydrogen versus bulk hydrogen gas.

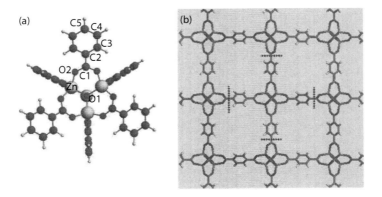

FIGURE 5.17   MOF-5. (a) Representative fragment and (b) segment of the crystal structure. (Reprinted with permission from Ji, M. et al. 2012. Luminescent properties of metal–organic framework MOF-5: Relativistic time-dependent density functional theory investigations. *Inorg. Chem.* 51 (22):12389–12394. Copyright 2012, American Chemical Society.)

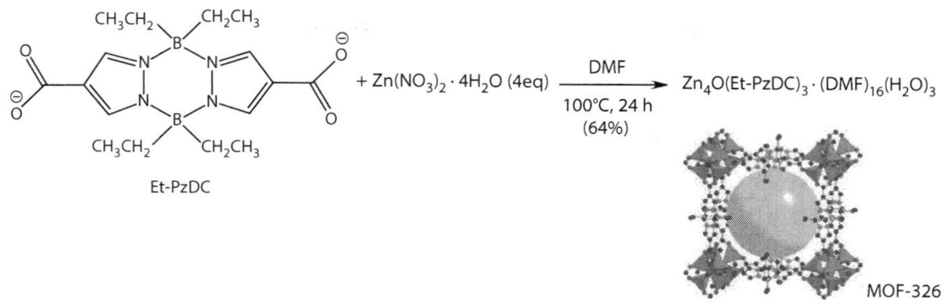

FIGURE 5.18 Synthesis of MOF-326. (Reprinted with permission from Tranchemontagne, D.J., K.S. Park, H. Furukawa et al. 2012. Hydrogen storage in new metal–organic frameworks. *J. Phys. Chem. C* 116 (24):13143–13151. Copyright 2012, American Chemical Society.)

The U.S. Department of Energy (DOE) has set a target for onboard hydrogen storage systems for light-duty vehicles of 5.5 wt.% hydrogen by 2015 (U.S. Department of Energy and Office of Energy Efficiency and Renewable Energy 2009). To achieve this target, research has focused on increasing the surface area of the MOFs and altering the structures to increase the strength of the interactions between H_2 and the MOF surface. One approach to achieving this is to increase the surface area of the MOF. A 2010 paper reported a hydrogen storage capacity of 176 mg/g in MOF-210 at 77 K and 1 atm, the highest at that time (Furukawa et al. 2010). However, there are several problems with MOF hydrogen storage, not the least of which is that the more promising results have been achieved at 77 K—an expensive proposition for practical application in the transportation sector. Less than 1% (by weight) of hydrogen can be stored at 298 K and 100 atm pressure (Jena 2011).

It should also be noted that MOFs are frequently hosts to the solvent molecules in which the synthesis was carried out. The removal of the guest solvent molecules can result in the collapse of the MOF framework and therefore, loss of porosity and decrease in surface area, impairing the material's ability to adsorb hydrogen.

TABLE 5.2
Selected Data for Heterocycle-Linked MOFs

MOF Identification	Metal	Linker	Pore Diameter (Å)	BET Surface Area (m²/g)
MOF-324	Zn	PyC	7.6	1600
MOF-325	Cu	PyC	19.6	<10
MOF-326	Zn	Et-PzDC	15.4	1380
IRMOF-61	Zn	EDB	12.2, 15.0	1410
IRMOF-62	Zn	BDB	5.2	2420

Source: Reprinted with permission from Tranchemontagne, D.J. et al. Hydrogen storage in new metal–organic frameworks. *J. Phys. Chem. C* 116 (24):13147. Copyright 2012, American Chemical Society.

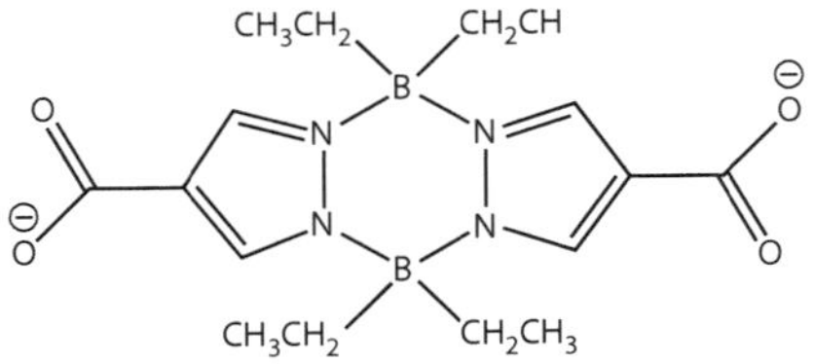

FIGURE 5.19 Ligands used in the synthesis of MOF variants shown in Figure 5.18.

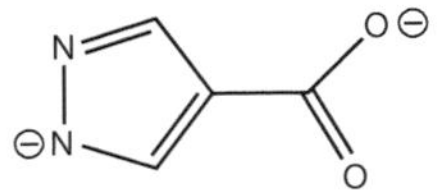

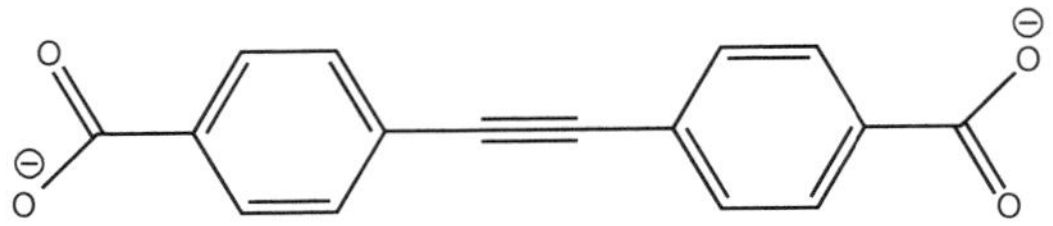

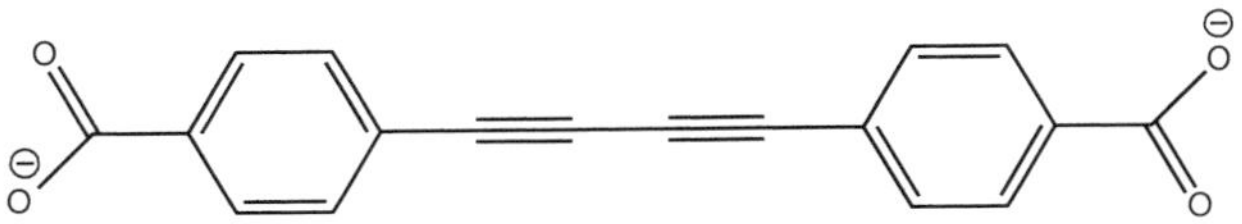

Overall, while the use of MOFs for hydrogen storage shows good promise, much improvement is needed to make them viable materials for this application.

5.3.2 METAL HYDRIDES

As has already been noted, the key problem with hydrogen gas is its volumetric energy density. Chemical methods of storing hydrogen as covalently bound *atomic* hydrogen can circumvent this issue. Metal hydrides release dihydrogen not by desorption but instead by a chemical reaction that may be catalyzed, should be reversible, and should, ideally, take place within the range of 1–10 atm and 298–393 K to meet the operating requirements of the fuel cell (Armaroli and Balzani 2011). Certainly, a simple example of this is the generation of hydrogen from metal hydrides as was shown in Equations 5.1 through 5.3 (Section 5.2). Thus, in contrast to the research focus on MOF/hydrogen storage where scientists are trying to *increase* the binding energy of H_2 to the MOF surface, success in the metal hydride realm will mean *decreasing* the hydrogen-binding energy. In any case, these solid materials hold the potential to provide compact, portable, and low-pressure hydrogen storage.

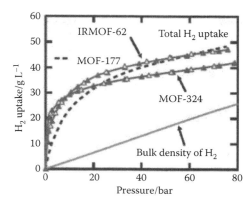

FIGURE 5.20   Hydrogen uptake performance of MOFs.

In principle, the use of metal hydrides for hydrogen storage would work as shown in Figure 5.21. The hydride is dehydrogenated in a *reversible* manner to provide hydrogen gas for the onboard fuel cell at a minimum pressure of about 0.3–0.5 MPa as required for fuel cell operation. Once depleted, the metal hydride storage is regenerated by hydrogenation. For this scheme to work, the initial weight percent of hydrogen in the material must be relatively high. In addition, both the thermodynamics and the kinetics of the forward and reverse reactions must be finely matched. In terms of kinetics, hydrogen must be generated at a rate sufficient to keep the vehicle powered (1.6 g H_2/s; Yang et al. 2010), and reactivation must favorably compare to refilling a tank with gasoline. As always, cost and sustainability are the key issues.

The most simple metal hydrides (MH_x) are either too stable or too unstable for use in hydrogen storage. For example, MgH_2, which has an enthalpy of formation of −75 kJ/mol H_2, would require a temperature of at least 290°C to release $H_2(g)$ at 0.1 MPa (Yang et al. 2010). At the other end of the spectrum is aluminum hydride, AlH_3, with $\Delta H = -7.6$ kJ/mol H_2—too unstable for use as a hydrogen storage medium. To find the elusive hydrogen storage material that can work with favorable kinetics and thermodynamics in a reasonable temperature range, researchers have turned to a variety of complex metal hydrides and light metal hydrides. Complex metal hydrides refer to those compounds that are made up of a metal cation and a complex anion such as BH_4^- (borohydrides) or AlH_4^- (alanates). Sodium alanate ($NaAlH_4$), with an attractive 7.4% hydrogen by weight, can decompose to give off hydrogen gas in the three-step process shown below (where T_{icat} refers to a titanium catalyst with which the material is doped):

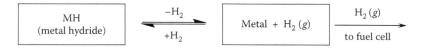

FIGURE 5.21   Scheme for metal hydride delivery of hydrogen gas.

$$NaAlH_4 + Ti_{cat} \rightleftarrows \tfrac{1}{3} NaAlH_6 + \tfrac{2}{3} Al + H_2 \qquad (5.20)$$

$$Na_3AlH_6 \rightleftarrows 3NaH + Al + \tfrac{3}{2} H_2 \qquad (5.21)$$

$$NaH \rightleftarrows Na + \tfrac{1}{2} H_2 \qquad (5.22)$$

The discovery that doping of the sodium alanate with titanium compounds can render Equation 5.20 thermodynamically favorable at temperatures above 33°C and 1 atm pressure was a breakthrough in metal hydride/hydrogen storage research (Bogdanovic and Schwickardi 1997; U.S. Department of Energy 2008). Without catalysis, bulk $NaAlH_4$ does not begin to decompose until it reaches its melting point (180°C) and then only generates an equilibrium pressure of 100 kPa H_2 as the NaH decomposes to $Na°$ and H_2 at 425°C.

Magnesium borohydride [$Mg(BH_4)_2$], too, has a high gravimetric density (14.8 wt.% hydrogen), but is not amenable to reversible hydrogenation except under bruising conditions of 950 bar and 400°C (Paskevicius et al. 2010; U.S. Department of Energy 2011). Sodium borohydride (10.8 wt.% hydrogen) does not release hydrogen until well above 500°C and is not reversible under reasonable conditions.

In addition to metal hydrides, alanates, and borohydrides, amides such as $LiNH_2$, $Ca(NH_2)_2$, and $Mg(NH_2)_2$ have been explored for their ability to deliver H_2. Lithium amide, for example, with a weight percent of 10% hydrogen, can dehydrogenate as shown in Equation 5.23 at 170–400°C (the enthalpy of dehydrogenation is calculated to be somewhere between 80.5 and 99 kJ/mol H_2).

$$LiNH_2 + 2LiH \rightleftarrows Li_3N + 2H_2 \qquad (5.23)$$

Alloying with magnesium lowered the dehydrogenation temperature to almost 200°C (Orimo et al. 2005, 2007).

Not surprisingly, nanotechnology has infiltrated research on the use of metal hydrides for hydrogen storage. Even simply reducing the particle size of the hydrides to the nanoscale results in improved thermodynamics and kinetics for hydrogen storage: the dehydrogenation/hydrogenation cycling for sodium borohydride was dramatically enhanced by preparing nanoscale $NaBH_4$ particles, lowering the temperature for the release of hydrogen by more than 100°C. Pursuing this material further, the application of the "core–shell" approach of reducing nickel chloride onto $NaBH_4$ nanoparticles yielded excellent improvements in performance (Christian and Aguey-Zinsou 2012).

The addition of carbon nanomaterials has generated similar improvements (Berseth et al. 2009). The incorporation of nanomaterials is often as simple as physical mixing (by ball milling) of the components. For example, mixing $NaAlH_4$ into a nanoporous carbon scaffold infused with $TiCl_3$ nanoparticles (*nanoconfinement*) gave a material with greatly improved hydrogen desorption kinetics compared to the material without $TiCl_3$ and to ball-milled samples of $NaAlH_4$–$TiCl_3$ (Nielsen et al. 2011). In an interesting integration of MOFs and complex metal hydrides, a magnesium–MOF "nanoreactor" preloaded with a titanium catalyst was infiltrated with

$NaAlH_4$. The resultant material ($NaAlH_4(Ti)@MOF-74(Mg)$) had a BET surface area of 407 m²/g but exhibited fast desorption kinetics and excellent reversibility. The authors attribute this success to several factors, including open metal sites to stabilize the titanium catalyst and charged atoms within the material that prevent diffusion or agglomeration by stabilizing decomposition products (Stavila et al. 2012). For magnesium hydride (which also has unfavorable dehydrogenation kinetics, *vide supra*), mixing with CNTs, fullerenes, graphite, and so on has improved the uptake of hydrogen, although the reasons why are not yet clear (Adelhelm and de Jongh 2011). There is no doubt that the small particle size helps with both diffusion and heat flow. Similar results are seen with sodium alanate (Zaluska et al. 2000). All in all, the outlook for further improvement in the use of complex metal hydrides—especially as they crossover into the nanoscale—is very promising.

5.3.3 OTHER CHS MATERIALS

It would be advantageous to be able to store hydrogen in a liquid instead of the less convenient solid materials. The so-called "liquid organic hydrides" or "liquid organic hydrogen carriers" such as methylcyclohexane (6.1 wt.% hydrogen) can be dehydrogenated to give toluene and hydrogen (Equation 5.24) in a highly endothermic reaction.

$$C_7H_{14} \rightleftarrows C_7H_8 + 3H_2 (\Delta H_{450°C} = 216\,kJ/mol) \tag{5.24}$$

Unfortunately, this reaction typically requires catalysis by a platinum group metal to be technologically feasible. Alcohols such as methanol and ethanol, too, can be considered as hydrogen carriers, since they can be reformed to H_2 via steam reforming as we saw in Section 5.2.1.

At 19.6 wt.% hydrogen, the Lewis acid–base complex ammonia–borane (AB) ($H_3N \cdot BH_3$) has been extensively studied as a potential CHS material. AB has a high gravimetric density and can undergo thermolysis to release hydrogen as shown in Equations 5.25a, 5.25b, and 5.25c. Note that $(HN \cdot BN)_x$ refers to any number of poorly characterized boron/nitrogen cross-linked polymeric materials.

$$H_3N \cdot BH_3 \rightarrow \tfrac{1}{x}(H_2N \cdot BH_2)_x + H_2 \tag{5.25a}$$

$$(H_2N \cdot BH_2) \rightarrow (HN \cdot BH)_x + H_2 \tag{5.25b}$$

$$(HN \cdot BH)_x \rightarrow BN + H_2 \tag{5.25c}$$

The overall reaction and a proposed partial mechanism showing the formation of the key intermediate (DADB, the *diammoniate of diborane*) is shown in Figure 5.22 (Staubitz et al. 2010).

There are numerous issues associated with AB as a hydrogen carrier, including the high temperatures (>150°C) needed for dehydrogenation, the unfavorable thermodynamics for regeneration (see thermodynamic data for AB and tetrahydroazaborine in Figure 5.23), formation of ammonia as a by-product (a problematic contaminant for subsequent use of the product hydrogen in fuel cells), and

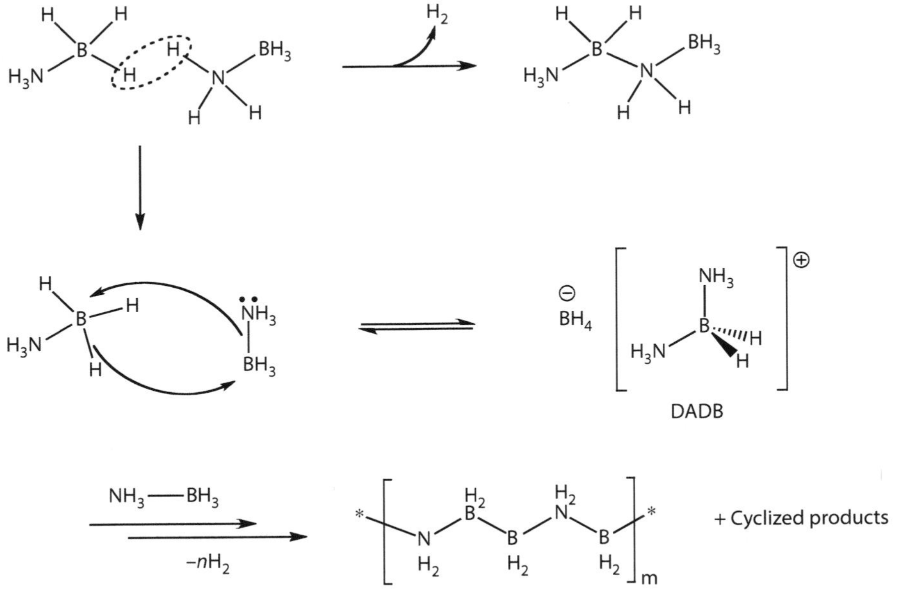

FIGURE 5.22 Proposed mechanistic steps in the decomposition of AB to produce hydrogen.

the eventual formation of boron nitride (BN), a highly stable compound. Another drawback of using AB is that it is a crystalline solid, not the desired liquid feed-stock. Attempts to catalyze the release of hydrogen from AB have shown excellent initial results, however (Denney et al. 2006) and the incorporation of additives such as ammonium chloride or copper(II) chloride has also resulted in promising improvements in its behavior as a hydrogen storage material (Staubitz et al. 2010). Another proposed improvement is to carry out the AB dehydrogenation in an ionic liquid such as 1-butyl-3-methylimidazolium chloride (Figure 5.24). This approach

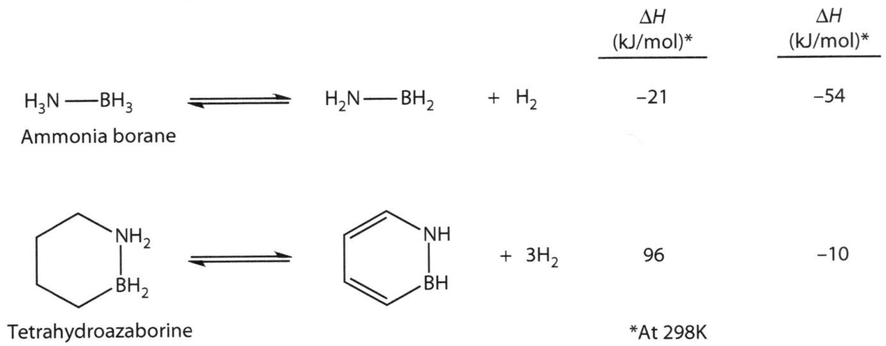

	ΔH (kJ/mol)*	ΔH (kJ/mol)*
$H_3N\!-\!BH_3 \rightleftharpoons H_2N\!-\!BH_2 + H_2$ Ammonia borane	−21	−54
Tetrahydroazaborine $\rightleftharpoons$ $+ 3H_2$	96	−10
	*At 298K	

FIGURE 5.23 Thermodynamic data for AB and tetrahydroazaborine.

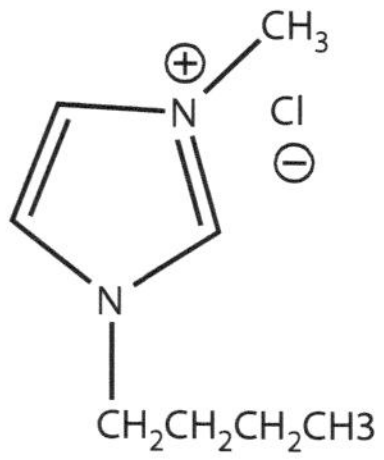

FIGURE 5.24   1-Butyl-3-methylimidazolium chloride.

improves the extent and the rate of hydrogen release, presumably by again promoting the formation of the DADB, the reactive intermediate in hydrogen release (Himmelberger et al. 2009).

Carbon–boron–nitrogen derivatives such as tetrahydroazaborine (Figure 5.22) are another approach since the free energy for release of hydrogen is closer to the optimum of $\Delta G = 0$ (ideal for reversibility; Figure 5.22). However, their promise is as yet unfulfilled for several reasons: they must be synthesized, they too are solids, and their dehydrogenation requires catalysis (Liu 2011). Thus, research into the development of viable hydrogen storage materials is ongoing with several promising results being reported. At this point, however, a viable hydrogen storage system does not exist (Yang et al. 2010).

OTHER RESOURCES

BOOKS

Bartholomew, C.H. and R.J. Farrauto. 2006. *Fundamentals of Industrial Catalytic Processes.* Hoboken, NJ: Wiley & Sons.

Centi, G. and R.A. van Santen, Eds. 2007. *Catalysis for Renewables: From Feedstock to Energy Production.* Weinheim, FRG: Wiley-VCH.

Higman, C. and M. van der Burgt. 2003. *Gasification.* Burlington, MA: Elsevier (Gulf Professional Publishing).

Okada, T. and M. Kaneko. 2009. *Molecular Catalysts for Energy Conversion.* Berlin: Springer.

Press, R.J., K.S.V. Santhanam, et al. 2009. *Introduction to Hydrogen Technology.* Hoboken, NJ: Wiley & Sons.

ONLINE RESOURCES

U.S. Department of Energy Hydrogen and Fuel Cells Program—http://www.hydrogen.energy. gov/index.html

U.S. Department of Energy|Energy Efficiency & Renewable Energy|Fuel Cell Technologies Office|Hydrogen Storage—http://www1.eere.energy.gov/hydrogenandfuelcells/storage/index.html

REFERENCES

Adelhelm, P. and P.E. de Jongh. 2011. The impact of carbon materials on the hydrogen storage properties of light metal hydrides. *J. Mater. Chem.* 21 (8):2417–2427.

An, H., T. Song, L. Shen, et al. 2012. Hydrogen production from a Victorian brown coal with *in situ* CO_2 capture in a 1 kWth dual fluidized-bed gasification reactor. *Ind. Eng. Chem. Res.* 51:13046–13053.

Armaroli, N. and V. Balzani. 2011. The hydrogen issue. *Chem. Sus. Chem.* 4:21–36.

Armstrong, F.A. 2004. Hydrogenases: Active site puzzles and progress. *Curr. Opin. Chem. Biol.* 8:133–140.

Bartholomew, C.H. and R.J. Farrauto. 2006. *Fundamentals of Industrial Catalytic Processes.* 2nd ed. Hoboken, NJ: Wiley & Sons.

Batten, S.R., N.R. Champness, X.-M. Chen, et al. 2012. Coordination polymers, metal–organic frameworks and the need for terminology guidelines. *Cryst. Eng. Comm.* 14 (9):3001–3004.

Berseth, P.A., A.G. Harter, R. Zidan, et al. 2009. Carbon nanomaterials as catalysts for hydrogen uptake and release in $NaAlH_4$. *Nano Lett.* 9:1501–1505.

Bogdanovic, B. and M. Schwickardi. 1997. Ti-doped alkali metal aluminium hydrides as potential novel reversible hydrogen storage materials. *J. Alloys Compd.* 253–254:1–9.

Carroll, M.E., B.E. Barton, T.B. Rauchfuss, et al. 2012. Synthetic models for the active site of the [FeFe]–hydrogenase: Catalytic proton reduction and the structure of the doubly protonated intermediate. *J. Am. Chem. Soc.* 134 (45):18843–18852.

Christian, M.L. and K.-F. Aguey-Zinsou. 2012. Core shell strategy leading to high reversible hydrogen storage capacity for $NaBH_4$. *ACS Nano* 6 (9):7739–7751.

Das, D. and T.N. Veziroglu. 2001. Hydrogen production by biological processes: A survey of literature. *Int. J. Hydrog. Energy* 26:13–28.

Denney, M.C., V. Pons, T.J. Hebden, et al. 2006. Efficient catalysis of ammonia borane dehydrogenation. *J. Am. Chem. Soc.* 128 (37):12048–12049.

Ewan, B.C.R. and R.W.K. Allen. 2005. A figure of merit assessment of the routes to hydrogen. *Int. J. Hydrog. Energy* 30:809–819.

Furukawa, H., N. Ko, Y.B. Go, et al. 2010. Ultrahigh porosity in metal–organic frameworks. *Science* 329:424–428.

Furukawa, H., M.A. Miller, and O.M. Yaghi. 2007. Independent verification of the saturation hydrogen uptake in MOF-177 and establishment of a benchmark for hydrogen adsorption in metal–organic frameworks. *J. Mat. Chem.* 17 (30):3197–3204.

Ghirardi, M.L., A. Dubini, J. Yu, et al. 2009. Photobiological hydrogen-producing systems. *Chem. Soc. Rev.* 38:52–61.

Gloaguen, F. and T.B. Rauchfuss. 2009. Small molecule mimics of hydrogenases: Hydrides and redox. *Chem. Soc. Rev.* 38:100–108.

Gordon, J.C. and G.J. Kubas. 2010. Perspectives on how nature employs the principles of organometallic chemistry in dihydrogen activation in hydrogenases. *Organometallics* 29 (21):4682–4701.

Graetz, J. 2009. New approaches to hydrogen storage. *Chem. Soc. Rev.* 38 (1):73–82.

Haryanto, A., S. Fernando, N. Murali, et al. 2005. Current status of hydrogen production techniques by steam reforming of ethanol: A review. *Energy Fuels* 19 (5):2098–2106.

Himmelberger, D.W., L.R. Alden, M.E. Bluhm, et al. 2009. Ammonia borane hydrogen release in ionic liquids. *Inorg. Chem.* 48 (20):9883–9889.

Jena, P. 2011. Materials for hydrogen storage: Past, present and future. *J. Phys. Chem. Lett.* 2:206–211.

Ji, M., X. Lan, Z. Han, et al. 2012. Luminescent properties of metal–organic framework MOF-5: Relativistic time-dependent density functional theory investigations. *Inorg. Chem.* 51 (22):12389–12394.

Lipman, T. 2011. An overview of hydrogen production and storage systems with renewable hydrogen case studies. *Clean Energy States Alliance*. Montpelier, VT.

Liu, S.-Y. 2011. Hydrogen storage by novel CBN heterocycle materials. 2011 Annual Progress Report, U.S. Department of Energy Hydrogen and Fuel Cells Program, 425–428.

Lloyd, L., D.E. Ridler, and M.V. Twigg. 1989. The water-gas shift reaction. In *Catalyst Handbook*, edited by M.V. Twigg, 283–338. London: Wolfe Publishing Ltd.

Logan, B.E., D. Call, S. Cheng, et al. 2008. Microbial electrolysis cells for high yield hydrogen gas production from organic matter. *Env. Sci. Tech.* 42 (23):8630–8640.

Magnuson, A., M. Anderlund, O. Johansson, et al. 2009. Biomimetic and microbial approaches to solar fuel generation. *Acc. Chem. Res.* 42 (12):1899–1909.

McCormick, T.M., B.D. Calitree, A. Orchard, et al. 2010. Reductive side of water splitting in artificial photosynthesis: New homogeneous photosystems of great activity and mechanistic insight. *J. Am. Chem. Soc.* 132:15480–15483.

Nielsen, T.K., M. Polanski, D. Zasada, et al. 2011. Improved hydrogen storage kinetics of nanoconfined $NaAlH_4$ catalyzed with $TiCl_3$ nanoparticles. *ACS Nano* 5 (5):4056–4064.

Nocera, D.G. 2012. The artificial leaf. *Acc. Chem. Res.* 45 (5):767–776.

Ohtsuka, Y., and K. Asami. 1997. Highly active catalysts from inexpensive raw materials for coal gasification. *Catal. Today* 39:111–125.

Orimo, S., Y. Nakamori, J.R. Eliseo, et al. 2007. Complex hydrides for hydrogen storage. *Chem. Rev.* 107 (10):4111–4132.

Orimo, S., Y. Nakamori, G. Kitahara, et al. 2005. Dehydriding and rehydriding reactions of $LiBH_4$. *J. Alloys Compd.* 404–406 (0):427–430.

Paskevicius, M., D.A. Sheppard, and C.E. Buckley. 2010. Thermodynamic changes in mechanochemically synthesized magnesium hydride nanoparticles. *J. Am. Chem. Soc.* 132 (14):5077–5083.

Rostrup-Nielsen, J.R. 2002. Syngas in perspective. *Catal. Today* 71:243–247. Elsevier.

Rulkens, W. 2008. Sewage sludge as a biomass resource for the production of energy: Overview and assessment of the various options. *Energy Fuels* 22:9–15.

Schulz, H. 1999. Short history and present trends of Fischer–Tropsch synthesis. *Appl. Catal. A Gen.* 186 (1–2):3–12.

Staubitz, A., A.P.M. Robertson, and I. Manners. 2010. Ammonia–borane and related compounds as dihydrogen sources. *Chem. Rev.* 110 (7):4079–4124.

Stavila, V., R.K. Bhakta, T.M. Alam, et al. 2012. Reversible hydrogen storage by $NaAlH_4$ confined within a titanium-functionalized MOF-74(Mg) nanoreactor. *ACS Nano* 6 (11):9807–9817.

Stranges, A.N. 2007. A history of the Fischer–Tropsch synthesis in Germany 1926–45. *Stud. Surf. Sci. Catal.*, 163: 1–27.

Tachibana, Y., L. Vayssieres, and J.R. Durrant. 2012. Artificial photosynthesis for solar water-splitting. *Nat. Photon.* 6:511–518.

Tard, C. and C.J. Pickett. 2009. Structural and functional analogues of the active sites of the [Fe]-, [NiFe]-, and [FeFe]-hydrogenases. *Chem. Rev.* 109:2245–2274.

Tranchemontagne, D.J., K.S. Park, H. Furukawa, et al. 2012. Hydrogen storage in new metal–organic frameworks. *J. Phys. Chem. C* 116 (24):13143–13151.

U.S. Department of Energy. *Metal Hydrides* [cited 18 December 2012] Available from http://www1.eere.energy.gov/hydrogenandfuelcells/storage/metal_hydrides.html.

U.S. Department of Energy. FY 2011. Progress report for the DOE hydrogen and fuel cells program. 2011.

U.S. Department of Energy, and Office of Energy Efficiency and Renewable Energy. 2009. Targets for onboard hydrogen storage systems for light-duty vehicles.

U.S. Energy Information Administration. 2008. The impact of increased use of hydrogen on petroleum consumption and carbon dioxide emissions. Report no. SR-OIAF-CNEAF/2008–04. U.S. EIA.

Vagnini, M.T., A.L. Smeigh, J.D. Blakemore, et al. 2012. Ultrafast photodriven intramolecular electron transfer from an iridium-based water-oxidation catalyst to perylene diimide derivatives. *Proc. Natl. Acad. Sci.* 109:15651–15656.

Verne, J. 1920. *The Mysterious Island (L'île Mystérieuse)*. New York: Charles Scribner's Sons.

Wang, H.-Y., G. Si, W.-N. Cao, et al. 2011. A triad [FeFe] hydrogenase system for light-driven hydrogen evolution. *Chem. Commun.* 47:8406–8408.

Wang, F. et al. 2012. Artificial photosynthetic systems based on [FeFe]–hydrogenase mimics: The road to high efficiency for light-driven hydrogen evolution. *ACS Catal.* 2 (3):407–416. American Chemical Society.

World Health Organization. 2012. *The International Decade for Action: Water for Life 2005–2015* [cited 1 November 2012]. Available from http://www.who.int/water_sanitation_health/decade2005_2015/en/

Xiang, Z. et al. 2011. CNT@Cu3(BTC)2 and metal–organic frameworks for separation of CO_2/CH_4 mixture. *J. Phys. Chem. C* 115 (40):19864–19871. American Chemical Society.

Yang, J., A. Sudik, C. Wolverton, et al. 2010. High capacity hydrogen storage materials: Attributes for automotive applications and techniques for materials discovery. *Chem. Soc. Rev.* 39 (2):656–675.

Yuhas, B.D., A.L. Smeigh, A.P.S. Samuel, et al. 2011. Biomimetic multifunctional porous chalcogels as solar fuel catalysts. *J. Am. Chem. Soc.* 133 (19):7252–7255.

Zaluska, A., L. Zaluska, and J.O. Ström-Olsen. 2000. Sodium alanates for reversible hydrogen storage. *J. Alloys Compd.* 298 (1–2):125–134.

6 Fuel Cells

6.1 INTRODUCTION

The greatest demand for petroleum is concentrated in the transportation sector, where the internal combustion engine's inherent inefficiency and the staggering number of vehicles in use have helped put an ever-increasing amount of CO_2 into the atmosphere. In the United States, transportation is responsible for 27% of our primary energy use, with the combustion of petroleum fuels being the largest source of carbon dioxide emissions (U.S. Energy Information Administration 2008). U.S. transportation, in fact, consumes almost 6% of the *global* primary energy supply (Armaroli and Balzani 2011). Given this unfortunate reality, one potential solution to the fossil fuel/CO_2 emission problem is to power cars with hydrogen fuel cells. The U.S. Department of Energy has targeted a reduction in cost for a portable fuel cell system to \$30/kW by 2017; as of 2011, the cost was \$49/kW—a noteworthy 82% decrease since 2002 (U.S. Department of Energy 2011a). In addition to decreasing cost, increasing the durability of fuel cells over a wide breadth of operating conditions and through thousands of startup/shutdown cycles is an essential objective of fuel cell research.

There are a number of different types of fuel cells as shown in Table 6.1. Fuel cells can be used for portable or stationary applications and a variety of different fuels can be used, depending upon the type of cell. We will primarily focus on hydrogen gas as the fuel of choice for fuel cell-powered transportation, with a brief examination of alcohol-based fuel cells.

6.1.1 FUEL CELL BASICS

While fuel cells have been known since the 1800s, it was modern space exploration that provided the impetus for their development. Electrochemical cells (also known as *voltaic* or *galvanic* cells) come in all sorts of shapes and sizes, and fuel cells are but one version (Figure 6.1). The reverse of a fuel cell is an electrolytic cell, using electricity to split water into H_2 and O_2, as discussed in Section 5.2.4. Electrochemical cells are basically devices that claim the energy hidden in chemical bonds and convert it directly into electrical energy. For example, batteries are a convenient and portable electrochemical cell that powers everything from our cell phones to our cars. A fuel cell is simply another type of electrochemical cell that is designed such that the chemicals required to power the cell (the fuel and the oxidant) can be continually replenished—it is, in essence, a "fuelable" battery.

The beauty of the fuel cell is that it is a streamlined energy conversion process. Like other energy conversion devices, a fuel cell takes a fuel and oxidizes it to produce energy, but a fuel cell accomplishes this feat *without* the intermediacy of thermal energy conversions. When coal is combusted to generate electricity, for

TABLE 6.1
Types of Fuel Cells[a]

Acronym	Name	Efficiency (%)[b]	Operating Range (°C)	Electrolyte	Oxidation of Fuel at Anode	Reduction at Cathode	Other
PEMFC, SPFC, PEFC	Polymer electrolyte membrane fuel cell	≈55	50–200	H^+ (sulfonic acid embedded in ionomer membrane)	$H_2(g) \rightarrow 2H^+(aq) + 2e^-$	$\frac{1}{2}O_2(g) + 2e^- \rightarrow H_2O$	Also known as solid polymer fuel cell or polymer electrolyte fuel cell. Quick startup; fewer corrosion and electrolyte management problems. Uses noble metal catalyst prone to poisoning by fuel impurities.
AFC	Alkaline fuel cell	60	50–200	6 M KOH	$H_2(g) + 2HO^-(aq) \rightarrow 2H_2O(l) + 2e^-$	$2e^- + \frac{1}{2}O_2(g) + H_2O(l) \rightarrow 2HO^-(aq)$	Well suited to space explorations where CO_2 contamination is not an issue $(CO_2 + HO^- \rightarrow HCO_3^-)$
DMFC	Direct methanol fuel cell	40	50–110	H^+ (sulfonic acid embedded in ionomer membrane)	$CH_3OH + H_2O \rightarrow CO_2(g) + 6H^+ + 6e^-$	$6H^+ + 6e^- + \frac{3}{2}O_2 \rightarrow 3H_2O$	Uses methanol for fuel without initial steam reforming to hydrogen; therefore, more cost effective, but low power density. Like PEMFC, uses an ionomer membrane electrolyte; "crossover" of methanol fuel an issue.
PAFC	Phosphoric acid fuel cell	>40	≈200	H_3PO_4			"First-generation" fuel cell. Stable, inexpensive electrolyte (H_3PO_4) adsorbed on SiC; produces high-quality "waste" heat. Insensitive to CO_2 but platinum catalyst is sensitive to fuel impurities (CO); corrosive. Typically used for stationary applications.

						Comments
MCFC	Molten carbonate fuel cell	≈46	630–650	CO_3^{2-}	$H_2(g) + CO_3^{2-} \rightarrow H_2O + CO_2(g) + 2e^-$ or $CO + CO_3^{2-} \rightarrow 2CO_2 + 2e^-$ $\frac{1}{2}O_2(g) + CO_2(g) + 2e^- \rightarrow CO_3^{2-}$	"Second-generation" fuel cell. Can use a variety of fuels, including H_2, CO, and CH_4. Electrolyte mixture is typically 50:50 Li_2CO_3:K_2CO_3 with trace other carbonates and additives to provide mechanical strength. High operating temperatures; therefore, can use less expensive electrodes. Anode is a porous nickel stabilized with about 10% Cr or Zr oxide to prevent sintering (Sandstede, 2000). High efficiency without noble metal catalysts, but slow startup, high cost, and intolerant to sulfur impurities.
SOFC	Solid oxide fuel cell	≈40	700–1000	Ceramic solid oxide (Y, Zr)	$2H_2(g) + 2O^{2-} \rightarrow 2H_2O + 4e^-$ and $2CO + 2O^{2-} \rightarrow 2CO_2 + 4e^-$ $O_2(g) + 4e^- \rightarrow 2O^{2-}$	Also known as a ceramic fuel cell; good efficiency with several multi-kilowatt stationary applications already in existence. Generates high-quality waste heat. High cost with slow startup and intolerance to sulfur.

Source: Adapted from Bartholomew, C.H. and R.J. Farrauto. 2006. *Fundamentals of Industrial Catalytic Processes*, 2nd ed. Hoboken, NJ: Wiley & Sons; Mekhilef, S., R. Saidur, and A. Safari. 2012. Comparative study of different fuel cell technologies. *Renew. Sustain. Energy Rev.* 16:981–989; Kirubakaran, A., S. Jain, and R.K. Nema. 2009. A review on fuel cell technologies and power electronic interface. *Renew. Sustain. Energy Rev.* 13:2430–2440; U.S. Department of Energy. 2008. Energy efficiency and renewable energy information center. *Comparison of Fuel Cell Technologies*. Available from www.hydrogen.energy.gov.

[a] Microbial fuel cell technology is not yet sufficiently developed to include in this table.

[b] Efficiency in this case refers to "electrical efficiency," that is, the efficiency calculated on the basis of the energy generated from the electrochemical reaction.

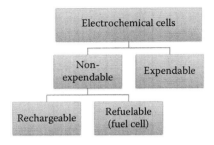

FIGURE 6.1   The categories of electrochemical cells.

example, it is a multistep process with several energy conversion steps, each of which has limited efficiency (Figure 6.2). Furthermore, since generation of electricity in a coal-fired plant is based on converting thermal energy into electrical energy, this step is ultimately limited by the Carnot efficiency (recall Equation 3.5), while a fuel cell is not. For these reasons alone, the use of fuel cells to power vehicles is an attractive alternative.

While there are a number of different types of fuel cell, they all consist of the same basic parts as shown in Figure 6.3. Of course, there is the *fuel* and the *oxidant*. All fuel cells require two electrodes: the negatively charged *anode*, at which the fuel is oxidized (releasing a flow of electrons to an external load), and an electron sink— the positively charged *cathode* (at which the oxidant is reduced). The cathode is required to take up the electrons exiting from the load. There must be a way for ionic conduction to occur to complete the electrochemical circuit; hence some sort of *electrolyte* is required. Early fuel cells used liquid electrolytes (concentrated alkaline or acid) but the current trend is in the direction of solids—mostly polymers or ceramics. We will discuss these individual components and how they work in a variety of fuel cells in greater detail later in this chapter.

6.1.2   AN ELECTROCHEMISTRY REVIEW

The overall operation of a fuel cell is straightforward: fuel and oxidant go into the cell, electricity is produced, and waste products and waste heat are removed (Figure 6.3). Managing these waste products is not a trivial matter, but treatment of this topic is beyond the scope of this text. But how do we get from fuel and oxidant to electricity? Since the generation of electrical energy by a fuel cell is based on electrochemistry, a very brief review is in order. As noted above, some electron-rich compound

FIGURE 6.2   The multiple conversions needed for conventional electricity generation. (Reprinted with permission from Li, X. 2006. *Principles of Fuel Cells*. New York: Taylor & Francis.)

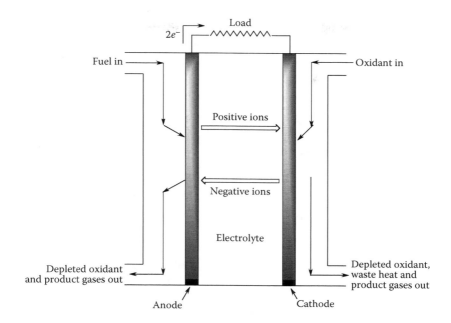

FIGURE 6.3   Schematic of a prototypical fuel cell.

(the fuel) is oxidized at the anode while something with a high potential for reduction (the oxidant) is, in turn, reduced at the cathode. The pairing of these two *half-cell* reactions leads to the overall electrochemical equation. For the typical hydrogen fuel cell, the two half-cell reactions are

$$H_2(g) \rightarrow 2H^+(aq) + 2e^- \quad (oxidation \text{ of hydrogen at anode}) \qquad (6.1)$$

and

$$\tfrac{1}{2}O_2(g) + 2H^+ + 2e^- \rightarrow H_2O \; (reduction \text{ of oxygen at cathode}) \qquad (6.2)$$

with the overall reaction being

$$H_2(g) + \tfrac{1}{2}O_2(g) \rightarrow H_2O \qquad (6.3)$$

The *cell potential* (E_{cell}, also referred to as the *electromotive force* (EMF) or *cell voltage*) is the force (in units of volts; 1 V = 1 joule/coulomb) that moves the electrical current through the circuit. The magnitude of E_{cell} is a reflection of the difference in *reduction potentials* of the chemical reactants involved. A compound that is easily reduced has a high, positive reduction potential while a compound that is strongly prone to oxidation has a highly negative reduction potential. A few representative values of standard reduction potentials are shown in Table 6.2 and more complete

TABLE 6.2
Selected Standard Half-Cell Potentials

Half-Reaction	E° (V)
$F_2(g) + 2e^- \rightarrow 2F^-(aq)$	2.87
$O_3(g) + 2H^+(aq) + 2e^- \rightarrow O_2(g) + H_2O$	2.07
$Cl_2(g) + 2e^- \rightarrow 2Cl^-(aq)$	1.36
$O_2(g) + 4H^+(aq) + 4e^- \rightarrow 2H_2O$	1.23
$NO_3^-(aq) + 4H^+(aq) + 3e^- \rightarrow NO(g) + 2H_2O$	0.96
$Ag^+(aq) + e^- \rightarrow Ag°(s)$	0.08
$I_3^-(aq) + 2e^- \rightarrow 3I^-(aq)$	0.53
$2H^+(aq) + 2e^- \rightarrow H_2(g)$	0.000
$Zn^{++}(aq) + 2e^- \rightarrow Zn°(s)$	−0.763
$2H_2O + 2e^- \rightarrow H_2(g) + 2OH^-(aq)$	−0.83
$Li^+(aq) + e^- \rightarrow Li°(s)$	−3.05

Note: Values are standard state (25°C, 1 M concentration, 1 atm pressure).

tables of standard reduction potentials can be found in many reference texts. Thus, of the examples shown, fluorine gas is most easily reduced (2.87 V) while lithium metal is the most powerful reducing agent (−3.05 V).

The output of an individual cell is quite small—on the order of 0.7–0.8 V—so fuel cells are constructed so that many cells can be linked together through electrically conducting bipolar plates to make a fuel cell *stack*. The anode of one cell is in electrical contact with the cathode of the next cell (and so on) and the entire assembly is designed so that the fuel and oxidant supply can flow through every cell. The overall voltage of the stack is a function of the number of individual cells. Some solid oxide fuel cell (SOFC) systems are able to generate power at levels up to multi-100 MW and are used for stationary power generation (U.S. Department of Energy National Energy Technology Laboratory (NETL) 2012).

6.2 THERMODYNAMICS AND FUEL CELLS

To understand how fuel cells work—and how *much* work we can obtain from fuel cells—we must return to thermodynamics. A detailed treatment of thermodynamics and fuel cells can be found in Li (2006); we will consider some generalities here. The cell potential is determined by the difference in the reduction potentials of the fuel and oxidant. Therefore, Gibbs free energy, enthalpy, entropy, and E_{cell} are all interrelated.

6.2.1 CALCULATION OF CELL POTENTIAL

It is simple to calculate $E°_{cell}$ (the cell potential at standard conditions of 1 atm, 298 K, and 1 M concentrations) using the half-cell potential values in Table 6.2. For the H_2/O_2 redox pair

$$E_{cell}^{\circ} = E_{cathode}^{\circ} - E_{anode}^{\circ} = 1.23 - 0.00 = 1.23 \text{ V} \tag{6.4}$$

This value represents the maximum possible amount of energy that can be generated by this fuel cell under standard conditions, provided that the cell is operating in a thermodynamically reversible fashion.

> *N.B.* Things can get quite confusing when talking about cell potential. There is E_{cell}, $E_{cell(reversible)}$, and E_{TN}, the thermoneutral cell potential. E_{cell}° is the cell potential at standard conditions. $E_{cell(reversible)}^{\circ}$ just spells out that this is the cell potential when the cell is operating under reversible conditions. It represents the thermodynamic maximum, but it is *not* the 100% *theoretically* possible maximum voltage. The theoretical maximum is called the "thermoneutral" cell potential (E_{TN}, also referred to as the "potential voltage" to make things *really* confusing), which does not take into account the energy losses that must result by the second law of thermodynamics, *vide infra*. The thermoneutral cell potential is the voltage a fuel cell would exhibit if all the chemical energy of the fuel and oxidant is converted perfectly into electric energy—a thermodynamic impossibility.

6.2.2 CELL POTENTIAL AND GIBBS FREE ENERGY

For a given redox couple, the Gibbs free energy, ΔG, corresponds to the net energy that this pair can produce. The relation between Gibbs free energy, enthalpy, and entropy is given by

$$\Delta G = \Delta H - T\Delta S \tag{6.5}$$

where ΔH amounts to the thermal energy that is generated and ΔS the entropic cost. These concepts are related in the equation*

$$\Delta G = -nFE_{cell} \tag{6.6}$$

where F is the Faraday constant of 9.65×10^4 C/mol and n is the number of mols of electrons transferred in the balanced electrochemical equation. Note that this equation illustrates the connection of sign to the spontaneity of the reaction: a negative ΔG relates to a *positive* E_{cell}; therefore, the cell potential must be positive in order for the electrochemical reaction to proceed as written. Since an electrochemical reaction is ultimately about transferring some quantity of electrons, ΔG is also related to the amount of charge, q. Since $q = nF$

$$\Delta G = qE_{cell} = -nFE_{cell} \tag{6.7*}$$

In the end analysis, ΔG is simply the cell potential recast in joules to encompass the molar amount of charge.

---

* You may be wondering "where did that negative sign come from?!" It all goes back to thermodynamics and the convention of sign, as discussed in Section 3.2. Free-energy change (ΔG) represents the maximum possible work that can be done. Because the work is done on the surroundings, the negative sign must appear.

6.2.2.1 State of Water

There is an important caveat in calculating the cell potential for the hydrogen fuel cell: the water produced in the cell may be a liquid or a gas depending upon the conditions. This has a profound effect on the thermodynamic values and, therefore, E°_{cell} because of the heat of vaporization of water (recall the discussion of lower heating value (LHV) and higher heating value (HHV) in Section 1.3). When calculating E°_{cell}, one must make sure to use the correct thermodynamic data (see Table 6.3). To illustrate, we will first *confirm* the value of E°_{cell} for H_2O (l) by using thermodynamic data (instead of the half-cell reduction potentials).

1. Calculation of ΔG° using $\Delta G = \Delta H - T\Delta S$. Referring to Table 6.3, at 1 atm and 25°C, the enthalpies for H_2 and O_2 are both zero by definition and the enthalpy of formation for H_2O (l) is −285,830 J/mol. Therefore, ΔH is calculated as

$$\Delta H^\circ = H_{prod} - H_{react} = -258{,}830 \text{ J/mol} - 0 \text{ J/mol} = -258{,}830 \text{ J/mol} \quad (6.8)$$

   The change in entropy is calculated in a similar fashion, taking into account the stoichiometry of the reaction:

$$\Delta S^\circ = S_{prod} - S_{react} = 69.92 \text{ J/mol} \cdot \text{K} - (130.68 \text{ J/mol} \cdot \text{K} + 0.5 \text{ mol } O_2/\text{mol } H_2$$
$$\times 205.14 \text{ J/mol} \cdot \text{K}) = -163.2 \text{ J/mol} \cdot \text{K} \quad (6.9)$$

   Substituting in our calculated values for ΔH° and ΔS°, we find ΔG°:

$$\Delta G^\circ = -285{,}830 \text{ J/mol} - 298 \text{ K}(-163.2 \text{ J/mol} \cdot \text{K}) = -237{,}196 \text{ J/mol} \quad (6.10)$$

2. Calculation of E°_{cell} from ΔG°. Since 2 mol of electrons are transferred for each mol of hydrogen fuel (recall Equations 6.1 and 6.2), calculation of the standard, reversible cell potential $E^\circ cell$ is straightforward by rearrangement of Equation 6.6:

TABLE 6.3
Thermodynamic Data for Hydrogen, Oxygen, and Water

Substance	ΔH (J/mol), 298 K, 1 atm	ΔS (J/mol · K), 298 K, 1 atm
H_2	0	130.595
O_2	0	205.14
H_2O (l)	−285,830	69.92
H_2O (g)	−241,845	188.83

Source: Adapted from Li, X. 2006. *Principles of Fuel Cells.* 1st ed. New York: Taylor & Francis.
Note: Enthalpy of vaporization for water = 44,010 J/mol.

$$E_{cell}^{\circ} = \frac{-237{,}196 \text{ J/mol}}{(2 \text{ mol } e^-/\text{mol } H_2)(96{,}487 \text{ C/mol } e^-)} = 1.2292 \text{ J/C} = 1.23 \text{ V}$$

As expected, the value matches that obtained from the measured half-cell potentials.

If the water produced is not in the liquid state but is instead a gas, a different value for E_{cell} will result given the different values for enthalpy and entropy in the formation of the product, in water. Referring again to the thermodynamic values provided in Table 6.3 and running through the calculations as described above, we find that the maximum energy produced is *less* when water is produced in the gaseous state (1.18 V < 1.23 V) (Equations 6.11 through 6.14). This is because energy is lost in the vaporization of water from liquid to gas.

Calculation of ΔH:

$$\Delta H = H_{prod} - H_{react} = -241{,}845 \text{ J/mol} - 0 \text{ J/mol} = -241{,}845 \text{ J/mol} \quad (6.11)$$

Calculation of ΔS:

$$\Delta S = S_{prod} - S_{react} = 188.83 \text{ J/mol} \cdot K - (130.68 \text{ J/mol} \cdot K + 0.5 \text{ mol } O_2/\text{mol } H_2$$
$$\times 205.14 \text{ J/mol} \cdot K) = -44.42 \text{ J/mol} \cdot K \quad (6.12)$$

Calculation of ΔG:

$$\Delta G = -241{,}845 \text{ J/mol} - 298 \text{ K}(-44.42 \text{ J/mol} \cdot K) = -228{,}608 \text{ J/mol} \quad (6.13)$$

Calculation of E_{cell}:

$$E_{cell} = \frac{-228{,}608 \text{ J/mol}}{(2 \text{ mol } e^-/\text{mol } H_2)(96{,}487 \text{ C/mol } e^-)} = 1.185 \text{ V} \quad (6.14)$$

The take-home message in these calculations is that it is important to pay close attention to the physical states and thermodynamic values of the compounds involved.

6.2.2.2 Effect of Temperature and Pressure

Because E_{cell} is ultimately a reflection of Gibbs free energy, it stands to reason that temperature will influence the operating potential of a fuel cell. The $T\Delta S$ term in the Gibbs free energy relationship (Equation 6.5) reveals that the influence of temperature is tied to entropic changes. Almost all fuel cell reactions have a negative change in entropy, so an increase in temperature means that E_{cell} decreases with temperature. This is true, for example, for the hydrogen fuel cell, where a decrease of about 0.3 V over a temperature range spanning 300–1300 K is observed (Li 2006).

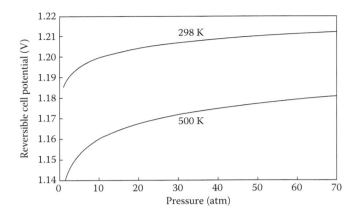

FIGURE 6.4   Impact of temperature and pressure on cell potential. (Reprinted with permission from Li, X. 2006. *Principles of Fuel Cells*. New York: Taylor & Francis.)

> An aside regarding temperature: the operating temperature of a fuel cell impacts considerably more than just the cell potential. Of course, there are advantages to operating at a higher temperature: the rate of the redox reaction increases leading to an increase in the current density and the power of the cell. Poisoning of noble metal catalysts decreases at higher temperatures as well. But with higher temperatures can come increased degradation of the fuel cell materials and additional mechanical problems, particularly with polymer components.

The effect of pressure on E_{cell} is significant for those fuel cell systems that have gaseous reactants and/or products. In this case, the E_{cell} is dependent upon the change in the number of moles of gas species in the electrochemical reaction. Again, most fuel cell reactions trend in the same direction—most see a decrease in the number of moles of gas species formed. This results in an increase in E_{cell} but at an even lower level than the impact of temperature. Figure 6.4 graphically illustrates the impact of temperature and pressure for the hydrogen/oxygen fuel cell (Li 2006).

6.3   EFFICIENCY AND FUEL CELLS

Our treatment of thermodynamics in Chapter 2 is revisited here as we consider the efficiency of fuel cells: even though fuel cells are *more* efficient than, say, a mechanical heat engine, they are still limited by the restrictions posed by the second law of thermodynamics. Some energy will be lost as waste heat even under ideal operating conditions, even for a fuel cell. In the following discussion of fuel cell efficiency, it should be noted that the focus is strictly on the electrical efficiency of the fuel cell: efficiency in terms of the energy generated by the electrochemical reaction only. Because fuel cells can generate high-quality heat that may be captured and used as part of the process, a separate value for efficiency—the *CHP efficiency* (for combined heat and power, also known as cogeneration power)—is sometimes reported. The potential for >80% CHP efficiency exists for fuel cells (Vogel et al. 2009).

The efficiency of a fuel cell operating under thermodynamically reversible conditions is

$$\eta_{rev} = \frac{\text{Electricity produced}}{\text{Heating value of fuel used}} = \frac{W_{max}}{-\Delta H} = \frac{\Delta G}{\Delta H} = \frac{E_{rev}}{E_{tn}} \times 100\% \quad (6.15)$$

Note that the efficiency is a ratio of free energy to enthalpy *or* of the reversible cell potential as it compares to the thermoneutral cell potential. The thermoneutral cell potential for the hydrogen/oxygen redox pair is 1.48 V; comparing that to the E°_{cell} of 1.23 V (liquid water) gives a reversible efficiency of 83% for the hydrogen/oxygen fuel cell. (A word of warning: this efficiency calculation is based on the HHV for hydrogen gas.)

It cannot be stressed enough that this "reversible efficiency" is an *ideal*, whereas practical, everyday fuel cells invariably operate under irreversible conditions. Incomplete reaction, by-product formation, voltage losses due to resistance, and other miscellaneous issues are reality. For comparison, the practical efficiency of fuel cells typically ranges from about 50% to 65% (based on the LHV of the fuel used), while steam and gas turbines have practical efficiencies in the realm of 40% (also based on LHV, Li 2006).

6.4 CELL PERFORMANCE: WHERE DO INEFFICIENCIES COME FROM?

6.4.1 VOLTAGE, CURRENT, AND POWER

Given that even the ideal, reversible fuel cell cannot operate with 100% efficiency, where do these losses in voltage come from? Before we can answer this question, we need to review the relationships between voltage, current, resistance, and power:

$$\text{Current (amps)} = \frac{\text{Potential (volts)}}{\text{Resistance (ohms)}} \quad (6.16)$$

$$\text{Power (watts} \equiv \text{J/s)} = \text{Current (amps)} \times \text{Potential (volts)} \quad (6.17)$$

These relationships allow us to understand the two extreme conditions in electrical circuits: an *open circuit* (zero current, maximum voltage) and a *short circuit* (maximum current, zero voltage). When the circuit is open, the anode and cathode are not connected and no electricity can flow, hence the voltage is at its maximum. This is the *open-circuit voltage* (V_{OC}). *The open circuit voltage is not a theoretical quantity*: it is the actual measurement of voltage when the circuit is open. Hence, in a perfect fuel cell at standard temperature and pressure, V_{OC} could, in theory,

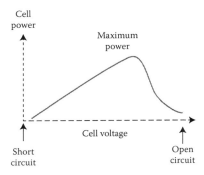

FIGURE 6.5   Relationship between power and voltage. (Reprinted with permission from Cracknell, J.A., K.A. Vincent, and F.A. Armstrong. 2008. Enzymes as working or inspirational electrocatalysts for fuel cells & electrolysis. *Chem. Rev.* 108:2439–2461. Copyright 2008, American Chemical Society.)

match $E°_{cell}$. In reality, however, V_{OC} is often less than E_{cell} due to the inefficiencies we are about to discuss. Clearly, the closer V_{OC} is to E_{cell}, the greater the integrity of the cell.

At the other extreme, when the circuit is closed with no load, a maximum current results. This is denoted the *short circuit current* (I_{SC}). Given the relationship between the short circuit and the open circuit, the maximum power output for a given cell will lie somewhere in between, as depicted in Figure 6.5 (Cracknell et al. 2008).

6.4.2   POLARIZATION

In an ideal reversible cell, the plot of voltage versus current would yield a straight line. However, as Figure 6.6 illustrates, the cell potential typically decreases nonlinearly at high and low current draw. A close examination of Figure 6.6 reinforces some important relationships regarding cell potentials:

- The thermoneutral cell potential (E_{tn}) is the absolute, perfect, maximum potential that can be obtained only in theory.
- The reversible cell potential ($E_{cell(reversible)}$) represents that maximum that can be obtained *in theory and* taking into account inevitable losses due to entropy (the second law of thermodynamics).
- The actual cell potential is diminished from E_{rev} by irreversible losses due to inefficiencies in mass transport, resistance, and kinetics (polarization).

These irreversible losses, generally referred to as *polarization* (or *overpotential* or *overvoltage*), fall into three basic categories and areas on a voltage/current curve: *activation polarization*, *ohmic polarization*, and *concentration polarization*. The sources of these losses are briefly described below.

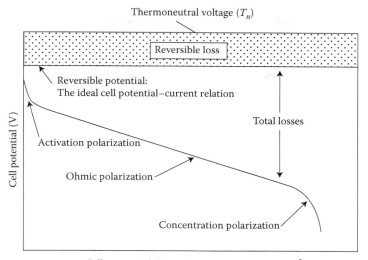

FIGURE 6.6   A typical fuel cell polarization curve. (Reprinted with permission from Li, X. 2006. *Principles of Fuel Cells*. New York: Taylor & Francis.)

6.4.2.1   Loss Due to Activation

Loss of potential due to activation polarization is a result of the lag time of a slow electrochemical reaction as the current is drawn. In essence, the reaction cannot "keep up" with the current demand, so excess energy is required to speed up the reaction.

6.4.2.2   Ohmic Losses

Ohmic losses are exactly what one would predict: there is invariably resistance in the cell, whether it is resistance to ionic or electronic conductivity. In either case, the efficiency of the cell is reduced.

6.4.2.3   Concentration Effects

Losses due to concentration differentials are related to mass transfer. When current density is high, accumulation of products or depletion of reactants at the relevant electrode can significantly hinder the transfer of the reacting species, leading to decreased rate of reaction relative to the current demand.

6.4.3   Exchange Current

One additional source of inefficiency is seen in the fact that even at nearly zero current, something is going on to reduce the potential significantly below $E_{cell(reversible)}$ (see Figure 6.6). This is a result of what is known as the *exchange current*, a phenomenon that amounts to "leakage" of electrons through the electrolyte. Electrons are transported to the cathode and ions to the anode, where more electrons are released to migrate back through the electrolyte to the cathode, continuing the exchange. This process can reduce the efficiency of a real fuel cell by as much as 8–16% from the reversible cell efficiency (Li 2006).

6.4.4 CELL PERFORMANCE AND NERNST EQUATION

Lastly, the actual cell potential obtained from an electrochemical cell must somehow take into account the actual concentration of the reacting species. For example, in the hydrogen fuel cell, the oxidant is typically not 100% oxygen but is instead air. The Nernst equation

$$E_{cell} = E_{cell}^{\circ} - \frac{RT}{nF} \ln Q \qquad (6.18)$$

allows the calculation of E_{cell} from the standard potential by taking into account concentration in the form of the reaction quotient, Q ($Q = $ [red]/[ox]):

$$E_{cell} = E_{cell}^{\circ} = \frac{RT}{nF} \ln \frac{(n_{H_2}/V_{H_2})}{(n_{O_2}/V_{O_2})} \qquad (6.19)$$

6.5 FUEL CELL ELECTROCATALYSTS

As noted in the introduction, there are many different types of fuel cells and all are made up of basically the same components. There are important differences, however, in terms of the makeup of the catalysts needed to carry out the electricity-generating reactions. In this section, we will focus on some important aspects of the electrodes and catalysts that carry out fuel cell electrochemistry.

6.5.1 ELECTROCATALYSIS

Certainly among the most crucial components of a fuel cell are the electrodes, and the *sine qua non* of the electrode is the electrocatalyst. Electrode construction and catalyst preparation is a combination of art and science that has a profound impact on fuel cell performance. All that needs to happen—adsorption of reactants, reaction, conduction of electrons and ions, and diffusion of products—must take place at the atomic level at the boundary of the catalysts and the reactants. Ultimately, mass transport of the reacting species to the electrodes is key to maximizing the improvement in kinetics provided by the catalyst. Reactants must be able to diffuse to the reaction sites, adsorb onto the electrode surface, react, and the products diffuse off—plus the electrons must be collected and allowed to flow through the circuit with minimization of the exchange current. Thus the catalyst material, surface areas and morphology all impact the overall efficiency of the cell, although a detailed treatment of transport phenomena in fuel cells is beyond the scope of this text.

At this point, we will consider only the H_2/O_2 polymer electrolyte membrane fuel cell (PEMFC), although many of the following concepts apply across the spectrum of fuel cell types. An electrocatalyst is needed for both hydrogen oxidation at the cathode and oxygen reduction at the anode, and the platinum group metals (PGM) are currently the most commonly used. The electrodes themselves must be porous so that the gases can diffuse through the cell to the electrocatalytic surface—hence the label *gas diffusion electrodes*. The usual preparation of the catalyst takes place by adsorption of an aqueous solution of the metal salt (e.g., $PtCl_2$) onto the support material, typically carbon black. This mixture is then treated with a reducing agent (hydrogen or a metal hydride, for example) to reduce the metal. Other materials are added and the resultant "catalyst ink" is coated onto a gas diffusion layer. Unfortunately, this method of electrode preparation leaves as much as 30% of the metal inaccessible—and therefore catalytically inactive—so that developing new methods for preparing the electrocatalyst is a vigorous area of research (Mitzel et al. 2012).

6.5.2 Oxygen Reduction Reaction

While efficient transformations at both the cathode and the anode are required for successful fuel cell performance, it is the oxygen reduction reaction (ORR) that is the prime target for improving efficiency. The ORR is notoriously slow, an Achille's heel that impacts any fuel cell that uses oxygen as the oxidant. Given this (and the acidic reaction conditions), platinum is largely considered indispensible as the ORR electrocatalyst, presenting a serious sustainability and cost issue.

The mechanism of the ORR is quite complex and dependent upon the specific electrode material. As a result, it is not well understood beyond the knowledge that there are several elementary steps and uncertain intermediates. Two simplified overall pathways were introduced by Wroblowa et al. in 1976 (Wroblowa et al. 1976): the two-electron peroxide pathway and the four-electron pathway:

$$\text{Four-electron pathway: } O_2 + 4e^- + 4H^+ \rightleftharpoons 2H_2O \qquad (6.20)$$

and

$$\text{Two-electron ("series") pathway: } (1)\ O_2 + 2e^- + 2H^+ \rightleftharpoons 2H_2O_2$$
$$(2)\ H_2O_2 + 2e^- + 2H^+ \rightleftharpoons 2H_2O \qquad (6.21)$$

The intermediacy of hydrogen peroxide and the specific role of the metal have yet to be fully understood. The peroxide/two-electron pathway predominates on most carbon materials, gold, and metal oxides (Erable et al. 2012). The four-electron pathway predominates when platinum is involved; a plausible role for the metal is shown in Figure 6.7. Molecular oxygen, pi-bound to the metal, accepts a large amount of electron density into its π^* orbital through back-bonding, thus weakening the oxygen–oxygen bond. Sequential protonation and reduction leads to a bis-hydroxyl species that can undergo reductive elimination to generate water (Okada and Kaneko 2009).

$$Pt \ + \ O = O \longrightarrow Pt - \overset{\displaystyle O}{\underset{\displaystyle O}{\|}} \quad or \quad Pt \overset{O}{\underset{O}{\diagdown}} \xrightarrow{H^+ + e^-} Pt \overset{O\text{----}H}{\underset{O}{\diagdown}} \xrightarrow{H^+ + e^-}$$

$$Pt \overset{OH}{\underset{OH}{\diagdown}} \xrightarrow{2H^+ + 2e^-} Pt \ + \ 2H_2O$$

FIGURE 6.7 Proposed intermediates in the oxygen reduction reaction. (Adapted from Toda, T. et al. 1999. *J. Electrochem. Soc. 1* 146 (10):3750–3756.)

While platinum is an effective catalyst for the ORR, it comes at a cost: about one-half of the expense of a fuel cell stack (even at high production volumes) is a result of its use (Garland 2008). Given this reality and the fact that platinum is a limited resource, reducing the amount of Pt (or finding reliable, efficient, sustainable, and less expensive alternatives) is a high priority. A multitude of approaches exist and several representative examples follow.

Platinum alloys. In a seminal discovery, $Pt_3Ni(111)$ was determined to be much more active than the state-of-the-art Pt/C catalysts for PEMFC (Stamenkovic et al. 2007). The notation "111" is a crystallographic descriptor that designates the type of exposure of the surface atoms, as indicated by the shaded area of a unit cell shown in Figure 6.8. While the reason for this dramatic increase in activity is not yet well understood (Sha et al. 2012), binary and ternary platinum alloys with 3d transition metals (Ni, Co, Mn, Fe, Cu) are among the most active ORR electrocatalysts to date, with $Pt_3Ni(111)$ the most active by far (van der Vliet et al. 2012).

Much research has been devoted to understanding the role of the bimetallic surface in catalytic activity. The presence of the metal heteroatoms somehow alters the electronic and chemical properties of the catalyst as the alloy shows activity different from that of either metal alone. Two factors responsible for this phenomenon are

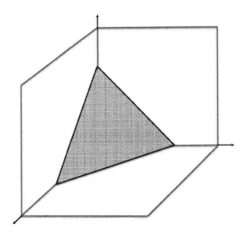

FIGURE 6.8 Representation of the {111} crystallographic plane.

postulated: (1) formation of heteroatom bonds at the metal surface, thus changing the electronic environment and modifying its electronic structure and (2) alteration of the geometry of the structure so that the average metal–metal bond length is changed. As a result, strain energy is introduced that modifies the electronic structure through changes in orbital overlap (Kitchin et al. 2004). These factors are believed to lessen the adsorption of spectator species on the alloy surface and, therefore, make additional active sites available to increase activity (Wang et al. 2012; Yu et al. 2012).

Inroads via nanotechnology. Controlling the synthesis of the electrocatalyst in terms of particle size, shape, stoichiometric composition, and homogeneity has greatly elevated the performance of fuel cell catalysts (Wang et al. 2012), and incorporation of nanoparticles has improved both the performance and durability of these catalysts. "Nano-segregated" PtNi catalysts synthesized at Argonne National Laboratories (US) have shown activity seven times higher than the Pt/C comparison for the ORR (Papageorgopoulos 2011) and a synthesis of PtNi nanoparticles using dimethylformamide as both solvent and reductant showed an enhancement almost 15 times higher than a "state-of-the-art" Pt/C catalyst (Carpenter et al. 2012). Nanostructured thin-film (NSTF) ternary electrocatalysts consisting of a $Pt_{68}Co_{29}Mn_3$ core have been "the workhorse cathode and anode of choice" according to a recent U.S. DOE report (U.S. Department of Energy 2011b) (U.S. DOE 2011). Overall, impressive progress has been made toward the reduction of platinum use in fuel cells, with PGM content being reduced to 0.05 mg/cm^2 of PGM on the anode and 0.1 mg/cm^2 on the cathode (U.S. Department of Energy 2011b) (U.S. DOE 2011).

Non-PGM electrocatalysts. As platinum remains the primary factor in fuel cell cost, research toward non-PGM or even metal-free catalysts is especially relevant. In the realm of non-PGM catalysts, iron, cobalt, and manganese have all shown promise (Morozan et al. 2011; Tan et al. 2012; Zhang and Shen 2012). Nitrogen-doped carbon nanotubes, porous carbon, and carbon nitride are among the metal-free alternatives that have been studied with some initial success (Yang et al. 2011; Yu et al. 2012). Treatment of graphene oxide (see Section 6.10.4) with ammonia in the presence of boric acid yields a boron- and nitrogen-doped material with better electrocatalytic activity than a commercially available Pt/C electrode (Wang et al. 2012). Enzymes are, potentially, the ultimate metal-free catalyst alternative for the ORR. We will examine their role in fuel cell chemistry in the section on microbial fuel cells (MFCs) (Section 6.8). While the performance of these alternatives to platinum is still quite low compared to the standard Pt/C catalyst, early results are encouraging.

Although most research has focused on the ORR at the cathode, reducing platinum loading at the anode (where hydrogen oxidation takes place) is also critical. Furthermore, given that most production methods for hydrogen result in some proportion of the by-product CO, development of an electrocatalyst that can effectively tolerate this poison in the H_2 gas stream has been a high priority. Incorporation of ruthenium (in the form of a $Pt_{0.5}Ru_{0.5}$ alloy) increases the tolerance of the anode to poisoning by CO (daRosa 2009). The presence of ruthenium–platinum bonding on the catalyst surface removes the CO from the platinum, allowing for the water gas shift reaction (Equation 5.5) to take place on the ruthenium surface (Takeguchi et al. 2012). The rate-determining step for oxidation of hydrogen is the adsorption of hydrogen onto the metal surface; thus, model studies on the kinetics and transport phenomena at the catalyst boundaries are helping to elucidate the electrochemical mechanisms (Zenyuk and Litster 2012).

6.5.3   CHARACTERIZATION OF CATALYSTS

It would be a serious omission to conclude this coverage of electrocatalysts for fuel cells without at least a brief mention of the experimental techniques used in their characterization. As should be clear, being able to investigate the surface properties as well as the electrochemical properties of potential electrocatalysts is critical to evaluation of their potential. As regards methods to evaluate the homogeneity, particle size, surface roughness, etc. on catalyst activity, techniques such as scanning electron microscopy (SEM), transmission electron microscopy (TEM), scanning transmission electron microscopy (STEM), extended x-ray absorption fine structure (XAFS), x-ray diffraction (XRD), x-ray photoelectron spectroscopy (XPS), and energy-dispersive x-ray spectroscopy (EDS) are routinely used to give an atomic-level view of the topography, particle size, and homogeneity of these materials. Cyclic voltammetry and other electrochemical studies (rotating disk electrode and rotating ring-disk electrode) provide a kinetic and mechanistic probe of the redox chemistry. Finally, theoretical methods (particularly density functional theory) have proven to be useful in the calculation of charge density and modeling the mechanism of the redox reactions on novel catalyst surfaces.

6.6   POLYMER ELECTROLYTE MEMBRANE FUEL CELL

6.6.1   INTRODUCTION

With a basic understanding of the role of the electrocatalysts and the redox chemistry of hydrogen and oxygen, we can now examine some specific fuel cells in depth. We will look at four: (1) the PEMFC (alternatively described as the proton exchange membrane fuel cell), (2) the *direct methanol fuel cell* (DMFC), (3) the SOFC, and (4) MFCs. While this selection is limited, it covers the most important fuel cells and, arguably, the most promising in terms of sustainable energy generation.

A detailed schematic of a hydrogen PEMFC is shown in Figure 6.9. (Note: it should not be assumed that a PEMFC is inevitably using hydrogen as a fuel. As we will see in Section 6.6.4, the DMFC is a PEMFC that uses methanol as the fuel. However, unless otherwise noted, in this text, PEMFC *does* presume that hydrogen gas is the fuel source.) The PEMFC is a fuel cell that is of particular interest in portable applications, that is, transportation. It is considered a low-temperature ($-20°C$ to $100°C$) fuel cell, although significant effort has been devoted to the development of higher-temperature PEMFCs that can operate above $100°C$. Exclusive use of hydrogen as a fuel in vehicles powered by a PEMFC would erase the carbon footprint of the transportation sector (neglecting, of course, the carbon cost of producing the

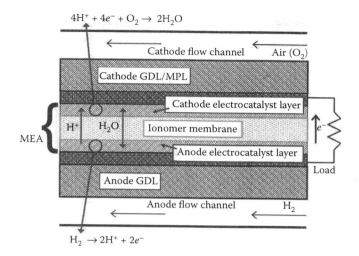

$$4H^+ + 4e^- + O_2 \rightarrow 2H_2O$$

$$H_2 \rightarrow 2H^+ + 2e^-$$

FIGURE 6.9 Schematic of a PEM fuel cell. GDL, gas diffusion layer; MPL, microporous layer; MEA, membrane electrode assembly. (Reprinted with permission from Borup, R. et al. 2007. Scientific aspects of polymer electrolyte fuel cell durability and degradation. *Chem. Rev.* 107 (10):3904–3951. Copyright 2007, American Chemical Society.)

hydrogen initially and the carbon cost in making the fuel cell materials). However, not only are technological challenges associated with the generation of the necessarily huge volume of hydrogen fuel, the development of a sufficiently robust PEMFC remains a substantial challenge.

What are the constraints faced by polymer chemists as they design a polymer for use as an electrolyte membrane in a fuel cell?

- Perhaps foremost, the polymer membrane must be mechanically stable, having a reasonably high T_g for operation at relatively high temperatures (up to 120°C; U.S. Department of Energy 2011a).
- Water is a critical component in the successful operation of a PEMFC so the polymer must perform well and be durable under conditions of high humidity.
- In addition to thermal/mechanical stability, the polymer must be robust with respect to its chemical reactivity.
- As with any electrolyte, it must act as an insulator to electrons.
- It must also be impervious to the fuel, preventing its *crossover* through the electrolyte membrane to the cathode which would lower the V_{OC}.

Because PEMs are used as the electrolyte in both the hydrogen/oxygen fuel cell and the DMFC, much polymer development has been motivated by trying to minimize the *methanol crossover* problem: diffusion of methanol through the PEM to the cathode. Crossover amounts of up to 10% can occur leading to a significant decrease in the open circuit voltage and requiring increased catalyst loadings. The thinner the PEM, the lower the ohmic losses, but the membrane then becomes more susceptible to crossover and mechanical failure. Finally, the electrocatalysts must be engineered into a PEM *membrane electrode assembly* (MEA, discussed below). Of course, it is also desirable that the PEM should be inexpensive, easy to fabricate, and synthesized in a sustainable manner! These challenges have led to (and continue to lead to) a multitude of approaches in the development of a better PEM; we describe some of the more important ones below.

6.6.2 GENERAL CONSIDERATIONS

6.6.2.1 Membrane Electrode Assembly

The heart of a PEMFC is the MEA consisting of the electrocatalysts and the electrolyte membrane (the "ionomer membrane" core shown in Figure 6.9, where an ionomer is simply an ion-conducting polymer). The design and improvement of MEAs is very much an empirical undertaking, and both chemical and mechanical variables play large roles in its effectiveness. Because the PEMFC may use gases for both the fuel and the oxidant (H_2 and O_2 or air), the electrodes typically consist of a porous carbon gas diffusion layer to which the catalyst layer is bonded. Fabrication of the MEA takes place either by bonding the catalyst layer (ca. 10 μm thick) to the ionomer membrane first, or by bonding it to the carbon electrode layer first. The ionomer membrane itself is also quite thin—usually around 10–175 μm. In either case, good bonding is essential so that the proper conductivity (and minimal loss due to ohmic resistance) can be achieved. We will look at MEAs in more detail in the context of the DMFC in Section 6.6.4.

6.6.2.2 Water Management

Water management is an important issue in PEMFCs and both the physical and chemical properties of the polymer play important roles. Moisture is necessary to provide the level of proton conductivity needed to sustain the electrochemical reaction. A dry membrane has low conductivity and increased resistance leading to a decreased current. Evaluation of PEM performance, therefore, must somehow reflect the conductivity, reported as σ in units of S/cm at specific conditions of temperature and relative humidity (RH). Exposure of the membrane to a sufficient amount of moisture allows for the development of networks of water (clusters, pools, channels, etc.) that provide a conduit for the conduction of protons by diffusion and *proton-hopping*, hence (as usual) the morphology of the polymer is a crucial variable in its performance as an ion conductor (Jorn et al. 2012). Dehydration can occur via *electro-osmotic drag*, where water molecules from the anode are literally dragged to the cathode by the proton current (Cheah et al. 2011). On the other hand, too much water "drowns" the electrodes, flooding the gas flow channels and interfering with mass transport. A hydrophobic agent such as PTFE (polytetrafluoroethylene, aka Teflon®, $-(CF_2)_n-$) is added to prevent the pores from being choked with water (Wang et al. 2011).

6.6.3 Polymer Development

Given the demands upon the polymer electrolyte, what kind of materials have been developed that can work effectively in a fuel cell? The ionomers used as electrolytes possess highly polar functionality that allow for the required proton conductivity through the channels of the membrane. Because the role of the ionomer in a fuel cell is to facilitate proton conductivity, these membranes consist of a polymer backbone with acidic functionality (typically sulfonic or phosphonic acids) scattered throughout. The backbone may vary from perfluorinated, partially fluorinated, or without fluorine; completely aliphatic to primarily aromatic; primarily made up of carbon and hydrogen or a polymer interspersed with a high proportion of heteroatoms. A special concern in the synthesis and fabrication of PEM ionomers is their unusually high polarity, making them particularly slow to dissolve in relatively nonpolar organic solvents. In any case, the development of a relatively inexpensive, easily fabricated ionomer with a high proton conductivity, low electron conductivity, and good resistance to degradation is a challenge. There is a tremendous variety of ionomers that have been investigated for use in PEMFCs but we will focus on four classes: perfluorosulfonic acids, sulfonated poly(arylene) ethers/ketones/sulfones, polyimides, and polybenzimidazoles (PBI).

6.6.3.1 Perfluorosulfonic Acid Membranes

If platinum on carbon is the standard for electrocatalyst composition in fuel cells, then Nafion® (DuPont Co.) can be considered the standard for the ionomer membrane material of PEMFCs. Nafion membranes are a series of perfluorosulfonic acid/PTFE copolymer membranes of different thicknesses and strengths (Figure 6.10). The common acronym for these membranes is PFSA or PFSI (for perfluorinated sulfonic acid or ionomer). The poly-CF_2 backbone presents excellent chemical stability and the sulfonic acid groups allow for good proton conductivity when hydrated (up to 0.10 S/cm; Li et al. 2003). PFSA membranes also offer low O_2 (g) and H_2 (g) permeability but, unfortunately, significant permeability for methanol, making them

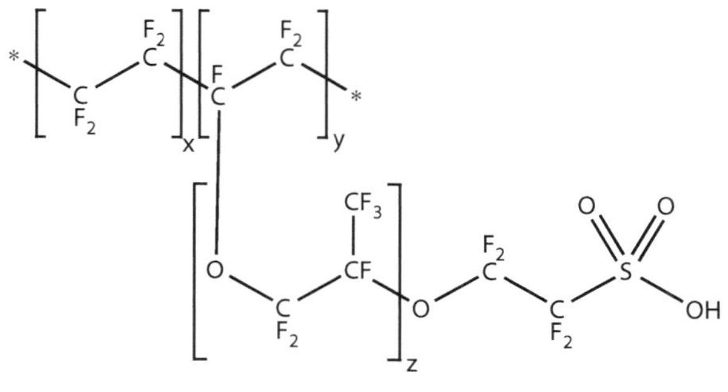

FIGURE 6.10   The Nafion copolymer structure.

poor candidates for the DMFC. In addition, not only is Nafion expensive, it also has a relatively low T_g and its operating temperature range is quite limited: PFSA membranes are effectively unusable below 0°C, at which point the water necessary for proton conductance is frozen, or above 100°C, where the stability of Nafion is poor (Nafion's ordinary operating temperature is ≈80°C). Nafion also presents serious problems in its environmental impact as it is not biodegradable and incineration generates SO_2, HF, and CO_2 (DuPont 2009). Thus, while they are widely used PEMs, much research effort has focused on modifying, or finding a replacement for, PFSA membranes.

Researchers have attempted to modify the Nafion membrane by impregnating inorganic materials (e.g., SiO_2, TiO_2, or zirconium phosphates) or functionalized fillers (carbon nanotubes, montmorillonite clays, or zeolites) into the polymer, primarily to improve thermal stability (Li et al. 2003; Zhang and Shen 2012). Doping of Nafion with an ionic liquid (triethylammonium trifluorosulfonate $(CH_3CH_2)_3NH^+CF_3SO_3^-$) was shown to match the proton conductivity of regular Nafion under *anhydrous* conditions, but its addition acted as a plasticizer, preventing its use at elevated temperatures (Sood et al. 2012). Attempts to modify the PFSA membrane by *radiation-grafting* (exposing the fluoropolymer to UV or gamma-ray radiation to prepare a reactive radical surface on the film, upon which a new monomer (e.g., styrene) can be grafted and further functionalized) have presented greatly mixed results in terms of durability and conductivity. Proton conductivities of up to 0.25 S/cm have been reported for these modified PFSA membranes, but with substantial accompanying degradation (Zhang and Shen 2012).

6.6.3.2 Poly(Arylene Ether) Membranes

Another prominent class of PEM membranes is based on sulfonated poly(arylene ether) ionomers of which there are several varieties: poly(arylene ether) ketones, poly(arylene ether) sulfones, poly(arylene ether)s with heterocyclic functionality, and so on. Poly(arylene ethers) offer more flexibility in terms of polymer synthesis and modification, are less expensive than PFSA membranes, and have shown good performance overall with proton conductivities that match or surpass PFSA membranes under similar conditions. One particularly well-known example is the semicrystalline sulfonated poly(ether)ether ketone SPEEK (Figure 6.11). SPEEK membranes, however, are prone to hydrolysis or radical-induced degradation and can become brittle at high temperatures. As with other ionomer classes, addressing the issue of

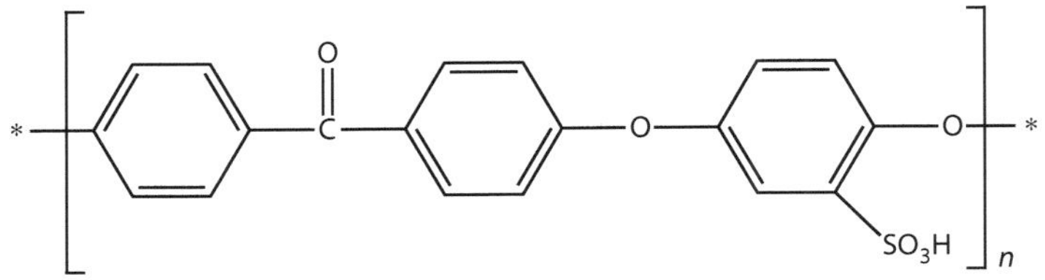

FIGURE 6.11 The sulfonated poly(ether)ether ketone repeating unit.

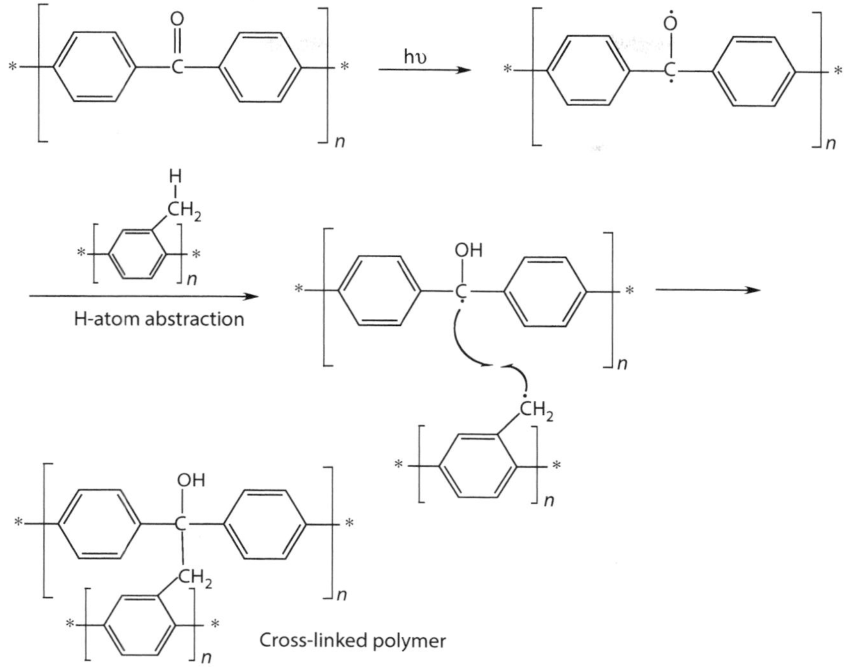

FIGURE 6.12   Radical-induced cross-linking of a SPEEK membrane.

methanol crossover is a high priority. Some modified poly(arylene ether) PEMs that have been prepared to address these concerns are discussed below.

Cross-linking of SPEEK membranes greatly increases the durability of the membrane but can have a detrimental impact on the proton conductivity. As shown in the light-induced radical coupling of Figure 6.12, cross-linking can take place between the benzophenone moiety and suitable substituent groups. As a result, this polymer showed improved selectivity for proton conduction relative to methanol permeability (Ishikawa et al. 2007). The synthesis of a cross-linked organic/inorganic hybrid poly(ether) ether membrane is shown in Figure 6.13. In this example, sulfonation of the PEEK precursor (1) provides for the development of a sulfonate-bridged SPEEK network (2). This network can then undergo ortholithiation with n-butyllithium followed by treatment with tetrachlorosilane and hydrolysis to give the silated/sulfonated ionomer product. This material demonstrated greatly improved mechanical properties (e.g., an elastic modulus of 260 ± 40 MPa, compared to 25 for Nafion), attributed to the silanol functionality and extra cross-linking. However, the enhanced mechanical rigidity lowered the proton conductivity relative to Nafion (DiVona et al. 2006). In another approach to a modified poly(ether)sulfone membrane, a bisazide was used as a coupling agent with a vinyl-appended polyethersulfone (PES) as shown in Figure 6.14. The resultant ionomer gave well-enhanced proton conductivity of 0.79 S/cm at 100°C, compared to 0.48 S/cm for Nafion 112 under identical conditions (Oh et al. 2008).

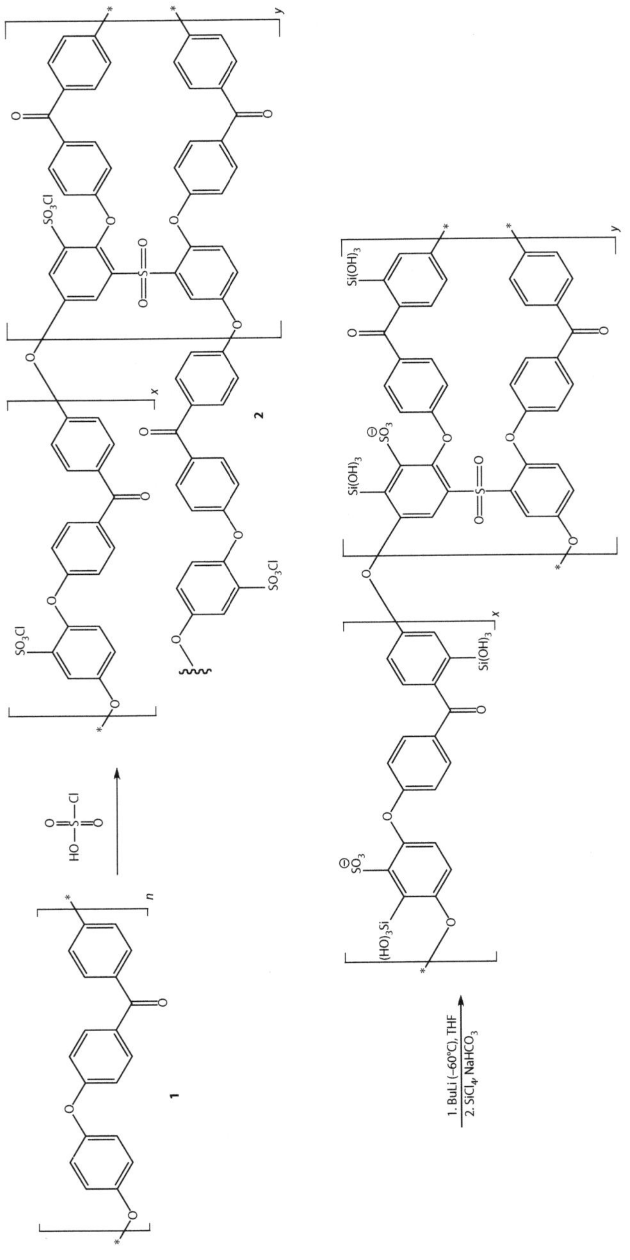

FIGURE 6.13 A silated, cross-linked SPEEK derivative.

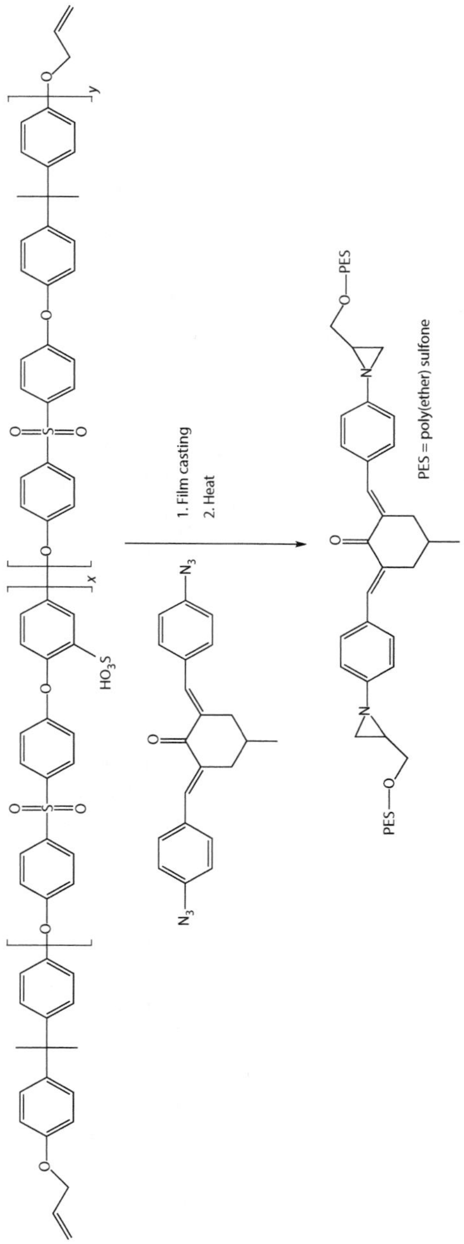

FIGURE 6.14 Poly(ether)sulfone coupling to an aromatic bisazide for enhanced mechanical stability.

While promising inroads have been made with respect to methanol crossover, the issue of PEM degradation by hydrolysis remains. Researchers have found that modification of the arene units by the incorporation of additional bulky substituents reduces the tendency toward hydrolysis. The poly(arylene ether sulfone) polymer electrolyte membrane shown in Figure 6.15 gave results comparable to Nafion in terms of both conductivity and durability (Miyatake et al. 2007). Stability to hydrolysis is even more improved when the polymer backbone includes heterocycles such as the oxazole- or triazole-containing copolymers shown in Figure 6.16; compound **b** also showed an increased proton conductivity relative to SPEEK at the same level of sulfonation (Li and Yu 2007; Ponce et al. 2008).

6.6.3.3 Polyimides and Imidazoles

Sufonated polyimides (SPI). Imides are not particularly resistant to hydrolysis, but SPI offer high mechanical strength, minimal crossover, and excellent thermal and chemical stability. Naphthalenic polyimides (Figure 6.17a) have been prepared and tested in order to address the hydrolysis issue. Ar_1 and Ar_2 represent any of a number of sulfonated oligomers, primarily of the arene ether variety. Binaphthyl SPIs (Figure 6.17b) appear to be even more resistant to hydrolysis and oxidation. For example, the naphthyl SPI (**A**, Figure 6.18) demonstrated a hydrolytic stability of 34 h at 90°C and the binaphthyl PEM **B** was stable for more than 1000 h at this temperature (Zhang and Shen 2012).

Polybenzimidazoles. Beyond concerns of methanol crossover and hydrolysis lies the issue of higher-temperature operation to take advantage of improved kinetics and reduced propensity to poisoning. PBI (Figure 6.19) are particularly attractive PEM candidates for higher-temperature PEMFC operation (≈150–200°C) due to their enhanced chemical and thermal stability. These membranes are doped with phosphoric acid to make the proton-conducting acid–base complex (Figure 6.19). Both *meta-* and *para-* PBI have been prepared, with the *para* isomer exhibiting better mechanical properties that persist even at a high level of doping with H_3PO_4 (Zhang and Shen 2012).

Like the SPI and PFSA membranes, much research has been devoted to modifying the basic PBI skeleton to improve the ionomer characteristics. Additionally, research has been done examining blends of PBI with other polymers. An SPI blended with a PBI membrane and doped with phosphoric acid (Figure 6.20) gave a proton conductivity of about 0.5 S/cm (120°C/45% RH), with no sign of decreased performance even after 1000 h of operation at 120°C and 0% RH (Suzuki et al. 2012). A 50:50 blend of PBI and poly(vinyl-1,2,4-triazole) doped with phosphoric acid (Figure 6.21) showed moderate conductivity at 120°C and no humidification (≈0.09 S/cm) compared to the conductivity of neat PBI under the same conditions of roughly 0.02 S/cm. The researchers postulated that the reason for the increased conductivity relative to PBI was due to the incorporation of more phosphoric acid in the blend, as well as to the fact that the blend includes both the triazole and imidazole rings. This material also exhibited excellent thermal stability (Hazarika and Jana 2012).

6.6.3.4 Metal–Organic Frameworks

In Chapter 5, we learned about how metal–organic frameworks are well suited for potential use as adsorbents for H_2 storage. Not only does the porosity of an MOF lend

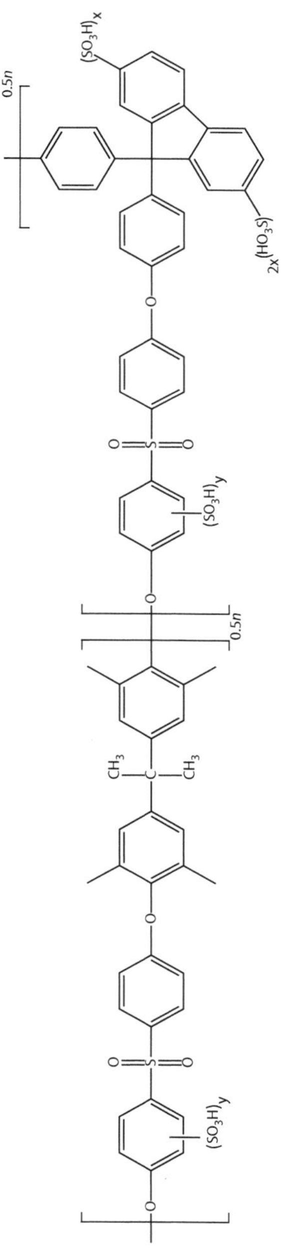

FIGURE 6.15 A hydrolysis-resistant PEM.

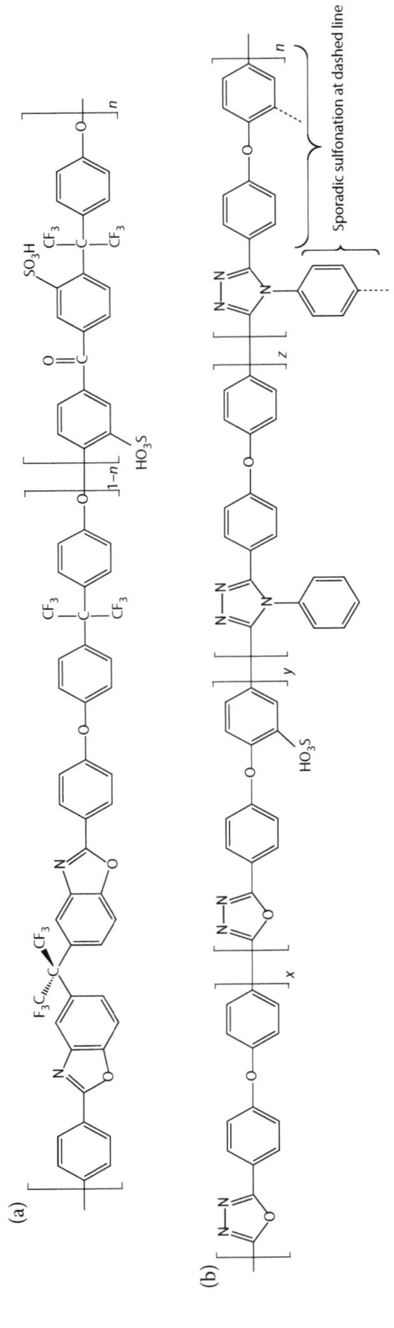

FIGURE 6.16 Polymers with improved hydrolytic resistance. (a) An oxazole-containing copolymer, (b) a triazole-containing copolymer.

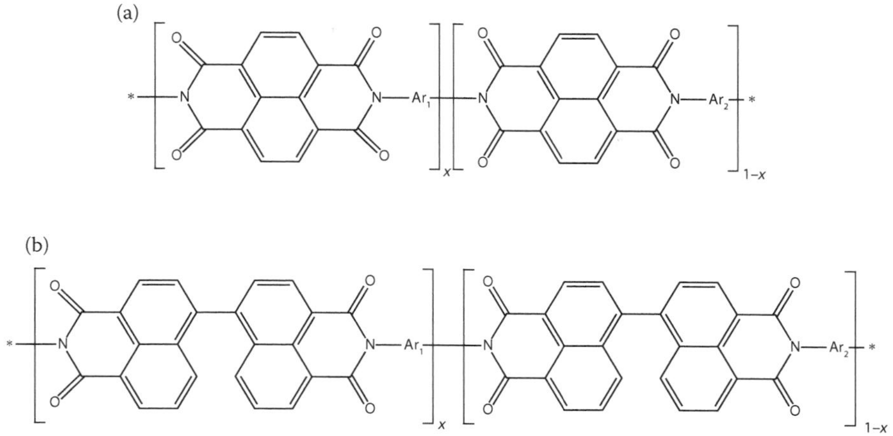

FIGURE 6.17 (a) Naphthalenic SPI. (b) Binaphthyl SPI.

itself to incorporation of H_2, but also the well-defined channels within MOFs make them excellent candidates for proton conductivity and use as a membrane electrolyte in a fuel cell. Three examples follow.

One approach to taking advantage of MOFs in PEMFCs is to chemically incorporate MOFs in the polymer membrane. For example, a sulfonated poly(2,6-dimethyl-1,4-phenylene oxide) membrane was coupled with an iron-aminoterephthalate MOF (Fe-MIL-101-NH_2) to give a "mixed matrix membrane" as depicted in Figure 6.22. This mixed matrix membrane showed a proton conductivity of 0.10 S/cm at 6% MOF loading, a value far superior to that of either the MOF-free polyphenylene oxide polymer or Nafion 117 under the same conditions. A maximum value of 0.25 S/cm was obtained at about 90°C. The proton conduction was attributed to increased acidity of the water molecule by coordination to an Fe(III) cation (Wu et al. 2013).

While integrating MOFs into the polymer electrolyte is one approach to taking advantage of the special properties they provide, there are advantages to using the MOF *alone* as the electrolyte membrane. For example, use of a metal–organic framework as the electrolyte membrane in a fuel cell should allow for an increase in operating temperature beyond that seen for typical organic polymers. Furthermore, MOFs are not necessarily limited to using water as the proton carrier (use of water limits the operating temperature of the PEM fuel cell). To that end, researchers examined the *anhydrous* proton-conducting capabilities of β-PCMOF2, the trisodium salt of 2,4,6-trihydroxy-1,3,5-benzenetrisulfonate. By entrapping 1H-1,2,4-triazole (Figure 6.23) in the channels of β-PCMOF2 to act as an organic proton carrier, researchers were able to demonstrate anhydrous proton conductivity of 5×10^{-4} S/cm at 150°C (Hurd et al. 2009). Incorporation of an MEA fabricated out of this β-PCMOF2-triazolium conglomerate and using it in an H_2/air fuel cell gave a V_{oc} value of 1.18 V at 100°C. The V_{oc} decreased, unfortunately, with increasing temperature, presumably due to fuel crossover. Thus, while this performance is still well below that shown by a hydrated Nafion membrane, this approach is an attractive alternative to hydrated fuel cells and their limitations.

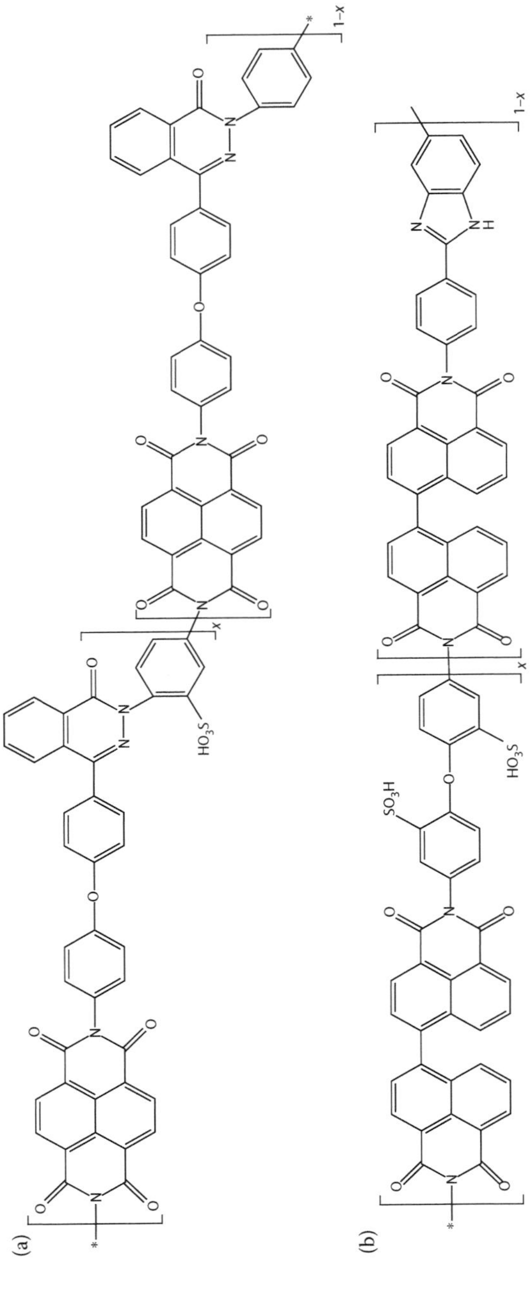

FIGURE 6.18   Naphthalenic (a) versus binaphthyl (b) SPI ionomers.

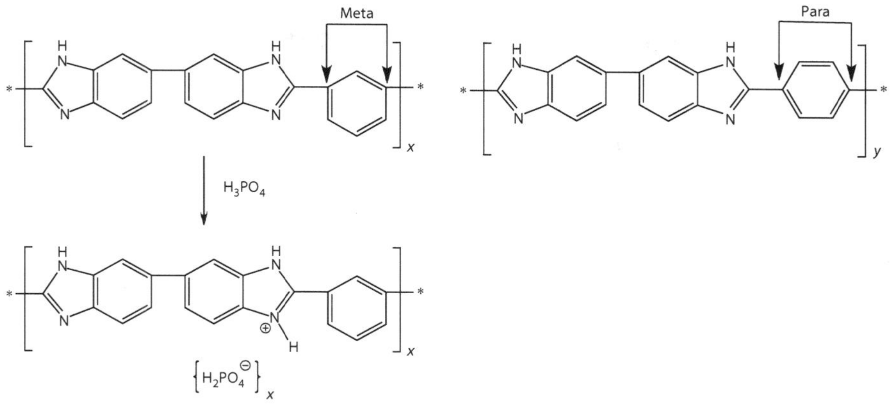

FIGURE 6.19 Phosphoric acid-doped polybenzimidazole.

In a similar approach, the doping of a chromium(III) terephthalate MIL-101 MOF with either sulfuric or phosphoric acid resulted in a framework that captured the acid in the pores, allowing for proton conductivity at both low humidity (as low as 0.13% RH) and high temperature (up to 150°C). Figure 6.24 presents the results of proton conductivity of the sulfuric acid-doped material (H_2SO_4@MIL-101) and phosphoric acid-doped sample (H_3PO_4@MIL-101) in comparison to the individual acid, Nafion, and the triazolium-doped MOF discussed above (Ponomareva et al. 2012). Clearly, this simple approach bodes well for further studies on the incorporation of MOFs into fuel cell systems.

6.6.4 DIRECT METHANOL FUEL CELLS

While excellent progress is being made in the development of the hydrogen/oxygen PEMFC, it is hard to set aside the multitude of problems with hydrogen fuel—its production from fossil fuels, its flammability, and its low energy density on a per volume basis. In Chapter 5, we saw the use of the so-called liquid organic hydrides as masked hydrogen storage and noted that alcohols can be considered hydrogen storage given that they can undergo steam reforming to generate H_2. Alcohols such as methanol, glycerol, and ethanol offer the significant advantage of being liquids and so conveniently fit into the current fuel delivery infrastructure. They also have a much higher energy density on a per volume basis than hydrogen. The research problem thus becomes "can we somehow bypass the large-scale production of hydrogen and reform the alcohol *directly* in the fuel cell, generating the hydrogen *in situ*, on demand?" The answer is yes, and the direct alcohol fuel cell (DAFC) is the result. At this point, methanol is the best alcohol fuel for a PEMFC since it shows sufficient power densities. The rest of this section, therefore, will be focused on the DMFC. (N.B. Just as we had to clarify that a PEMFC does not necessarily rely on hydrogen as its fuel, similarly, the use of methanol as a fuel does not automatically mean that the fuel cell is a PEMFC. Methanol can be used as a fuel in alkaline fuel cells,

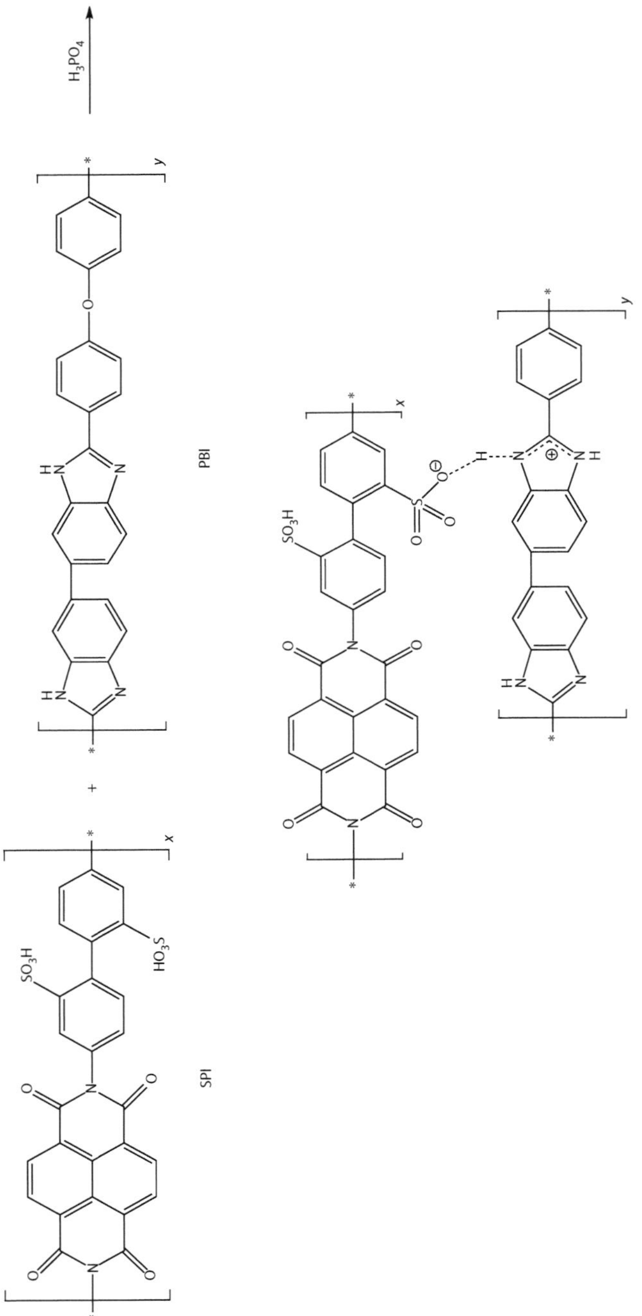

FIGURE 6.20 SPI–PBI blend.

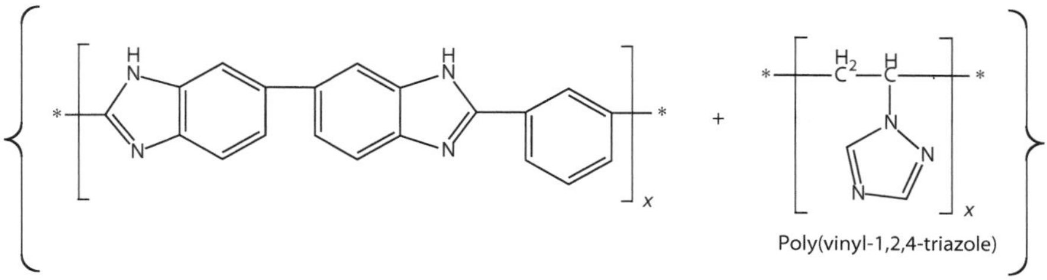

FIGURE 6.21 PBI/polytriazole blend.

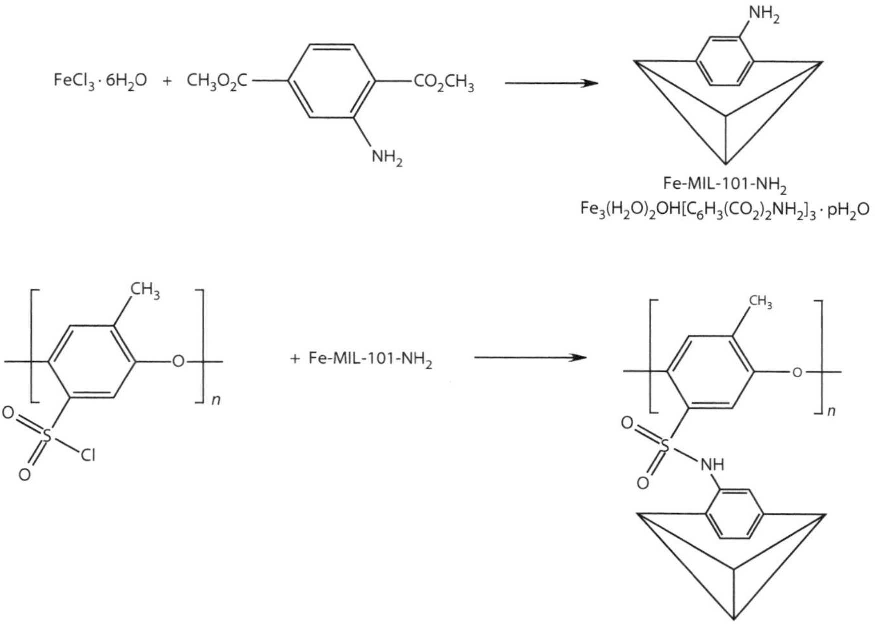

FIGURE 6.22 Cartoon depicting the Fe-MIL-101-NH$_2$/polymer conglomerate.

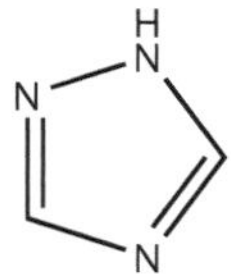

FIGURE 6.23 1H-1,2,4-triazole.

SOFCs, and others. In this section, however, we will restrict our discussion to the use of methanol in a PEMFC and refer to this combination as a DMFC.)

The use of methanol as the fuel in a DMFC brings with it its own host of challenges. We are already familiar with the considerable problem of methanol crossover. We already know that the kinetics at the cathode are sluggish (the dreaded ORR). In

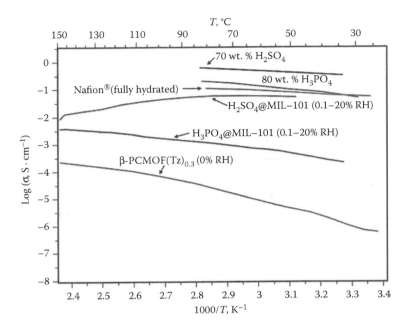

FIGURE 6.24 Proton conduction performance of various materials. (Adapted with permission from Ponomareva, V.G. et al. 2012. Imparting high proton conductivity to a metal–organic framework material by controlled acid impregnation. *J. Am. Chem. Soc.* 134 (38):15640–15643. Copyright 2012, American Chemical Society.)

the DMFC, the kinetics at the anode—where the methanol must be oxidized—are also problematic. Because we are now operating at the MEA with both liquids and gases, a good catalyst layer requires an effective *triple*-phase boundary—the solid of the catalyst layer, the liquid methanol, and the gaseous oxygen (Wang et al. 2006). The importance of morphology is magnified in terms of mass transport. Finding a good electrocatalyst for the DMFC is further complicated by the fact that, at some point in the oxidation of methanol, CO is formed as an intermediate (*vide infra*), which can poison the electrocatalyst. Then there is the ironic detail that the most economical source of methanol is natural gas (although production from biomass has been demonstrated and developing alternative sources of methanol continues to be a key area of research). Nonetheless, the advantages to using this liquid fuel in a portable fuel cell application for transportation mean a great deal of research is focused on making this a viable sustainable energy solution.

6.6.4.1 Half-Cell Reactions
The redox reactions of the DMFC are shown in Table 6.1. Methanol is oxidized at the anode in the presence of water (forming carbon dioxide) while the cathode reaction is, as usual, the reduction of oxygen to water. The necessity for water in the oxidation reaction means that the actual fuel in the DMFC is a mixture of methanol and water, contributing to the problem of methanol crossover. As we saw in the thermodynamics of the H_2/O_2 fuel cell, the physical state (liquid or gas) of the products

TABLE 6.4
Thermodynamic Values for Methanol and Water at 25°C and 1 atm

Methanol	ΔH (kJ/mol)	ΔG (kJ/mol)
Gas	200.670	−162.000
Liquid	238.660	−166.360
Water	ΔG (kJ/mol)	ΔG (kJ/mol)
Gas	241.826	−228.590
Liquid	285.826	−237.180

Source: Adapted from Black W.Z. and J.G. Hartley, 1991. *Thermodynamics—2,* New York: HarperCollins.

and reactants in the DMFC will impact all thermodynamic calculations as well. The relevant thermodynamic data are given in Table 6.4.

The mechanism of the alcohol oxidation reaction on a platinum catalyst has been extensively investigated and a postulated mechanism (for the electrooxidation of *ethanol*) is shown in Figure 6.25 (Simões et al. 2007). Without going into much detail, it can be seen that the key role of the noble metal is to adsorb the alcohol, activate the C–H bond and then oxidize the alcohol by a multistep dehydrogenation. As Figure 6.25 shows, several possible pathways to different organic products exist;

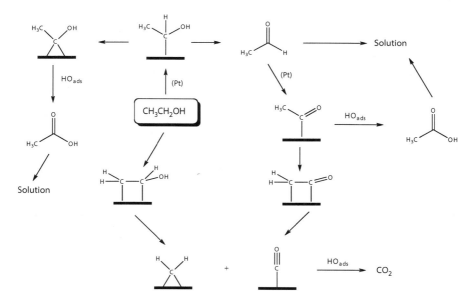

FIGURE 6.25 Proposed intermediates in the ethanol oxidation on Pt surface (as indicated by the heavy black line) in acidic media. (Reprinted from *J. Power Sources* 167 (1), Simões, F.C. et al. Electroactivity of tin modified platinum electrodes for ethanol electrooxidation. 1–10. Copyright 2007, With permission from Elsevier.)

for methanol oxidation, the presumed organic intermediates are formate and/or CO. It is important to note that these electrochemical steps are driven by the concurrent low free energy oxidation of water, as noted above and as shown below in Equations 6.22 through 6.24 where $(OH_2)_{ads}$ means that water is adsorbed on the metal surface; Bianchini and Shen 2009.

$$(OH_2)_{ads} \rightarrow (OH)_{ads} + H^+ + e^- \tag{6.22}$$

$$(OH)_{ads} \rightarrow (O)_{ads} + H^+ + e^- \tag{6.23}$$

$$(CO)_{ads} + (O)_{ads} \rightarrow CO_2 \tag{6.24}$$

It is the production of the adsorbed oxygen atom in this oxidation scheme that propels the oxidation of CO (from methanol) to CO_2.

6.6.4.2 DMFC Electrocatalysts

The best catalysts for the alcohol oxidation reaction are binary, ternary, or quaternary platinum alloys since the combination of metals enhances the kinetics of both C–H and O–H activation. Furthermore, alloys once again address the issue of CO poisoning, as seen in the case of the Pt–Ru alloy used in the PEMFC (Section 6.5.2) (Kamarudin and Hashim 2012; Petrii 2008). Many other metals have been tested in Pt alloys for the methanol oxidation reaction, including tin, rhenium, cobalt, tungsten, iridium, palladium, and nickel, among others—all driven by the need to minimize the amount of platinum used. In a recent example, an electrocatalyst prepared by depositing a Pt monolayer on a gold(111) surface showed enhanced catalytic activity for the ethanol oxidation reaction and a vastly enhanced current density relative to platinum alone (Li et al. 2012). This phenomenon is attributed to surface strain: the stretched Pt–Pt interactions enhance the ability of the platinum atoms to bind CO and OH.

Efforts to improve the efficiency of the DMFC center on the MEA and include both chemical modifications and improvements in fabrication. Needless to say, these efforts are complicated by the triple-phase boundary of the DMFC. The MEA for a DMFC is a complex assembly consisting of the gas diffusion layer, the catalyst layers (for both anode and cathode), and the PEM (Figure 6.26; adapted from Xu et al. 2006). The methanol/water mixture is transported through the diffusion layer into the catalyst layer at the anode, where the methanol is oxidized to CO_2 and the electrons harvested to generate the current. Reduction of oxygen takes place simultaneously at the cathode. As always, every component of the MEA must be optimized to minimize resistance, maximize mass transport, and in this case, limit methanol crossover. Increasing the thickness of the PEM limits the crossover but also increases resistance through the MEA. Extensive research effort has led to many improvements in the polymer membrane (*vide supra*).

While the PEM and the electrocatalysts may seem like obvious contributors to the overall efficiency of the DMFC, even the catalyst supports and diffusion layers are prime targets for improved efficiencies, particularly as they influence catalyst efficiency and therefore decrease catalyst loading. These components are usually made

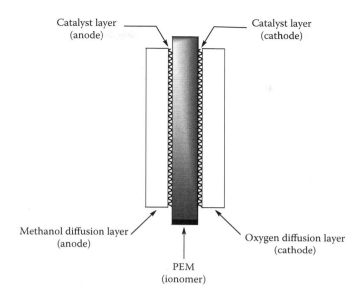

Catalyst layer
(anode)

Catalyst layer
(cathode)

Methanol diffusion layer
(anode)

Oxygen diffusion layer
(cathode)

PEM
(ionomer)

FIGURE 6.26 Schematic of the membrane electrode assembly for a direct methanol fuel cell.

up of some kind of porous carbon, as noted before, but these materials are prone to corrosion. Instead of carbon black (an amorphous form of carbon), nanostructured carbon materials (nanotubes, nanofibers, or graphene), as well as nanostructured noncarbon materials such as metal oxides (titanium oxide, tin oxide, indium–tin oxide, plus many others), have all been recently explored for use in the catalyst layer to improve both catalyst efficiency and resistance to corrosion (Sharma and Pollet 2012). This "nanostructuring" of materials is an approach we will see recurring throughout the areas of sustainable energy generation.

DAFCs have the foremost advantage of a liquid fuel, thus requiring no major change in the current fuel storage and delivery infrastructure. However, their lower efficiencies (relative to the PEMFC) mean that further improvements in the effectiveness of the alcohol electrocatalysts is required before the DAFC can be considered commercially viable.

6.7 SOLID OXIDE FUEL CELLS

6.7.1 INTRODUCTION

SOFCs are reliable, relatively inexpensive, and fairly efficient sources of electricity—at around 40% efficiency, SOFC are on par with DAFC. SOFC stacks are used as stationary power sources for small applications (e.g., residential) to commercial-scale power plants. While the DMFC and PEMFC are considered low-temperature fuel cells, the SOFC is at the other end of the operating temperature range (700–1000°C). Given that the SOFC is, as its name suggests, a *solid*, elevated temperatures are necessary to achieve the necessary level of conductivity. Furthermore, the high-temperature exhaust from an SOFC can be coupled to another power generation source (e.g., a turbine), boosting the efficiency of a CHP system as described previously. While scientists work to develop more durable polymer electrolyte membranes that can operate at *higher* temperatures (for improved kinetics and therefore efficiency in a PEMFC), scientists working on SOFCs strive to find materials that will allow the operation of the SOFC at *lower* temperatures (ca. 500–700°C) for enhanced safety, durability, and lower operating costs. However, the higher operating temperature of the SOFC allows for more flexibility in fuel use: as long as the fuel can be reformed *in situ* to generate syngas (the typical fuel for SOFC), it is acceptable for use in the SOFC. Another significant feature that distinguishes the SOFC from PEMFCs is that the higher operating temperature range allows for the use of less expensive (non-PGM) electrocatalysts, a particularly important consideration with respect to the sustainability of resources.

SOFC stacks that are linked to coal gasification are known as *IGFC* systems (integrated gasification fuel cell). These systems fall in the realm of "clean coal" technology, in which coal is gasified in the presence of steam to generate the syngas feedstock for an SOFC. In the SOFC, the CO_2 generated from the electrochemical oxidation of CO from the syngas (see Equation 6.26) would, in theory, be sequestered to make this a carbon-neutral process.

There are two general types of SOFCs—those with an oxide-conducting electrolyte and those that are proton-conducting. Proton-conducting SOFCs (naturally abbreviated PC-SOFC) consist of an oxide electrolyte that is hydrated to allow for the proton to hop from one stationary oxide to the next. An advantage of these PC-SOFCs is that the fuel is not diluted with H_2O and there is therefore no need for water management or fuel recirculation. In addition, PC-SOFCs can operate at somewhat lower temperatures ($\approx$400–800°C) because of the lower E_{act} of proton mobility. Mixed proton-oxide-conducting solid ion fuel cells are currently under development as the next generation of SOFCs, but we will focus exclusively on the oxide-conducting fuel cell as the dominant SOFC in current use.

6.7.2 REACTIONS

The reactions that take place in an SOFC are shown below (Equations 6.25 and 6.26).

$$2H_2 + 2O^{2-} \rightarrow 2H_2O + 4e^- \tag{6.25}$$

$$2CO + 2O^{2-} \rightarrow 2CO_2 + 4e^- \tag{6.26}$$

A detailed theoretical study using density functional theory revealed the plausible steps whereby CO is oxidized on the metal oxide surface as shown in Figure 6.27. While this study was focused on an ultrathin film of FeO/Pt(111), the binding of the CO, release of CO_2, and capture of O_2 on the metal oxide surface are all likely germane to the operation of an SOFC electrolyte (Sun et al. 2010). The ion conductor is the O_2^- ion that migrates through vacancies in the solid electrolyte. This oxide pathway is a result of interstitial oxide ion defects formed from doping the electrolyte material with other

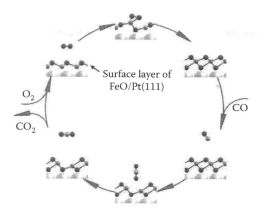

FIGURE 6.27 (**See color insert.**) Proposed steps in the oxidation of CO on a FeO/Pt(111) surface. (Reprinted with permission from Giordano, L. and G. Pacchioni. 2011. Oxide films at the nanoscale: New structures, new functions, and new materials. *Acc. Chem. Res.* 44 (11): 1244–1252. Copyright 2011, American Chemical Society.)

cations (*vide infra*). Because these cells operate at such a high temperature, the water is formed at the anode as steam with the concomitant decrease in the thermodynamic free energy of formation and 100 mV loss in cell potential relative to other types of fuel cells (Bartholomew and Farrauto 2006). However, as noted above, the waste heat generated in an SOFC makes for improved efficiency in a combined heating and power system. Given the configuration of a typical stationary SOFC stack installation, the waste streams—water and CO_2—can be separated with the pure water recycled (or used, for example, for drinking) and the CO_2 for sequestration (Adams et al. 2012).

6.7.3 ELECTRODE AND ELECTROLYTE MATERIALS

SOFCs are no different from other electrochemical cells in that an electrolyte is sandwiched between the two electrodes (Figure 6.28). However, given that all of these materials are solids, the distinction between the electrode, electrocatalyst, and the electrolyte is not always clear. Electrodes are porous mixed composites consisting of a conducting oxide and an oxide electrolyte. Thus, the electrolyte is often the medium for the electrode/electrocatalyst, and the specific electrolyte can dramatically impact the performance of the electrode. Furthermore, the "barrier" between the electrode and the electrolyte must be highly conductive, but the electrolyte must not react with the electrode. In any case, the compatibility of the electrode materials with the electrolyte is a big issue, especially given the high operating temperatures at which unwanted reactions can take place or materials can delaminate.

Many of the considerations we have already examined apply to SOFCs as well: improving the efficiency and lowering the cost of SOFCs is primarily empirical and relies upon optimizing the performance of every component in terms of both chemical and mechanical stability. Fabrication of the materials in an SOFC, however, is wildly different than that seen in PEMFCs since we are now working with inorganic materials, not more easily processed organic materials. Instead of chemical

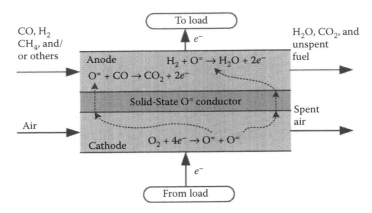

FIGURE 6.28 Schematic of a solid oxide fuel cell. (From Adams, T.A. et al. 2013. Energy conversion with solid oxide fuel cell systems: A review of concepts and outlooks for the short- and long-term. *Ind. Eng. Chem. Res.* 52 (9):3089–3111. Copyright 2012, American Chemical Society.)

substances that dissolve in solvents that can be layered as catalyst inks, for example, instead we are dealing with crystalline materials that must be able to be sintered (heated to a point at which the particles stick together without fusion or melting) at elevated temperatures in the fabrication process. Incorporation of nanotechnology to the improvement of SOFC materials has further led to remarkably creative fabrication methods with simultaneous improvement in the efficiencies of ion conduction and/or catalytic activity, as we will see.

6.7.3.1   Electrolytes

Oxide-conducting electrolytes in SOFCs, like other electrolytes, must optimize ionic conductivity. Increasing the conductivity of solid oxide materials can be accomplished by increasing the temperature or by minimizing the thickness of the electrolyte layer. However, extremely high temperatures are unfavorable (as mentioned above) and if the electrolyte layer is too thin it will not be impermeable to gas and will lead to crossover problems. As a result, development of improved electrolyte performance for SOFC electrolytes is, as usual, a balancing act.

The solid materials typically used as SOFC electrolytes are of the AO_2 (fluorite-type) or ABO_3 (perovskite-type) structure. An example of a fluorite structure can be seen in Figure 6.29, where the larger spheres are the oxide anions with the smaller spheres representing the tetravalent metal cation. In the realm of SOFC, the prototypical fluorite electrolyte is zirconia (ZrO_2) that has been doped with yttrium oxide to give $(ZrO_2)_{0.92}(Y_2O_3)_{0.08}$ (*yttrium-stabilized zirconia* or YSZ). An yttrium(III) ion displaces a zirconium(IV) ion in the zirconia lattice to create the oxygen vacancies necessary for conduction by the oxide ion, O^{2-}. YSZ shows good mechanical and chemical stability as well as measurable oxide ion conductivity at temperatures above 700°C (at 1000°C the conductance is 0.1 S/cm). (Bartholomew and Farrauto 2006). Zirconia may also be stabilized with scandium oxide to yield the "SSZ" electrolyte that has a higher conductivity but is considerably more expensive.

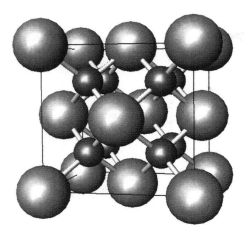

FIGURE 6.29 The fluorite crystal structure; the dark gray interior spheres represent the tetravalent metal ions and the light gray spheres the oxygen anions. (Reprinted with permission from Orera, A. and P.R. Slater. 2010. New chemical systems for solid oxide fuel cells. *Chem. Mater.* 22:675–690. Copyright 2010, American Chemical Society.)

Another popular fluorite electrolyte for the SOFC is ceria (CeO_2), which when doped with gadolinium oxide (Gd_2O_3) or samarium oxide (Sm_2O_3) results in higher oxide conductivities at lower (≈ 500–$700°C$) temperatures (Malavasi et al. 2010). The ceria-gadolinia combination is referred to as CGO. However, complications with competing redox processes suggest that the best approach for the electrolyte may be to blend doped ceria materials (with better conductivity) with doped zirconia materials (with better mechanical and chemical properties).

Perovskite materials (ABO_3, Figure 6.30) consist of a six-coordinate "B" cation plus a 12-coordinate "A" cation. These materials have also shown good performance as SOFC electrolytes. The classic example of a perovskite material used in SOFC

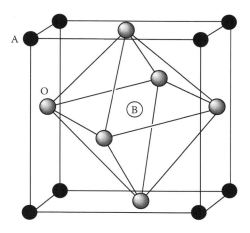

FIGURE 6.30 The perovskite unit cell.

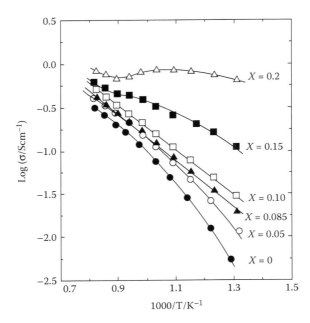

FIGURE 6.31   Arrhenius plots showing the electrical conductivity of $La_{0.8}Sr_{0.2}Ga_{0.8}Mg_{0.2-x}Co_xO_3$ with varying amounts of cobalt. (Reprinted with permission from Ishihara, T. et al. 1999. Improved oxide ion conductivity in $La_{0.8}Sr_{0.2}Ga_{0.8}Mg_{0.2}O_3$ by doping Co. *Chem. Mater.* 11 (8):2081–2088. Copyright 1999, American Chemical Society.)

is $LaGaO_3$ that has been doped with strontium or magnesium to give systems of the general formula $La_{1-x}Sr_xGa_yMg_{1-y}O_{3-\delta}$ (LSGM). Small amounts of other dopants (Co, Fe or Ni) can dramatically improve the conductivity, as illustrated in Figure 6.31 for a series of $La_{0.8}Sr_{0.2}Ga_{0.8}Mg_{2-x}Co_xO_3$ materials (Ishihara et al. 1999).

6.7.3.2   Electrodes

6.7.3.2.1   Cathode Materials

As is the case for cathodes in other fuel cells, the cathode for the SOFC is tasked with the challenge of reducing oxygen, but the elevated temperatures mean that this sluggish reaction is more facile in the SOFC. Perovskite materials are most widely used, with lanthanum strontium manganite composites (LSM; $La_{1-x}Sr_xMnO_{3-\delta}$) performing well at high operating temperatures (ca. 1000°C) both in terms of conductivity and in having a thermal expansion coefficient that matches well with the electrolyte YSZ, a crucial consideration in fabricating SOFCs. It is the presence of the mixed valence manganese ions—Mn(III) and Mn(IV)—that results in high electronic conductivity, enhanced even more by doping with strontium. The performance of LSM declines precipitously, however, at lower temperatures as polarization resistance increases almost 2000-fold over 500 degrees leading to increased efforts to find a useful electrode material for lower temperatures (Jacobson 2009). Cobaltites such as $PrBaCo_2O_{5+x}$ (PBC) and $GaBaCo_2O_{5+x}$ (GBC) have been studied because they are believed to provide enhanced reactivity for the oxidation reduction reaction. A lanthanum/strontium material doped with cobalt and iron ($La_{0.6}Sr_{0.4}CO_{0.2}Fe_{0.8}O_{3-\delta}$) has been shown to work at a lower temperature (<750°C) (Liu et al. 2011). Newer materials being investigated for use at the cathode include Ln_2NiO_{4+x} (where $Ln = La$, Pr, Nd) (Jacobson 2009; Orera and Slater 2010) and strontium-cerium mixtures. A nickel/samarium-doped ceria cell fabricated with a $Sr_{0.95}Ce_{0.05}CoO_{3-\delta}$ cathode gave a peak power density of 0.625 W/cm² at 700°C—the low end of the SOFC operating temperature spectrum. These are very encouraging results for the continued investigation of cathode materials that provide reasonable kinetics and conductivity in an SOFC (Yang et al. 2013).

6.7.3.2.2 Anode Materials

The most common SOFC anode material is a composite between nickel and the ceramic electrolyte known as a *nickel cermet*. Thus, an anode made for a fuel cell using the YSZ electrolyte would be a Ni-YSZ cermet. These Ni cermets—usually about 30–35% nickel—show excellent catalytic activity and electronic conductivity, while the YSZ of the composite provides the oxide conductivity that allows the O^{2-} to diffuse into the anode. However, the Ni cermets work best with hydrogen fuel—sulfur contaminants in hydrocarbon fuels can lead to poisoning by formation of nickel sulfides and deactivation by coking. Thus, the major research push in anode development for SOFCs is to develop an electrocatalyst that is less prone to poisoning and carbon formation. Much like the pathway seen in the development of electrocatalysts for the PEMFC, several researchers have taken the approach of using more than one metal. To this end, strontium titanate ($SrTiO_3$) doped with niobium or lanthanum has been examined. Another approach has been to add copper, which does not catalyze carbon formation. It does not catalyze the oxidation reaction either, but it is a good electronic conductor. The results have been mixed: electronic conductivity, stability, polarization resistance, and ion conductivity all must be balanced in the continuing search for the optimal material for the anode (McIntosh and Gorte 2004).

6.7.4 FABRICATION AND CHARACTERIZATION

Characterization of the components and materials for SOFCs may include FT-IR, Raman spectroscopy, and powder x-ray diffraction to identify functionality and chemical structure, and for the analysis of elemental composition. The particle sizes and morphology of the materials can be examined using transmission electron microscopy or scanning electron microscopy, and, ultimately, the performance of the components is tested.

Fabrication of SOFC components, as noted before, is not trivial and the challenges are many. Cost is, as always, an issue, as is the availability of the elements used in the electrolytes. Lanthanum is considered a near-critical risk in the short term and nickel and strontium a future risk to supply (Bauer et al. 2011; Knowledge Transfer Network 2010). The elevated operating temperature is especially demanding in terms of long-term stability. Many methods involve sintering; however, these high-temperature (≈1400°C) processes can lead to coarse materials that are less effective, so low-temperature alternatives are being sought. An "ion impregnation" method consists of preparing a metal ion solution, dropping the solution onto a porous framework, then firing the sample at 800°C in air. The resultant layered nanostructured electrocatalyst was shown to have low polarization resistance and a good peak power density at 600°C (Wang et al. 2012). A plethora of other techniques have been tried to yield a good, dense electrolyte sandwiched between porous electrodes, including the use of advanced deposition techniques such as sputtering, pulsed laser deposition, spark plasma sintering, and spray pyrolysis. Even "sintering aids" have been added to reduce the sintering temperature (Liu et al. 2010). Improved methods in the fabrication of SOFCs that will be amenable to mass production at low cost and high efficiency are ongoing area of study. Overall, future growth in the area of SOFC is likely, especially in the realm of stationary combined heat and power applications.

6.8 MICROBIAL FUEL CELLS

6.8.1 INTRODUCTION

Biofuel cells take advantage of bio-electrical systems to generate electricity. There are two subsets of biofuel cells: microbial fuel cells and *enzymatic fuel cells* (Cracknell et al. 2008). Like the microbial electrolysis cell discussed in Section 5.2.4, in MFCs the electrocatalyst(s) are actually living cells in the form of a biofilm of bacteria or algae affixed to an electrode. In contrast, in enzymatic fuel cells, the catalysts are inert, isolated enzymes attached to either or both electrodes. There are several advantages and disadvantages to each type of biofuel cell. The substrate specificity of enzymes means that an enzymatic fuel cell can be constructed more simply, since a fuel/oxidant separating membrane is not necessary. Enzymes can provide higher current density provided enough enzyme can be layered onto the electrode. At the same time, MFCs—being catalyzed by whole, living cells—can carry out redox reactions on a wider variety of nutrients, making them more energy efficient. Enzymes are fragile and difficult to make adhere to an electrode surface, whereas microbial electrodes, being composed of living matter, last longer and are self-adhering (enzymatic electrodes last only a few days under operating conditions; Erable et al. 2012). Because of the particularly attractive application of MFC to electrical generation from municipal solid waste (Rulkens 2008), we will focus on MFCs in this chapter.

Nature possesses an impressive array of enzymatic electrocatalysts. Microorganisms use a wide variety of fuels, oxidants, and chemical intermediates (formates and nitrates, carbon monoxide, hydrogen and oxygen, quinone/hydroquinone redox mediators, among many others) in *in vivo* energy-transforming electron transport chains. The birth of the MFC is considered to be 1911, when the electrochemical activity of microorganisms was first reported (Schröder 2012). In recent years, the pace of research activity on MFCs has increased exponentially, but they present some unique challenges. Obviously, a successful MFC electrode must not only conduct electrons, but it must also support life in the form of a biofilm of bacteria or algae. Successful MFCs require a neutral pH, whereas the ORR is more favorable under acidic conditions: the $E°$ for the four-electron reduction of oxygen under acidic conditions is 1.23 V while in alkaline solution, it is 0.40 V; at 25°C and neutral pH, the O_2/H_2O reduction potential is 0.82 V (Wiberg 2001). Changes in pH at the electrode surface can also lead to biofouling and loss in efficiency of the MFC. The solubility of the oxidant, O_2, is low in aqueous solution, so the challenge of mass transport at the catalyst/substrate boundary is amplified. Furthermore, while the detailed mechanism is not well understood, generation of reactive oxygen species such as the superoxide radical anion $O_2^{\bullet-}$ or the hydroxyl radical $HO^{\bullet}$ can mean cell damage and death to the microbial catalyst.

6.8.2 COMPONENTS

The schematic of an MFC is not unlike that of other fuel cells: anode, cathode, electrolyte, separator (membrane), circuit, and load (Figure 6.32). Oxygen- and nutrient-rich fuel from, for example, municipal wastewater is oxidized by microorganisms at the anode, releasing protons and carbon dioxide. The freed electrons travel the circuit to generate the electrical current while at the cathode the oxygen reduction reaction takes place, capturing the protons and electrons to produce water. The cation-exchange separator membrane plays the same role in an MFC as the polymer

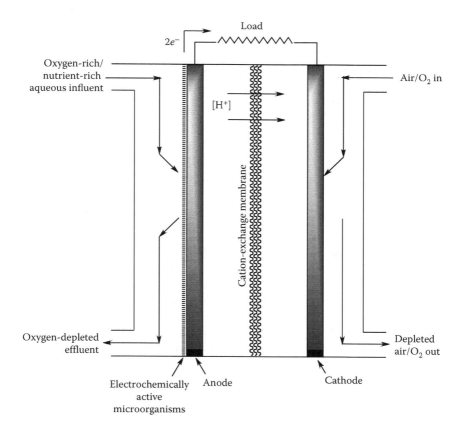

FIGURE 6.32 Schematic of a microbial fuel cell.

electrolyte membrane in a PEMFC, and is often the same material: Nafion®. Note that it is the anode in the MFC that qualifies it as an MFC; the cathode may contain a typical transition-metal electrocatalyst, although research is being carried out to develop microbial catalysts that will efficiently reduce oxygen in an MFC (Erable et al. 2012).

6.8.2.1 Anode Fabrication

Electron transfer is not quite trivial in an MFC. Some bacteria have the ability to transfer electrons directly to the anode material, typically some form of carbon, while others require a shuttle or mediator. This adds another layer of complexity to the electrode design for the MFC. As the anode is the electrode upon which the biological community exists, preparation of anodes for MFC is radically different from that for other fuel cells. The anode material is inoculated with a source of bacteria, often a colony from an already active MFC. All manner of different forms of carbon (brush, felt, fiber, mesh, granular graphite, and more) have been used as the scaffold for the anode. As expected, materials that provide a larger surface area have demonstrated better performance, with respect to both electron transfer and bacterial growth. Because of the variety of inoculation sources and reactor types, it is otherwise difficult to draw conclusions about the performance of these various types of carbon materials.

Metals such as stainless steel, titanium, and gold have also been used for MFC anode fabrication, with the requirement being that the metal must not corrode under the operating conditions of the cell. Because bacteria do not adhere well to a smooth metal surface, metal-based anodes have generally shown lower power densities than comparable graphite anodes (Wei et al. 2011). It is interesting to note that some sort of surface treatment is helpful even for carbonaceous materials, and several approaches (primarily chemical modifications such as coatings or surface changes) have been tried with positive results. In addition, composite materials (e.g., a metal–graphite pair) have been investigated. However, all of these alterations must take into account the impact on the growth and health of the bacterial electrocatalysts.

6.8.2.2 Cathode Materials

It is no surprise that the oxygen reduction reaction is the barrier to improved efficiency for the MFC. Not only is the oxygen reduction reaction the problem reaction in terms of kinetics, the limited solubility of oxygen in the aqueous media further hinders the cell efficiency. There are three general types of cathodes for MFCs: air cathodes, aqueous-air cathodes, and biocathodes, with the materials for their fabrication generally the same as for the anodes (primarily carbon). As noted above, much research has focused on the development of a functional biocathode for the ORR (Erable et al. 2012), although much work remains to be done.

A recent approach to the development of larger-scale MFCs has taken advantage of a poly(vinyl alcohol) (PVA) membrane separator placed against a carbon cloth/Pt cathode in order to mitigate oxygen diffusion to the carbon-brush anode. While various other separator materials have been tried their use has often led to increased ionic resistance in the cell among other drawbacks. By incorporating a neutral, porous PVA membrane in the cell, both ion transport and pH gradients are avoided. Comparison of this PVA-separated MFC to one with no separator showed that both oxygen diffusion to the anode and bacterial growth on the cathode were decreased in the presence of the separator. In addition, although the presence of the separator did increase the internal resistance in the cell (a result of slowed mass transfer to the cathodes), the maximum power (P_{max}) of the PVA-separated cell (1220 mW/m^2) was about 10% higher than the standard comparison with no separator. Similarly, the coulombic efficiency (a measure of the efficiency of electron collection at the anode) was significantly higher than the standard. Thus, this PVA approach resulted in both greater power production and higher coulombic efficiency, a very promising finding for the development of high-performing MFCs (Chen et al. 2012).

6.9   FUEL CELL SUMMARY

The development of fuel cells captures well the research process: empirical results contribute to a better understanding of the molecular basis of macromolecular phenomena. Creation of a better, more efficient fuel cell that can be used for transportation using a sustainably provided fuel is dependent upon the rational discovery of new electrolytes and electrocatalysts, green syntheses of robust but renewable polymers, cost-effective and sustainable pathways to hydrogen and methanol, wise choices with respect to scarce resources, and smart design of the fabricated whole so that, when the fuel cell life cycle is complete, waste is minimal. Progress must be based on a thorough understanding of the mechanisms of the redox processes involved and breakthroughs in computational research have contributed to this basic understanding. Fuel cell efficiencies are improving at an impressive rate such that their increasing use in a sustainable energy scenario for stationary or portable applications is a given.

6.10   ELECTROCHEMICAL ENERGY STORAGE

As noted in Chapter 1, the need for electrical energy storage (EES) is strongly yoked to cleaner energy producers such as wind and solar because of supply variability: electricity generated during peak production times is stored and delivered "off peak" to level the load on the grid. In addition, EES—in the form of rechargeable batteries—is key to reducing our reliance on fossil-fueled transportation via the development of electric vehicles (EV). Therefore, EES, whether stationary or mobile, is germane to the study of sustainable energy. While *supercapacitors* are broadly used for EES and large-scale physical modes of EES exist in the form of compressed air, flywheel, or pumped hydrodynamic storage, our emphasis on chemistry takes us directly to *electrochemical* EES, a huge and expanding area of research that has progressed more slowly than other areas of energy research.

Rechargeable batteries are the prototypical examples of EES. They must meet stringent requirements including high energy density, safety and reliability, and the ability to undergo hundreds of charge–discharge cycles. It is the goal of the U.S. Department of Energy to achieve a battery performance of 300 Wh/L and 250 Wh/kg—primarily for transportation purposes—but currently, the cost of electricity provided by these systems is far too high (> $700/kWh) (Liu et al. 2013). As our needs for storage of electrical energy continue to grow, we need both smaller (for EV and electronic devices) and larger EES with greater energy density and storage capacity.

Our focus in this section will be on a few of the more important and recent developments in the field.

6.10.1 LITHIUM ION BATTERIES

A lithium ion battery is likely one of the most familiar high-density EES devices in that it is found as the rechargeable battery in a multitude of electronic devices such as smart phones and laptops. It may also be familiar from the negative exposure received after a fire and explosion in the electronics bay of a Japan Airlines 787 caused the entire fleet of Boeing 787 Dreamliners to be grounded in early 2013, the result of a failed Li-ion battery (Clark 2013). Because the electrode is made up of graphite (or some carbonaceous material) intercalated with lithium metal, and because it operates at a potential nearly matching that of metallic lithium, dendrites (long fingers) of lithium metal tend to grow in the battery and, potentially, cause a short circuit and overheat (a *thermal runaway*). This poses a risk of fire given the organic solvents used in the cell (*vide infra*). Thus, the chemistry behind the early Li-ion batteries is "inherently unsafe" (Yang et al. 2011), but the advantages of Li-ion batteries make them widely used. The target for improvements is not only to improve their safety and reliability but also to increase the voltage and specific capacity so that an improved energy density results. This is particularly important for transportation applications and Li-ion batteries are just beginning to make inroads in the transportation sector, displacing the nickel–metal hydride batteries in use today (Girishkumar et al. 2010).

The Li-ion battery works by shuttling the Li^+ ion back and forth between host materials at the anode and cathode during discharge and recharge. Thus, as the battery is delivering electricity to its load, Li^+ ions migrate through an electrolyte from the anode to the cathode (Figure 6.33). The various materials making up the cell are summarized in Table 6.5. Both electrodes contain lithium and the electrolyte is typically a lithium salt dissolved in an alkyl carbonate solvent (e.g., ethylene carbonate, dimethyl carbonate, diethyl carbonate, or ethyl methyl carbonate) (Yang et al. 2011). The "rocking horse" redox chemistry of the Li^+ ion in a Li-ion battery is given in Equations 6.27 through 6.29, below, where C_6 indicates some form of graphite.

$$\text{Reaction at anode: } Li_xC_6 \xrightarrow{\text{discharge}} xLi + xe^- + C_6 \qquad (6.27)$$

$$Li_{1-x}CoO_2 + xLi^+ + xe^- \xrightarrow{\text{discharge}} LiCoO_2 \qquad (6.28)$$

$$\text{Cell reaction: } LiC_6 + CoO_2 \xrightarrow{\text{discharge}} C_6 + LiCoO_2 (E_{cell} = 3.7 \text{ V at } 25°C) \ (6.29)$$

These cells can deliver power greater than 200 Wh/kg with a capacity of 150 Ah/kg (Tarascon and Armand 2001).

Anode materials. Research focused on anode materials has addressed both capacity and safety issues. An anode made solely of lithium metal provides very high-specific capacity but is rarely used. Instead, graphite is interpenetrated with lithium

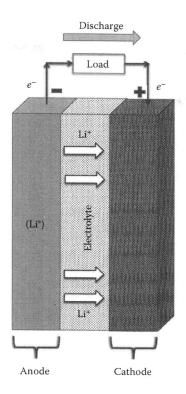

FIGURE 6.33 Schematic of a generic lithium-ion battery.

(an *intercalation compound*) and, while less than that of metallic lithium alone, capacities of up to 330 Ah/kg can be observed depending on the type of graphite used (e.g., well-ordered, heat-treated graphite or natural graphite) (Yang et al. 2011).

The formation of dendrites, as explained above, is the biggest issue surrounding the lithium anode. Upon onset of the electrochemical reaction, a thin layer of

TABLE 6.5
Typical Materials Used in Lithium-Ion Batteries

Anode Material	Cathode Material	Electrolyte Salt
Li° intercalated in graphite (Li_xC_6)	$LiCoO_2$	$LiPF_6$
		$LiBF_4$
	$LiMn_2O_4$	$LiClO_4$
	$LiFePO_4$	$LiBC_4O_8$
		$Li[PF_3(C_2F_5)_3]$

Source: Adapted from Yang, Z. et al. 2011. Electrochemical energy storage for green grid. *Chem. Rev.* 111 (5):3577–3613.

lithium salts forms at the interface between the anode and the electrolyte, impeding the diffusion of the lithium ions and impacting the reaction kinetics. It is this layer that wreaks havoc on the performance of the battery and accelerates dendrite formation. Two approaches to address the problem of dendrite formation have been to place an artificial protective layer on the anode or to use a polymer electrolyte that can allow ion conductivity while providing enough stability to block the formation of dendrites. To that end, a copolymer consisting of two distinct domains—one structural and one conductive—was developed and patented. The structural polymeric material is polystyrene, polymethacrylate, polyvinylpyridine or the like, while the conductive domain is made up of a polyether, polyamine, or polyethylene oxide (Singh et al. 2008). However, in general, solid-state lithium-ion batteries are far from viable.

Carbon-free alloys of lithium with tin, silicon, or tin oxide have also been explored for use as anodes. These alloyed electrodes exhibit a much higher capacity but are not always structurally stable since they show significant volume expansion during alloying with lithium leading to a fast loss in capacity with repeated cycling. The mechanism behind the poor performance is not yet understood. Nevertheless, these alloys show enough promise to warrant further exploration. For example, a nanostructured carbon–tin anode alloyed with lithium has been shown to give good results, provided that the tin nanoparticles are evenly distributed throughout the amorphous carbon matrix. The electrode so prepared had a maximum capacity of 300 Ah/kg (Hassoun et al. 2011).

Metal oxides have also been considered for possible use as the anode material in Li-ion batteries. Lithium titanate ($Li_4Ti_5O_{12}$) alleviates safety concerns due to the high potential of the titanate (1.55 V) versus Li^+/Li. However, the improvements in safety come at the expense of energy density: the theoretical capacity is a mere 175 Ah/kg. Titanium oxide, too, has been investigated, with the advantage of lower cost and the disadvantage of decreased electronic conductivity (Yang et al. 2011).

Cathode materials. The cathode has seen little development away from the metal oxide $LiCoO_2$, a problem because of its cost and toxicity. In addition, this standard cathode provides around 155 Ah/kg capacity, far below that of the theoretical capacity of the standard graphite-based anode (372 Ah/kg). As a result, the sheer mass of the cathode is about twice that of the anode, leading to increased costs (Liu et al. 2013). Several materials with symmetries based on the α-NaFeO₂ structure such as $LiMn_xNi_yCo_zO_2$ (Figure 6.34a), ($LiMn_2O_4$) (Figure 6.34b), and $LiFePO_4$ (Figure 6.34c) have been investigated to improve stability, service life, and cost, but thermal runaways are still a problem, especially when the battery is overcharged (Yang et al. 2011). A notable achievement is a $Li[Ni_{0.45}Co_{0.1}Mn_{1.45}]O_4$ cathode, prepared by a coprecipitation method. Figure 6.35 shows field emission scanning electron microscopy images of the material at two different magnification levels, clearly illustrating the excellent uniformity of the micrometer-sized particles. This morphology provides excellent electrode performance. Coupling this cathode with the carbon–tin anode described above resulted in a battery that gave good cycling performance and energy density (roughly 170 Wh/kg) (Hassoun et al. 2011).

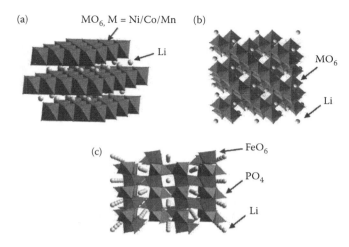

FIGURE 6.34 Cathode materials for Li-ion batteries. (a) Layered structure ($LiMn_xNi_yCo_zO_2$); (b) spinel structure ($LiMn_2O_4$); (c) olivine structure ($LiFePO_4$). (Reprinted with permission from Yang, Z. et al. 2011. Electrochemical energy storage for green grid. *Chem. Rev.* 111 (5):3577–3613. Copyright 2011, American Chemical Society.)

6.10.1.1 Lithium–Sulfur Batteries

The technology described above arguably suffers in the context of sustainability by using scarce and environmentally hazardous resources. A possible alternative is the lithium–sulfur battery in which sulfur—a relatively abundant and environmentally benign element—is used in the cathode. As the cell is discharged, the S_8 allotrope of sulfur (Figure 6.36) is reduced by $Li°$, reductively cleaving the ring until the final product, Li_2S, is formed (see Equation 6.30; Evers and Nazar 2012b).

$$S_8 + 16Li \rightleftarrows 8Li_2S \quad (2.15V \text{ vs. } Li^+/Li) \tag{6.30}$$

A particular advantage of using sulfur is the theoretical energy density of the Li–S cell: 2500 Wh/kg. On the negative side, sulfur conducts neither ions nor electrons; thus novel approaches are required to overcome this unfavorable property so that the diffusion of lithium ions and electrons is more feasible. By impregnating the sulfur within a nanoscale host (e.g., carbon nanospheres), researchers have been able to achieve high, stable capacities of up to 1000 Ah/kg (Evers and Nazar 2012b). Graphene, too, has been used to encapsulate sulfur for use as a cathode in a Li-ion battery with promising results (see Section 6.10.4).

The key problems with the Li–S battery are the short life cycle and low efficiency attributed to the use of liquid electrolytes that result in detrimental side reactions. Recent development of solid electrolytes for Li–S batteries has shown excellent promise. Lithium polysulfidophosphates (LPSP) are compounds of the general formula Li_3PS_{4+n} ($0 < n < 9$) that are formed from the reaction of sulfur with lithium thiophosphate ($LiPS_4$). LSPSs were found to have lithium ion conductivity roughly

FIGURE 6.35   FESEM images of Li[Ni$_{0.45}$Co$_{0.1}$Mn$_{1.45}$]O$_4$ cathode material. (Reprinted with permission from Hassoun, J. et al. 2011. An advanced lithium ion battery based on high performance electrode materials. *J. Am. Chem. Soc.* 133 (9):3139–3143. Copyright 2011, American Chemical Society.)

eight times higher than Li$_2$S at 25°C. A device fabricated with Li$_3$PS$_{4+5}$ used as a cathode material showed excellent cyclability (300 charge–discharge cycles at 60°C) and capacity (over 1200 Ah/kg). The proposed mechanism shown in Figure 6.37 illustrates that the cleavage and formation of the polysulfide S–S bonds is the basis of the redox behavior. Raman spectroscopy was used to confirm the electrochemical insertion and removal of the lithium ion between Li$_2$PS$_{4+5}$ and a mixture of Li$_2$S and Li$_3$PS$_4$ (Lin et al. 2013). While these results are quite preliminary and the relatively low ionic conductivity remains a hindrance, this serves as an encouraging illustration of the potential of an all-solid-state Li–S battery.

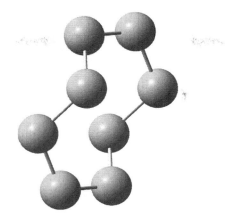

FIGURE 6.36   The S_8 allotrope of sulfur.

$$\underset{\substack{Li\\ \oplus\ominus}}{S}\!-\!\!\left(\!S\!\right)_{\!x}\!\!-\!\!\underset{(\,S\!\!\!+\!\!\!\!\underset{\ominus\ \oplus}{S}_{y}\ Li)}{\overset{\overset{S}{\|}}{P}}\!\!-\!\!\left(\!S\!\right)_{\!z}\!\!-\!\!\underset{\substack{Li\\ \ominus\ \oplus}}{S}\ \ +\ 2(x+y+z)\,Li\ \ \underset{\text{Charge reaction}}{\overset{\text{Discharge reaction}}{\rightleftharpoons}}\ \ \underset{\substack{Li\\ \oplus\ominus}}{S}\!-\!\!\underset{\underset{Li\ \oplus\ominus S}{\|}}{\overset{\overset{S}{\|}}{P}}\!\!-\!\!\underset{\substack{Li\\ \ominus\ \oplus}}{S}\ \ +\ (x+y+z)\,Li_2S$$

FIGURE 6.37   Charge and discharge reactions in the LPSP battery.

6.10.1.2   Lithium–Air Batteries

Another emerging—and very attractive—technology in the lithium battery field is the Li–air battery. Certainly, air is an environmentally and sustainably ideal redox partner and the *theoretical* gravimetric energy density for the Li–air battery compares very favorably to other battery systems, as shown in Figure 6.38 (Girishkumar et al. 2010). Furthermore, safety concerns are minimized because one of the redox partners—oxygen—is not stored in the battery, thus limiting the potential for a thermal runaway. Yet, the Li–air battery has several significant hurdles to overcome, not the least of which is the poor *practical* energy density (Figure 6.38).

Obviously, O_2/air is the oxidizing agent in the Li–air battery and lithium is the reducing agent so that the discharge half-reactions can be written as shown in Equations 6.31 and 6.32 for an anhydrous system.

$$\text{Reaction at anode: } 2Li \xrightarrow{\text{discharge}} 2Li^+ + 2e^- \tag{6.31}$$

$$\text{Reaction at cathode: } 2Li^+ + O_2 + 2e^- \xrightarrow{\text{discharge}} Li_2O_2 \tag{6.32}$$

Four potential electrolyte systems have been explored for the Li–air battery: aprotic, aqueous, mixed aqueous/aprotic, and solid-state electrolytes. At this point,

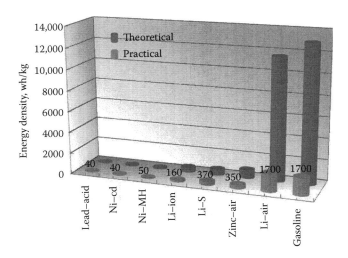

FIGURE 6.38 Comparison of energy storage densities. (Reprinted with permission from Girishkumar, G. et al. 2010. Lithium–air battery: Promise and challenges. *J. Phys. Chem. Lett.* 1 (14):2193–2203. Copyright 2010, American Chemical Society.)

only the liquid aprotic electrolyte (e.g., a lithium salt dissolved in carbonates, esters, or ethers such as tetrahydrofuran or dioxolane) is viable and our focus will be on this system, although the redox chemistry at the electrodes is not completely understood.

A schematic of a lithium–air battery is shown in Figure 6.39. The anode of the Li–air battery is often metallic lithium and the cathode is a porous, high-surface-area carbon imbedded with some sort of catalyst (e.g., α-MnO_2 nanorods) bound to a nickel current collector (Girishkumar et al. 2010). The working voltage of 2.96 V is significantly lower than the standard potential given by thermodynamic values and is indicative of one of the challenges that the Li–air battery faces: there is a large discharge overpotential, likely due to the plugging of the porous cathode by the formation of the solid discharge product Li_2O_2. Plugging the cathode not only reduces the diffusion of O_2/air through the system; Li_2O_2 is also an electron insulator. Furthermore, a charge overpotential exists as well—a high potential ($\approx$4 V) is required to reverse the reaction and charge the cell with the result that oxidation of the electrolyte and/or the graphite can take place, jeopardizing the durability of the cell. A further complication is the use of air versus pure oxygen: moisture, CO_2, N_2—all the contaminants in ordinary air can reduce the power and efficiency of the cell. Overall, the Li–air battery has a long way to go before large-scale implementation is even a remote possibility.

Ultimately, lithium-ion batteries have already proven to be valuable components in the EES arena, but they have a significant carbon footprint (Ishihara 2002). Again, the matter of scale is apparent in that up to 30% of the world's reserves of lithium would be used if all of the world's vehicles were converted into lithium-ion battery-powered EVs or hybrid plug-ins (Armand and Tarascon 2008). While innovations

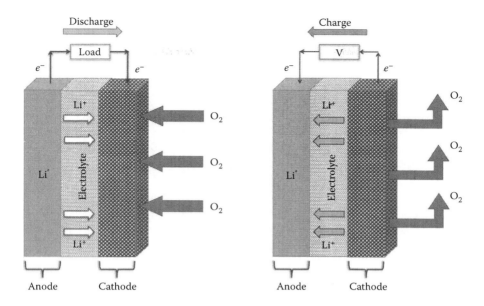

FIGURE 6.39 Schematic of a Li–air battery in discharge and charge modes. (Adapted with permission from Girishkumar, G. et al. 2010. Lithium–air battery: Promise and challenges. *J. Phys. Chem. Lett.* 1 (14):2193–2203. Copyright 2010, American Chemical Society.)

improving energy density continue, this sustainable energy approach, too, can only be a small part of the overall solution.

6.10.2 SODIUM-BASED BATTERIES

(N.B. This section on sodium-based batteries is largely summarized from the excellent review by Yang (Yang et al. 2011)). A logical leap for EES storage is the move from lithium to sodium. The standard reduction potential of this easily oxidized metal is –2.71 V, making it a good candidate for the redox processes of EES and, like lithium-ion batteries, sodium-based batteries offer high power and energy densities. However, its larger van der Waals radius presents issues with respect to ion migration and, like lithium, it raises real concerns about safety, especially as sodium EES devices are operated at high temperature (300–350°C). Nonetheless, sodium batteries have shown reasonable voltages (in the range of ≈1.8–2.6 V) and a large-scale sodium-based EES system is in use for load leveling of wind farm energy at the Rokkasho-Futamata Wind Farm in northern Japan. This facility, made up of seventeen 2-MW battery units, with each battery unit made up of forty 50-kW modules, has the capacity to store 238 MWh of energy (Clean Energy Action Project).

A simplified schematic of the sodium battery is shown in Figure 6.40. The fundamental processes underlying the operation of sodium batteries have many parallels

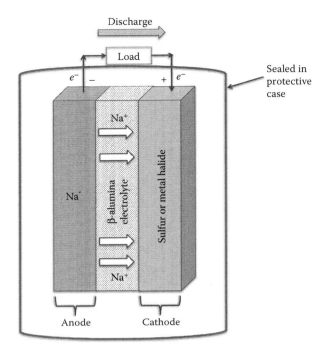

FIGURE 6.40   Schematic of a generic sodium-ion battery.

to those of SOFC given that the electrolyte is a solid oxide: beta alumina (β-Al$_2$O$_3$). The redox partner at the cathode is usually either sulfur (NAS batteries) or a metal halide. In either case, the anode material is liquid sodium, which is oxidized during discharge (Equation 6.33) and the resultant sodium ion transported through cavities/vacancies in the β-Al$_2$O$_3$ electrolyte separator.

$$\text{Reaction at anode: } 2Na \xrightarrow{\text{discharge}} 2Na^+ + 2e^- \tag{6.33}$$

As the volume of the molten sodium decreases during discharge, the liquid sodium is wicked to the β-Al$_2$O$_3$ surface to maintain physical contact with the electrolyte.

6.10.2.1   Electrolyte

Loosely packed layers of the solid electrolyte allow the migration of the mobile sodium ions through the cell under the influence of an electric field. The beta-alumina electrolyte has a variable formula of $(Na_2O)_{1+x}Al_2O_3$, where x can reach 0.57 for undoped β-Al$_2$O$_3$ (Sudworth and Tilley 1985). Doping of the electrolyte with lithium or magnesium ions improves the stability and/or the sodium ion conductivity, with $Na_{1.67}Al_{10.33}Mg_{0.67}O_{17}$ or $Na_{1.67}Al_{10.67}Li_{0.33}O_{17}$ being favorable stoichiometries. Preparing the electrolyte is no trivial matter, however, and solid-state/mechanical, solution, and vapor-phase processes have all been used, each with advantages and disadvantages. For example, sintering of the material above 1600°C is usually required to obtain a material with the appropriate density, mechanical strength, and electronic conductivity for cell operation. While fabrication of batteries with thinner electrolyte layers would result in improved conduction (and, therefore, the potential to operate the cells at a lower temperature), decreasing the thickness beyond the current 1–2 mm is done at the expense of safety as the strength of the membrane is diminished.

6.10.2.2 Cathode

As noted above, the material at the cathode is typically either sulfur or a metal halide. In the case of sulfur, the discharge reaction (Equation 6.34) forms sodium polysulfides that are electrical insulators.

$$\text{Reaction at cathode (sulfur): } xS + 2Na^+ + 2e^- \xrightarrow{\text{discharge}} Na_2S_x$$
$$\text{(1.78–2.21 V at 350°C)} \tag{6.34}$$

In order to promote the necessary electron transfer, a carbon felt is often inserted into the molten sulfur cathode. Because sulfur and the polysulfides are fairly corrosive and their reactivity is enhanced by the high temperatures, the materials that can be used for cell construction are limited and costly. What is more, given that both the anode and cathode materials are in the liquid state at high temperature, failure in the beta-alumina separator could result in their contact with potentially explosive results.

Metal halide cathode materials offer some advantages to sodium-based batteries. The metal halides are typically solids even at the elevated operating temperatures (e.g., the melting point of $NiCl_2$ is 1001°C), so a coelectrolyte such as $NaAlCl_4$ is often added to aid in ionic conductivity ($NaAlCl_4$ is the eutectic formed between $NaCl$ and $AlCl_3$ and is molten at the battery's operating temperature). The cathode reaction for a sodium/nickel chloride cell is

$$\text{Reaction at cathode (nickel chloride): } NiCl_2 + 2Na^+ + 2e^- \xrightarrow{\text{discharge}}$$
$$Ni^\circ + 2NaCl \tag{6.35}$$

with the overall cell reaction being

$$NiCl_2 + 2Na^\circ \xrightarrow{\text{discharge}} Ni^\circ + 2NaCl \quad \text{(2.58 V at 300°C)} \tag{6.36}$$

Iron(II) chloride has also been widely investigated as a less expensive alternative to nickel chloride, but overcharging during the recharge cycle can lead to the formation of Fe(III) and degradation of the beta alumina electrolyte. In either case, the metal halide cathodes are considerably less corrosive than the sulfur/polysulfide materials; hence their use is attractive from a material and safety viewpoint.

With any sodium batteries surface features again play a large role in battery performance, with the formation of insulating films leading to an increase in resistance

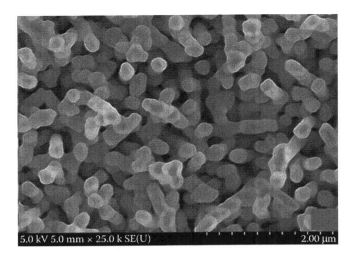

FIGURE 6.41   A SEM image of the tin/nickel anodes templated on the tobacco mosaic virus. (Reprinted with permission from Liu, Y. et al. 2013. Tin-coated viral nanoforests as sodium-ion battery nodes. *ACS Nano* 7 (4):3627–3634. Copyright 2013, American Chemical Society.)

and decrease in the battery capacity and performance. Improvements to remedy this issue and overall performance continue to be made, again particularly in the area of nanotechnology. For example, by growing a "forest" of tin nanorods for use as an anode in a sodium-ion battery, an initial capacity of 722 Ah/kg was achieved and a level of 405 Ah/kg retained after 150 cycles. This morphology was developed on a template derived from the tobacco mosaic virus that was genetically engineered to contain numerous cysteine residues (TMV1cys) so that the exposed thiol groups would readily bind nickel. The tin layer was plated on top of the nickel to result in a three-dimensional carbon/tin/nickel/TMV1cys nanorod (Figure 6.41) (Liu et al. 2013). Again, the application of nanotechnology in sustainable energy research is widespread with impressive accomplishments.

6.10.3   REDOX FLOW BATTERIES

Like the NaS batteries in use at the Rokkasho-Futamata Wind Farm, flow batteries are amenable to large-scale applications, making them a likely technology for storage of surplus energy from intermittent sources. The "flow" refers to the nature of the battery setup: the electrolytes are stored outside of the cell and flow through the cell. Unlike conventional cells, the electrolytes at the anode and cathode are typically different and are referred to as the *anolyte* and *catholyte*. Each side of the cell requires an electrocatalyst, as usual, and an ion-selective membrane allows for ionic conductivity (see Figure 6.42). Flow batteries have a long history with some of the more recent redox pairings listed below (Yang et al. 2011).

- Vanadium/vanadium (V^{3+}/V^{2+} vs. V^{4+}/V^{5+})
- Polysulfide/bromine (S/S^{2-} vs. Br^-/Br_2)

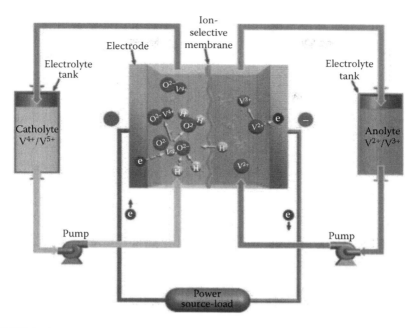

FIGURE 6.42 A vanadium redox flow battery. (Reprinted with permission from Yang, Z. et al. 2011. Electrochemical energy storage for green grid. *Chem. Rev.* 111 (5):3577–3613. Copyright 2011, American Chemical Society.)

- Iron/chromium (Fe^{3+}/Fe^{2+} vs. Br^-/Br_2)
- Zinc/bromine (Zn^{2+}/Zn vs. Br^-/Br_2)
- Vanadium/cerium (V^{3+}/V^{2+} vs. Ce^{3+}/Ce^{4+})

The amount of power a particular flow battery can deliver will depend upon the size of the electrodes in the cell and the number of cells in the stack, just as for fuel cells. The capacity is stoichiometric, that is, it depends upon the concentration and volume of the electrolyte. The largest vanadium flow battery to date is installed at the Tomamae Wind Villa power plant in northern Japan and is rated at 4.0 MW battery storage (Clean Energy Action Project; Yang et al. 2011).

All flow batteries use two redox couples, but we will focus on the vanadium flow battery because it is the simplest with only one redox-active element, vanadium, present in the cell. This is possible because of the fact that vanadium can exist in the 2^+, 3^+, 4^+, or 5^+ oxidation state. As can be seen from Figure 6.42, the composition of the vanadium flow battery is very similar to that of a fuel cell: two electrochemical compartments separated by a membrane. The catholyte consists of a solution of vanadium(IV)/vanadium(V) ions; the anolyte consists of vanadium (II)/(III) in solution. Sulfuric acid is the usual electrolyte medium, but controlling the concentrations of the active species is a challenge due to the varying solubility properties of the vanadium sulfate compounds. Temperature, the ratio of V(V) to V(IV) ions, and sulfate and vanadium ion concentrations all impact the electrochemical activity of the cell. It is a precarious

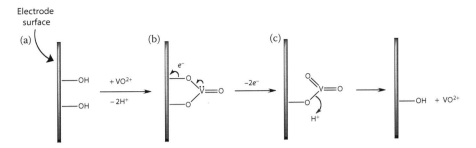

FIGURE 6.43 Proposed steps in the charging mechanism at the cathode of the vanadium flow battery (see text).

balance between precipitating out vanadium oxides and maximizing the concentrations for increased capacity of the cell. As a result, vanadium ion concentrations are typically limited to 2 M and the energy density is a modest 25 Wh/kg (Liu).

The half-reactions taking place in each compartment are shown below (Equations 6.38 and 6.39), with the overall cell reaction being

$$VO_2^+ + V^{2+} + 2H^+ \xrightarrow{\text{discharge}} VO^{2+} + V^{3+} \tag{6.37}$$

$$\text{Reaction at anode: } V^{2+} \xrightarrow{\text{discharge}} V^{3+} + e^- \tag{6.38}$$

$$\text{Reaction at cathode: } VO_2^+ + 2H^+ + e^- \xrightarrow{\text{discharge}} VO^{2+} + H_2O \tag{6.39}$$

On the cathode side, vanadium(V) is reduced to vanadium(IV), and on the anode side, vanadium(II) is oxidized to vanadium(III). The actual mechanism of electron transport is thought to be dependent upon the formation of a vanadium–oxygen–carbon bond at the electrode surface that is activated by heat treatment or chemical oxidation of the graphite electrode. In the example of the cathode electrochemistry (as shown in Figure 6.43 for the *charging* portion of the cycle), VO^{2+} reacts with neighboring hydroxyl groups on the cathode surface to form a coordinated vanadate (a). It is at this point that electron transport can take place to the cathode from vanadium to generate the electrons necessary for the current as vanadium is oxidized from (IV) to (V) (b). Acid-catalyzed hydrolysis releases the VO^{2+} from the cathode surface (c) (Sun and Skyllas-Kazacos 1992). The reverse sequence of steps takes place during discharge.

Because the chemistry is carried out under strongly acidic conditions, the choices of materials for the battery components are limited. It will come as no surprise that Nafion is the standard choice for the membrane, although much research has been devoted to improving its resistance to vanadium ion and water crossover by testing a variety of hybrids of Nafion with SPEEK, pyrroles, polyethylenimines, and so on. It is also noteworthy that flow batteries require materials with even greater mechanical stability because of their intended use in large-scale applications. Finally, vanadium, like Nafion, is relatively expensive, making the operating cost of a vanadium redox flow battery too high for broad penetration into the marketplace.

While flow batteries offer high capacity, they also exhibit low energy density relative to other electrochemical energy storage options. There are many active approaches to improving the performance of flow batteries and lowering their cost, including using additives to improve the electrolyte, making modifications to the ion membrane, and continuing to explore different redox couples, for example. In the end, a clearer understanding of the very complex redox chemistry involved should contribute to the rational development of a more effective system.

6.10.4 GRAPHENE

Graphene (Figure 6.44a) made a surprise entry into the scientific world when it was serendipitously discovered in 2004 (Novoselov et al. 2004). It is a fascinating, one-atom-thick sheet of carbon with an enormous surface area 2630 m^2/g), extraordinary electrical conductivity and charge mobility, good thermal conductivity, great mechanical strength and flexibility, as well as the ability to stack via strong π–π interactions (Dai 2012). As a result, it is a molecule of intense research interest in virtually all areas of electrochemistry, including fuel cells and energy storage. In order to make the most of graphene's properties, some modification is

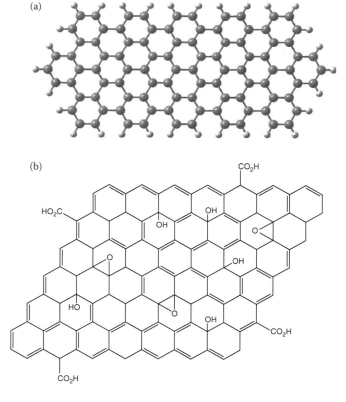

FIGURE 6.44 Representations of graphene (a) and graphene oxide (b).

usually undertaken, often starting with the oxidation of the material to make graphene oxide (GO, Figure 6.44b). As is evident from this approximate structure, the extended pi system is interrupted upon oxidation, making graphene oxide nonconductive and less able to layer due to the out-of-plane epoxides and hydroxyl functionality (Marcano et al. 2010). A careful balance of properties intermediate between graphene and graphene oxide can be achieved by reducing some of the oxygenated functionality in graphene oxide.

Improving the vanadium redox flow battery has been realized through the modification of graphene oxide to make a positive electrode with excellent catalytic activity. In this case, the superior electronic conductivity of graphene was balanced against the formation of the nucleophilic active sites on the graphene surface required for reaction with VO^{2+} (as shown in the mechanism of Figure 6.43). Researchers annealed graphene oxide at a wide range of temperatures (700–1050°C) with urea ($H_2NC(O)NH_2$) under an argon atmosphere, then characterized the nitrogen-doped product with SEM, TEM, XRD, Raman spectroscopy, and x-ray spectroscopy to confirm the morphology and composition. A material containing as much as 10 atom percent nitrogen was formed (Figure 6.45). At higher temperatures, less nitrogen was incorporated but it was found that the amount of nitrogen content was not related to the catalytic activity: instead, it is the *type* of nitrogen that is responsible. The researchers proposed that those nitrogens less likely to be protonated were responsible for the catalytic activity via the formation of a C–N–V bond, presumably like the related oxygen-based mechanism of Figure 6.43 (Jin et al. 2013).

Graphene is also a material of interest in the realm of lithium storage as well: the astonishing surface area of graphene makes it very attractive for increasing the energy density of the Li-ion battery. As a result of a density functional theory study on lithium storage in graphite or graphene, researchers determined that graphene's capacity for Li^+ is, essentially, zero. However, the results confirmed that doping with

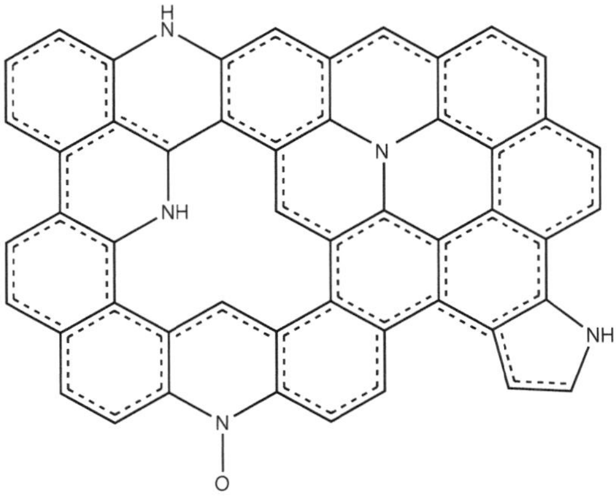

FIGURE 6.45 Plausible structure for N-doped graphene illustrating the different types of nitrogen functionality.

boron should result in greatly enhanced lithium storage by graphene, a theoretical result that has been supported experimentally (Liu et al. 2013). Graphene has also been a material of exploration in the development of the Li–S battery (see Section 6.10.1.1). For example, researchers enveloped sulfur in graphene using a ridiculously straightforward one-pot synthesis technique consisting of mixing together graphene oxide and NaOH with an aqueous NaOH solution of sulfur and sodium sulfide nonahydrate. After four hours of stirring a stoichiometric amount of 5% HCl was added and the graphene–sulfur composite was filtered and dried. In so doing, "intimate contact between the sulfur particles and the graphene sheets" was achieved resulting in a cathode material that provides a 705 Ah/kg discharge capacity (Evers and Nazar 2012a). However, higher energy and power densities are still required.

6.11 SUMMARY

Fuel cells have had a long technological evolution and their performance is quite advanced. There are many different types of fuel cells for a wide variety of applications, and the use of hydrogen (or syngas) as a fuel greatly minimizes our carbon emissions. But the move toward a hydrogen/fuel cell-based transportation system presents a perfect example of why we must not ignore life cycle analyses. In such an analysis of various fuels for transportation, researchers noted the following concerns with respect to fuel cell-powered vehicles (MacLean and Lave 2003):

1. The reliance on resource-scarce materials in the fuel cell itself
2. The heavy reliance on carbon-containing feedstocks in the production of hydrogen gas
3. The cost of a fuel cell vehicle (estimated to be roughly one million dollars in 2003!)

The science and technology of fuel cells is impressive and continues to evolve. New research findings in this area of electrochemistry are applied to EES with similarly impressive results. However, sustainability remains very much in doubt and large-scale implementation of fuel cell-powered vehicles is far in the future.

OTHER RESOURCES

BOOKS

Barbir, F. 2013. *PEM Fuel Cells. Theory and Practice*. Waltham, MA: Academic Press/Elsevier.
Li, X. 2006. *Principles of Fuel Cells*. New York: Taylor & Francis.

ONLINE RESOURCES

U.S. Department of Energy/Energy Efficiency and Renewable Energy/Fuel cells: http://www.eere.energy.gov/topics/hydrogen_fuel_cells.html.
U.S. Department of Energy Hydrogen and Fuel Cells Program: http://www.hydrogen.energy.gov/index.html.

REFERENCES

Adams, T.A., J. Nease, D. Tucker et al. 2013. Energy conversion with solid oxide fuel cell systems: A review of concepts and outlooks for the short- and long-term. *Ind. Eng. Chem. Res.* 52 (9):3089–3111.

Armand, M. and J.-M. Tarascon. 2008. Building better batteries. *Nature* 451 (February):652–657.

Armaroli, N. and V. Balzani. 2011. *Energy for a Sustainable World.* Weinheim, FRG: Wiley-VCH.

Bartholomew, C.H. and R.J. Farrauto. 2006. *Fundamentals of Industrial Catalytic Processes*, 2nd ed. Hoboken, NJ: Wiley & Sons.

Bauer, D., D. Diamond, J. Li et al. Critical materials strategy, U.S. Department of Energy. 2011. U.S. Department of Energy, http://energy.gov/sites/prod/files/DOE_CMS2011_FINAL_Full.pdf.

Bianchini, C. and P.K. Shen. 2009. Palladium-based electrocatalysts for alcohol oxidation in half cells and in direct alcohol fuel cells. *Chem. Rev.* 109:4183–4206.

Carpenter, M.K., T.E. Moylan, R.S. Kukreja, et al. 2012. Solvothermal synthesis of platinum alloy nanoparticles for oxygen reduction electrocatalysis. *J. Am. Chem. Soc.* 134:8535–8542.

Cheah, M.J., I.G. Kevrekidis, and J. Benziger. 2011. Effect of interfacial water transport resistance on couple proton and water transport across Nafion. *J. Phys. Chem. B* 115:10239–10250.

Chen, G., B. Wei, Y. Luo et al. 2012. Polymer separators for high-power, high-efficiency microbial fuel cells. *ACS Appl. Mater. Interfaces* 4 (12):6454–6457.

Clark, N. 2013. Boeing begins modifying 787 batteries. *New York Times*, 22 April 2013.

Clean Energy Action Project. *Energy Storage Case Studies.* Clean energy action project [cited 13 May 2013]. Available from http://www.cleanenergyactionproject.com/CleanEnergyActionProject/Energy_Storage_Case_Studies.html.

Cracknell, J.A., K.A. Vincent, and F.A. Armstrong. 2008. Enzymes as working or inspirational electrocatalysts for fuel cells & electrolysis. *Chem. Rev.* 108:2439–2461.

da Rosa, A.V. 2009. *Fundamentals of Renewable Energy Processes.* Burlington, MA: Academic Press/Elsevier.

Dai, L. 2012. Functionalization of graphene for efficient energy conversion and storage. *Acc. Chem. Res.* 46 (1):31–42.

Di Vona, M.L., D. Marani, C. D'Ottavi et al. 2006. A simple new route to covalent organic/inorganic hybrid proton exchange polymeric membranes. *Chem. Mater.* 18 (1):69–75.

E.I. du Pont de Nemours and Company. Dupont fuel cells safe handling and use of perfluorosulfonic acid products. dfc301.pdf. DuPont. 2009. Wilmington, DE, fuelcells.dupont.com.

Erable, B., D. Féron, and A. Bergel. 2012. Microbial catalysis of the oxygen reduction reaction for microbial fuel cells: A review. *ChemSusChem* 5:975–987.

Evers, S. and L.F. Nazar. 2012a. Graphene-enveloped sulfur in a one pot reaction: A cathode with good coulombic efficiency and high practical sulfur content. *Chem. Commun.* 48 (9):1233–1235.

Garland, N. 2008 Fuel Cell Seminar Paper GHT 33–1. http://www.fuelcellseminar.com/2008_presentations; cited in Martin, K.M., J.P. Kopasz, and K.W. McMurphy. 2010. Status of fuel cells and the challenges facing fuel cell technology today. In *Fuel Cell Chemistry and Operation*, 1–13. American Chemical Society (reference 7).

Garland, N. 2012. New approaches for high energy density lithium–sulfur battery cathodes. *Acc. Chem. Res.* 46 (5):1135–1143.

Girishkumar, G., B. McCloskey, A.C. Luntz et al. 2010. Lithium–air battery: Promise and challenges. *J. Phys. Chem. Lett.* 1 (14):2193–2203.

Hassoun, J., K.-S. Lee, Y.-K. Sun, et al. 2011. An advanced lithium ion battery based on high performance electrode materials. *J. Am. Chem. Soc.* 133 (9):3139–3143.

Hazarika, M. and T. Jana. 2012. Proton exchange membrane developed from novel blends of polybenzimidazole and poly(vinyl-1,2,4-triazole). *ACS Appl. Mater. Interfaces* 4 (10):5256–5265.

Hurd, J.A., R. Vaidhyanathan, V. Thangadurai et al. 2009. Anhydrous proton conduction at 150°C in a crystalline metal–organic framework. *Nat. Chem.* 1 (9):705–710.

Ishihara, K. 2002. In *5th International Conference on Ecobalance*. Tsukuba, Japan: Society of Non-Traditional Technology.

Ishihara, T., H. Furutani, M. Honda et al. 1999. Improved oxide ion conductivity in $La_{0.8}Sr_{0.2}Ga_{0.8}Mg_{0.2}O_3$ by doping Co. *Chem Mater.* 11 (8):2081–2088.

Ishikawa, J.I., S. Fujiyama, K. Inoue et al. 2007. Highly sulfonated poly(aryl ether ketone) block copolymers having a cross-linking structure. *J. Memb. Sci.* 298 (298):48–55.

Jacobson, A.J. 2009. Materials for solid oxide fuel cells. *Chem. Mater.* 22 (3):660–674.

Jin, J., X. Fu, Q. Liu, et al. 2013. Identifying the active site in nitrogen-doped graphene for the VO^{2+}/VO_2^+ redox reaction. *ACS Nano* 7 (6):4764–4773.

Jorn, R., J. Savage, and G.A. Voth. 2012. Proton conduction in exchange membranes across multiple length scales. *Acc. Chem. Res.* 45 (11):2002–2010.

Kamarudin, S.K. and N. Hashim. 2012. Materials, morphologies and structures of MEAs in DMFCs. *Renew. Sustain. Energy Rev.* 16 (5):2494–2515.

Kirubakaran, A., S. Jain, and R.K. Nema. 2009. A review on fuel cell technologies and power electronic interface. *Renew. Sustain. Energy Rev.* 13:2430–2440.

Kitchin, J.R., J.K. Nørskov, M.A. Barteau, et al. 2004. Role of strain and ligand effects in the modification of the electronic and chemical properties of bimetallic surfaces. *Phys. Rev. Lett.* 93 (15):156801–1 to 156801–4.

Knowledge Transfer Network. 2013. *Minerals and Elements Review.* Chemistry innovation Ltd. 2010 [cited 8 May 2013 2013]. Available from http://www.chemistryinnovation. co.uk/stroadmap/files/dox/MineralsandElementspages.pdf.

Li, J. and H. Yu. 2007. Synthesis and characterization of sulfonated poly(benzoxazole ether ketone)s by direct copolymerization as novel polymers for proton-exchange membranes. *J. Polym. Sci. A Polym. Chem.* 45 (11):2273–2286.

Li, M., P. Liu, and R.R. Adzic. 2012. Platinum monolayer electrocatalysts for anodic oxidation of alcohols. *J. Phys. Chem. Lett.* 3 (23):3480–3485.

Li, Q., R. He, J.O. Jensen et al. 2003. Approaches and recent development of polymer elec-trolyte membranes for fuel cells operating above 100°C. *Chem. Mater.* 15:4896–4915.

Lin, Z., Z. Liu, W. Fu et al. 2013. Lithium polysulfidophosphates: A family of lithium-conducting sulfur-rich compounds for lithium-sulfur batteries. *Angew. Chem. Int. Ed.* 52 (29):7460–7463.

Liu, J., J.-G. Zhang, Z. Yang et al. 2013. Materials science and materials chemistry for large scale electrochemical energy storage: From transportation to electrical grid. *Adv. Funct. Mater.* 23:929–946.

Liu, M., M.E. Lynch, K. Blinn, et al. 2011. Rational SOFC material design: New advances and tools. *Mater. Today* 14 (11):534–546.

Liu, Y., V.I. Artyukhov, M. Liu et al. 2013. Feasibility of lithium storage on graphene and its derivatives. *J. Phys. Chem. Lett.* 4 (10):1737–1742.

Liu, Y., Y. Xu, Y. Zhu et al. 2013. Tin-coated viral nanoforests as sodium-ion battery anodes. *ACS Nano* 7 (4):3627–3634.

Liu, Z., Z.-W. Zheng, M.-F. Han et al. 2010. High performance solid oxide fuel cells based on tri-layer yttria-stabilized zirconia by low temperature sintering process. *J. Power Sources* 195:7230–7233.

Li, X. 2006. *Principles of Fuel Cells*. New York: Taylor & Francis.

MacLean, H.L. and L.B. Lave. 2003. Life cycle assessment of automobile/fuel options. *Env. Sci. Technol.* 37 (23):5445–5452.

Malavasi, L., C.A.J. Fisher, and M.S. Islam. 2010. Oxide-ion and proton conducting electrolyte materials for clean energy applications: Structural and mechanistic features. *Chem Soc. Rev.* 39 (11):4370–4387.

Marcano, D.C., D.V. Kosynkin, J.M. Berlin et al. 2010. Improved synthesis of graphene oxide. *ACS Nano* 4 (8):4806–4814.

McIntosh, S. and R.J. Gorte. 2004. Direct hydrocarbon solid oxide fuel cells. *Chem. Rev.* 104:4845–4865.

Mekhilef, S., R. Saidur, and A. Safari. 2012. Comparative study of different fuel cell technologies. *Renew. Sustain. Energy Rev.* 16:981–989.

Mitzel, J., F. Arena, H. Natter et al. 2012. Electrodeposition of PEM fuel cell catalysts by the use of a hydrogen depolarized anode. *Int. J. Hydrogen Energy* 37:6261–6267.

Miyatake, K., Y. Chikashige, E. Higuchi et al. 2007. Tuned polymer electrolyte membranes based on aromatic polyethers for fuel cell applications. *J. Am. Chem. Soc.* 129 (13):3879–3887.

Morozan, A., P. Jégou, B. Jousselme et al. 2011. Electrochemical performance of annealed cobalt-benzotriazole/CNTs catalysts towards the oxygen reduction reaction. *Phys. Chem. Chem. Phys.* 13:21600–21607.

Novoselov, K.S., A.K. Geim, S.V. Morozov et al. 2004. Electric field effect in atomically thin carbon films. *Science* 306 (5696):666–669.

Oh, Y.S., H.J. Lee, M. Yoo et al. 2008. Synthesis of novel crosslinked sulfonated poly(ether sulfone)s using bisazide and their properties for fuel cell application *J. Membrane Sci.* 323:309–315.

Okada, T. and M. Kaneko. 2009. *Molecular Catalysts for Energy Conversion*. Berlin: Springer.

Orera, A. and P.R. Slater. 2010. New chemical systems for solid oxide fuel cells. *Chem. Mater.* 22:675–690.

Papageorgopoulos, D. 2011. PEMFC R&D at the DOE Fuel Cell Technologies Program. Edited by N. M. Markovic. Arlington, VA.

Petrii, O.A. 2008. Pt-Ru electrocatalysts for fuel cells: A representative review. *J. Solid State Electrochem.* 12:609–642.

Ponce, M.L., D. Gomes, and S.P. Nunes. 2008. One-pot synthesis of high molecular weight sulfonated poly(oxadiazole-triazole) copolymers for proton conductive membranes. *J. Memb. Sci.* 319 (1–2):14–22.

Ponomareva, V.G., K.A. Kovalenko, A.P. Chupakhin et al. 2012. Imparting high proton conductivity to a metal–organic framework material by controlled acid impregnation. *J. Am. Chem. Soc.* 134 (38):15640–15643.

Rulkens, W. 2008. Sewage sludge as a biomass resource for the production of energy: Overview and assessment of the various options. *Energy Fuels* 22:9–15.

Schröder, U. 2012. Microbial fuel cells and microbial electrochemistry: Into the next century! *ChemSusChem* 5:959–961.

Sha, Y., T.H. Yu, B.V. Merinov et al. 2012. Mechanism for oxygen reduction reaction on Pt_3Ni alloy fuel cell cathode. *J. Phys. Chem. C* 116 (40):21334–21342.

Sharma, S. and B.G. Pollet. 2012. Support materials for PEMFC and DMFC electrocatalysts. A review. *J. Power Sources* 208 (0):96–119.

Simões, F.C., D.M. dos Anjos, F. Vigier et al. 2007. Electroactivity of tin modified platinum electrodes for ethanol electrooxidation. *J. Power Sources* 167 (1):1–10.

Singh, M., I. Gur, H.B. Eitouni et al. 2008. Solid electrolyte material manufacturable by polymer processing methods. US Patent 8,268,197, filed November 14, 2008, and issued September 18, 2012.

Sood, R., C. Iojoiu, E. Espuche et al. 2012. Proton conducting ionic liquid doped Nafion membranes: Nano-structuration, transport properties and water sorption. *J. Phys. Chem. C* 116 (46):24413–24423.

Stamenkovic, V.R., B. Fowler, B.S. Mun et al. 2007. Improved oxygen reduction activity on $Pt_3Ni(111)$ via increased surface site availability. *Science* 315 (26 Jan 2007):493–497.

Sudworth, J.L. and A.R. Tilley. 1985. *The Sodium Sulphur Battery*. London: Chapman & Hall.

Sun, B. and M. Skyllas-Kazacos. 1992. Modification of graphite electrode materials for vanadium redox flow battery application—I. Thermal treatment. *Electrochim. Acta* 37 (7):1253–1260.

Sun, Y.-N., L. Giordano, J. Goniakowski et al. 2010. The interplay between structure and CO oxidation catalysis on metal-supported ultrathin oxide films. *Angew. Chem. Int. Ed.* 49 (26):4418–4421.

Suzuki, K., Y. Iizuka, M. Tanaka et al. 2012. Phosphoric acid-doped sulfonated polyimide and polybenzimidazole blend membranes: High proton transport at wide temperatures under low humidity conditions due to new proton transport pathways. *J. Mater. Chem.* 22 (45):23767–23772.

Takeguchi, T., T. Yamanaka, K. Asakura et al. 2012. Evidence of nonelectrochemical shift reaction on a CO-tolerant high-entropy state Pt–Ru anode catalyst for reliable and efficient residential fuel cell systems. *J. Am. Chem. Soc.* 134 (35):14508–14512.

Tan, Y., C. Xu, G. Chen et al. 2012. Facile synthesis of manganese-oxide-containing mesoporous nitrogen-doped carbon for efficient oxygen reduction. *Adv. Funct. Mater.* 22 (21):4584–4591.

Tarascon, J.M. and M. Armand. 2001. Issues and challenges facing rechargeable lithium batteries. *Nature* 414 (6861):359–367.

U.S. Department of Energy. 2008. Energy efficiency and renewable energy information center. *Comparison of Fuel Cell Technologies*. Available from www.hydrogen.energy.gov.

U.S. Department of Energy. 2011a. Fuel Cell Technology Challenges: Technical Plan—Fuel Cells. Edited by U.S. Department of Energy.

U.S. Department of Energy. 2011b. FY 2011 Progress Report for the DOE Hydrogen and Fuel Cells Program. U.S. Government. http://www.hydrogen.energy.gov/annual_progress11.html.

U.S. Department of Energy National Energy Technology Laboratory (NETL). 2012. A Primer on SOFC Technology. Edited by NETL Solid State Energy Conversion Alliance.

U.S. Energy Information Administration. 2008. *Environment/Emissions of Greenhouse Gases Report (DOE/EIA-0573(2008))*. Available from http://www.eia.gov/oiaf/1605/ggrpt/carbon.html-transportation.

van der Vliet, D.F., C. Wang, D. Tripkovic et al. 2012. Mesostructured thin films as electrocatalysts with tunable composition and surface morphology. *Nat. Mater.* 11:1051–1058.

Vogel, J., J. Marcinkoski, R. Tyler et al. 2009. V.D.4 FC40 International Stationary Fuel Cell Demonstration. *DOE Hydrogen Program FY2008 Annual Progress Report*: 912–915.

Wang, C., N.M. Markovic, and V.R. Stamenkovic. 2012. Advanced platinum alloy electrocatalysts for the oxygen reduction reaction. *ACS Catal.* 2 (5):891–898.

Wang, S., L. Azhang, Z. Xia et al. 2012. BCN graphene as efficient metal-free electrocatalyst for the oxygen reduction reaction. *Angew. Chem. Int. Ed.* 124:4285–4288.

Wang, Y., K.S. Chen, J. Mishler et al. 2011. A review of polymer electrolyte membrane fuel cells: Technology, applications, and needs on fundamental research. *Appl. Energy* 88:981–1007.

Wang, Y., H. Zhang, F. Chen et al. 2012. Electrochemical characteristics of nano-structured $PrBaCo_2O_{5+x}$ cathodes fabricated with ion impregnation process. *J. Power Sources* 203:34–41.

Wang, Z., Y. Liu, and V.M. Linkov. 2006. The influence of catalyst layer morphology on the electrochemical performance of DMFC anode. *J. Power Sources* 160 (1):326–333.

Wei, J., P. Liang, and X. Huang. 2011. Recent progress in electrodes for microbial fuel cells. *Bioresour. Technol.* 102 (20):9335–9344.

Wiberg, N. 2001. *Inorganic Chemistry*. San Diego: Academic Press.

Wroblowa, H.S., Y.-C. Pan, and G. Razumney. 1976. Electroreduction of oxygen. A new mechanistic criterion. *J. Electroanal. Chem.* 69:195–201.

Wu, B., X. Lin, L. Ge et al. 2013. A novel route for preparing highly proton conductive membrane materials with metal-organic frameworks. *Chem. Commun.* 49 (2):143–145.

Xu, C., T.S. Zhao, and Q. Ye. 2006. Effect of anode backing layer on the cell performance of a direct methanol fuel cell. *Electrochim. Acta* 51:5524–5531.

Yang, W., T.-P. Fellinger, and M. Antonietti. 2011. Efficient metal-free oxygen reduction in alkaline medium on high-surface-area mesoporous nitrogen-doped carbons made from ionic liquids and nucleobases. *J. Am. Chem. Soc.* 133 (2):206–209.

Yang, W., T. Hong, S. Li et al. 2013. Perovskite $Sr_{1-x}Ce_xCoO_{3-d}$ ($0.05 \leq x \leq 0.15$) as superior cathodes for intermediate temperature solid oxide fuel cells. *ACS Appl. Mater. Interfaces* 5 (3):1143–1148.

Yang, Z., J. Zhang, M.C.W. Kintner-Meyer et al. 2011. Electrochemical energy storage for green grid. *Chem. Rev.* 111 (5):3577–3613.

Yu, D., Y. Xue, and L. Dai. 2012. Vertically aligned carbon nanotube arrays co-doped with phosphorus and nitrogen as efficient metal-free electrocatalysts for oxygen reduction. *Phys. Chem. Lett.* 3:2863–2870.

Yu, W., M.D. Porosoff, and J.G. Chen. 2012. Review of Pt-based bimetallic catalysis: From model surfaces to supported catalysts. *Chem. Rev.* 112 (11):5780–5817.

Zenyuk, I.V. and S. Litster. 2012. Spatially resolved modeling of electric double layers and surface chemistry for the hydrogen oxidation reaction in water-filled platinum–carbon electrodes. *J. Phys. Chem. C* 116 (18):9862–9875.

Zhang, H. and P.K. Shen. 2012. Recent development of polymer electrolyte membranes for fuel cells. *Chem. Rev.* 112 (5):2780–2832.

7 Solar Photovoltaics

7.1 INTRODUCTION

Of all the approaches to meeting our energy needs sustainably, utilizing the energy from the sun is arguably the most viable and solar is a growing source of renewable energy. Cumulative global solar photovoltaic (PV) capacity reached 101 GW in 2012 and growth is expected to continue at a fantastic pace. According to the European Commission's *PV Status Report 2012*, "Since 2000, total PV production increased almost by two orders of magnitude, with annual growth rates between 40% and 90%. The most rapid growth in annual production over the last five years could be observed in Asia, where China and Taiwan together now account for more than 65% of world-wide production" (Jäger-Waldau 2012, p. 9). As world energy consumption continues to grow, solar photovoltaics will undoubtedly play an increasingly important role in the generation of low-carbon energy.

Generated through nuclear fusion, the Sun's energy strikes the outer atmosphere of the Earth with about 1370 W/m^2 of power, a value known as the *solar constant* (Armaroli and Balzani 2011). This radiation is then attenuated through reflection and absorbance by clouds, the atmosphere and the Earth's surface (Figure 7.1). The end result is the heating of our planet to a temperature that currently supports life and provides us with solar energy—about 3.7 million Quad per year (McElroy 2010). Given that the 2010 global energy consumption was roughly 344 Quad (IEA/ International Energy Agency 2013), the Earth receives enough energy from the Sun to fulfill the yearly world demand in *less than one hour* (Koster et al. 2009)—more than enough to sate our voracious appetite. The caveat to capturing and utilizing some of this solar energy is the usual one: it must be converted into a convenient, useful, cost-effective form in a sustainable manner.

The solar spectrum (Figure 7.2) ranges in wavelength from about 250 to 2500 nm corresponding to photons with energies that range from about 0.5 to 3.5 eV (Nozik 2002). The maximum photon flux density is located at approximately 700 nm (at the far end of the visible range; Cheng et al. 2009). Photovoltaic devices (also commonly known as *solar cells*) convert this breadth of energy from sunlight directly into that most convenient of energy forms, electricity. This direct conversion of a primary energy source (in this case, solar radiation) to electricity accentuates the parallel between solar cells and fuel cells. One can consider solar radiation—in the form of photons—not unlike the energy tied up in the chemical bonds that make up a fuel. Also like fuel cells, individual solar cells need to be interconnected in order to generate enough power, just like fuel cell stacks. Individual solar cells typically produce only 2–3 W, but when interconnected into panels, and panels into arrays, megawatts of power can be produced (Figure 7.3).

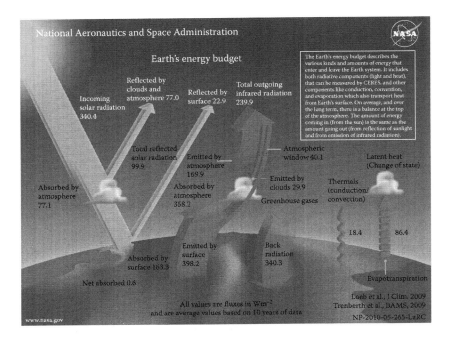

FIGURE 7.1   (**See color insert.**) The Earth's energy budget. (From National Aeronautic and Space Administration. "Earth's Energy Budget." Retrieved May 23, 2013, from http://www. nasa.gov/audience/foreducators/topnav/materials/listbytype/Earths_Energy_Budget.html.)

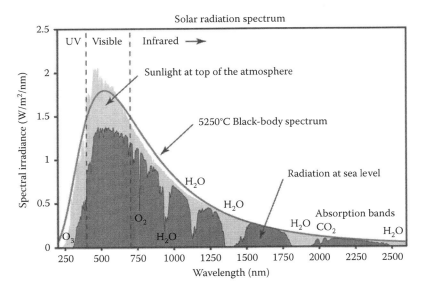

FIGURE 7.2   (**See color insert.**) The solar radiation spectrum. (From Rohde, R. A. Solar radiation spectrum, Wikimedia Commons/Global Warming Art. Retrieved May 23, 2013, from http://commons.wikimedia.org/wiki/File:Solar_Spectrum.png.)

FIGURE 7.3   A solar panel array. (From Shutterstock 61884094.)

Another way of taking advantage of the Sun's energy is by focusing the Sun's rays to intensify them and heat an appropriate material, then use the concentrated thermal energy. This is known as CSP (**c**oncentrating **s**olar **p**ower) and the solar tower shown in Figure 7.4 exemplifies this type of energy generation. This chapter will focus solely on solar photovoltaics, although the field of CSP is also a significant player in

FIGURE 7.4   (**See color insert.**) A solar thermal concentrator tower. (From Shutterstock 96405053.)

the renewable energy arena (Mehos 2008). In fact, CSP and photovoltaics have been coupled to concentrate the light striking the PV device, resulting in some of the highest reported solar cell efficiencies to date (Hanna et al. 2012; Siemens 2012).

One way to classify solar cells is by their basic chemical makeup, namely inorganic versus organic, although this is a fairly arbitrary and obsolete distinction. Inorganic solar cells based on silicon are firmly established as the most widely used commercial PV with megawatt arrays of flat silicon solar panels familiar worldwide. Intense research effort has led to the development of other types of inorganic PVs, a plethora of organic PVs, and hybrid cells that take advantage of the properties of both inorganic and organic materials. We will first examine silicon solar cells to present a basic understanding of photovoltaics, band theory, semiconductors, and the photoelectric effect. We will then present the more complex processes behind a survey of different types of photovoltaics in turn. Given the breadth and extraordinary growth of research in the field of photovoltaics, this chapter will be limited to silicon, thin-film inorganic, organic, dye-, and quantum dot (QD)-sensitized solar cells, although impressive new discoveries in this field are reported daily.

7.2   SOLAR PV BASICS

7.2.1   BAND THEORY AND THE PHOTOELECTRIC EFFECT

The fundamental material required for a PV device is a semiconductor, and to understand how semiconductors work we need to review the electronic structure of solids based on *band theory*. One can picture a nonionic solid like silicon as a single, unbounded molecule awash in a sea of electrons: the valence electrons supplied by each atom in the solid are dispersed throughout the entire solid. Given an almost infinitely large number of atoms that contribute atomic orbitals to this giant "molecule," the combinations of these atomic orbitals result in molecular orbitals for the solid that are very closely spaced in energy. In fact, their nearness in energy level means that these molecular orbitals lie together in *bands* (Figure 7.5). The lower energy band corresponds to the group of molecular orbitals (mostly filled) that contain the valence electrons (the *valence band*, VB). Higher in energy lies the band of molecular orbitals (mostly empty) that correspond to the *conduction band* (CB). The energy level between these two bands is known as the Fermi level. Above the Fermi level there is a 50% probability that, at room temperature, an empty electron state (a *hole*) will be filled.

Electrons promoted from the valence band into the conductive band are mobile and contribute to conductivity, leaving behind a *hole*. Holes, too, are mobile and are similarly conductive through the solid. If the valence band and the conduction band overlap (as shown in Figure 7.5a), the material is a *conductor*. However, if there is a large gap in energy between the valence and conductive bands (known as the bandgap, E_g), the material is an *insulator* (Figure 7.5c). In-between these two extremes lies the realm of *semiconductors* (Figure 7.5b). The bandgap energies for some common semiconductor materials are given in Table 7.1.

How does solar energy fit into all this? It was 1887 when Hertz serendipitously discovered the *photoelectric effect* when studying the wave nature of light. The

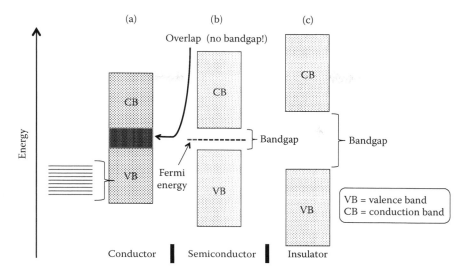

FIGURE 7.5 Molecular orbital bands for different types of materials. (a) Conductor; (b) semiconductor; (c) insulator.

photoelectric effect amounts to the ejection of electrons from a material when struck by light, a finding further illuminated by Lenard in 1900 (Eisberg 1961). The photoelectric effect was deciphered and Planck demonstrated that electromagnetic radiation must exist in discrete levels equivalent to $h\upsilon$ (where h is Planck's constant and υ is the frequency of the electromagnetic radiation). These findings threw classical physics on its head. Einstein pulled it all together, recognizing that the light striking the material delivers a "packet" (quantum) of energy in the form of a photon and that the particular amount of energy is dependent upon the *frequency* (not the intensity) of the radiation. Thus, an electron is promoted from the valence band to the conduction band only if the energy absorbed is *at least* as large as the bandgap—firmly

---

TABLE 7.1
Bandgaps for Selected PV Materials

Material	Si	Ge	GaN	GaP	GaAs	ZnSe	CdTe	P3HT	CZTS	CZTSe
E_g (eV)	1.11	0.67	3.4	2.26	1.43	2.7	1.45[a]	1.9	1.45[b]	0.94[c]

Source: Kitai, A. 2011. *Principles of Solar Cells, LEDs and Diodes.* West Sussex, UK: John Wiley & Sons, p. 26.

[a] Bonnet, D. 2012. CdTe thin-film PV modules. In *Practical Handbook of Photovoltaics*, edited by A. M. McEvoy, T. Castañer and L. Waltham. MA/Oxford UK, Academic Press/Elsevier, p. 284.

[b] Katagiri, H., N. Sasaguchi, S. Hando et al. 1997. Preparation and evaluation of Cu_2ZnSnS_4 thin films by sulfurization of E-B evaporated precursors. *Sol. Energy Mater. Sol. Cells* 49:407–414.

[c] Zoppi, G., I. Forbes, R.W. Miles et al. 2009. $Cu_2ZnSnSe_4$ thin film solar cells produced by selenisation of magnetron sputtered precursors. *Prog. Photovolt.: Res. Appl.* 17:315–319.

tying the frequency of the light to the energy needed to promote an electron. This minimum amount of energy is dependent upon the semiconductor material; for example, for crystalline silicon the bandgap energy is 1.1 eV at room temperature. (It is worth noting that it is actually the *photovoltaic* effect that applies to PVs in that the electrons aren't actually *ejected* in a PV, they are merely transferred to another material, *vide infra*.)

Tuning bandgaps to most efficiently harvest the Sun's energy is one of the fundamental challenges of PV research. The key principle, then, for the operation of semiconductors in PV devices is that an amount of energy at least as great as the *bandgap energy* E_g must be provided to promote the electrons from the valence band to the conductive band in order for electrical conduction to occur. This is known as the *cell threshold* and in PVs that energy is provided by photons. Those photons with energy $< E_g$ fail to promote an electron while all those with energy $\geq E_g$ do successfully promote an electron to the conduction band, with the excess energy potentially lost as heat.

7.2.2   ELECTRICAL CONDUCTION IN A PV DEVICE

As we saw in Chapter 6, in fuel cells an electrochemical redox reaction generates electrons and ions (typically protons) that are then transported through a circuit to generate electricity. In PVs, electrons are promoted to the conductive band of the semiconductor as a result of the photovoltaic effect, leaving behind a region of positive charge (the hole). In a good solar cell material, these charge carriers (the electron and the hole) will be swept away within hundreds of picoseconds to generate an electric current, more quickly than they can *recombine* to return to the ground state. There are several different modes of recombination (e.g., radiative or nonradiative), but the end result is the same: the electron and hole reunite before they make it into the circuit to generate the current. Thus, the longer the hole and electron can exist without recombining, the greater the current density and the better the material for use in a solar cell.

> *N.B.* The absorption of light actually generates an *electron–hole pair* (EHP), also known as an *exciton*. Whether it is a separate electron and hole that are generated or an exciton depends upon the particular PV device, as we will see. It is also crucial to understand that creation of the electron–hole pair is but the first step in the photovoltaic process. The bound pair must then be separated to free charge carriers and move out of the light-absorbing material, finally exiting the device by the appropriate electrical contact.

Pure semiconductor materials exist that can be made to be conductive by simple input of energy, but very low conductivity results. These are known as *intrinsic* semiconductors. However, a material can be altered to enhance the concentration of the holes and electrons (and therefore increase the current density) by a process known as *doping*—adding an impurity into the solid-state structure that has either an extra electron or one fewer electron. Such a material is considered an *extrinsic* semiconductor. How does doping work? We will look at silicon, a tetravalent element, to illustrate. By allowing a few parts per million of pentavalent phosphorus atoms to diffuse into the top few nanometers of the silicon layer, "extra" nonbonding electrons from the phosphorus are available, allowing for this now impure material to act as a *donor* of electrons. Given that an electron is negatively charged, this is known as *n-type doping*. In a similar fashion, doping Si with a trivalent atom (e.g., boron) results in fewer electrons in the material, or, more specifically, a material that possesses a vacancy where one of silicon's four electrons once was. This is *p-type doping* and this material would be considered an *acceptor*. Note that the Fermi energy levels change upon doping, with the Fermi level for p-type materials lying closer to the valence band and the Fermi level for n-type materials being raised so that it is closer to the conduction band.

The p–n junction. The *n* (donor) material and the *p* (acceptor) material interface is known as the *p–n junction* in an extrinsic semiconductor.

A reminder and word of warning: this discussion of the p–n junction is based on the architecture of silicon solar cells only. As we discuss other types of solar cells, the materials and architecture of the cell may be considerably different.

The p–n junction possesses an internal electric field by virtue of the fact that the p-region holes and the n-region electrons are attracted to one another. As incident light is absorbed by the semiconductor material and electrons are promoted, the electrons are attracted to the positive region and the holes to the negative region. The extra electrons from the n-region flow toward the p-material. A *depletion zone* builds up setting up a difference in electrical potential (Figure 7.6). Charge carriers diffuse to the p–n junction and, with any luck, are separated by the internal electric field before recombining. After separation these free charge carriers are swept to the appropriate contact to generate an electric current. Figure 7.7 illustrates this conduction process. The "bend" in the diagram reflects the differential in the n/p energy levels due to the distribution of charge and resultant internal electric field. *The p–n junction's internal electric field is the sine qua non of an inorganic semiconductor PV device.*

Types of p–n junctions. When one singular material (as in silicon) is doped in two different ways to make p-type material and n-type material, the resultant device is

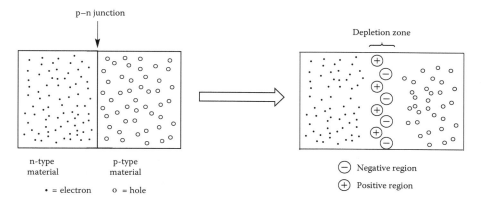

FIGURE 7.6 The p–n junction and evolution of the depletion zone.

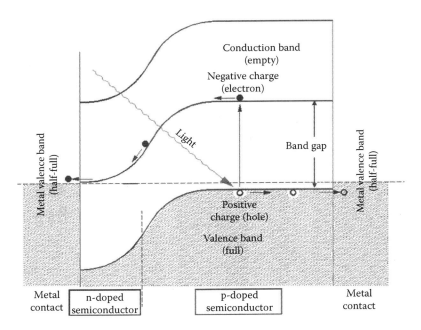

FIGURE 7.7   The conduction process at the p–n junction. (Adapted from http://en.wikipedia. org/wiki/File:bandDiagramSolarCell-en.gif. Accessed January 9, 2013.)

said to have a *homojunction*. For a homojunction solar cell with only one E_g, the optimum bandgap energy is about 1.3 eV, corresponding to the area of the solar spectrum at which the greatest density of photon flux can be absorbed (Kirchartz and Rau 2011). Only that portion of the solar spectrum corresponding to wavelengths meeting this minimum bandgap energy will contribute to the cell current, since energy from those photons with energy $<E_g$ are insufficiently energetic to promote an electron into the conduction band. Therefore, for a homojunction solar cell, the 1.3 eV bandgap means that light of all wavelengths ≤950 nm will contribute to the photocurrent, as demonstrated below in Equation 7.1.

$$\lambda(nm) = \frac{1240}{E(eV)} = \frac{1240}{1.3} \approx 950 \text{ nm}$$

$$(7.1)$$

Cells may also contain multiple junctions (*multijunctions*) with two or more different E_g. This allows the semiconductor materials to play different roles, namely, one material with a relatively large bandgap will absorb only a small part of the light, allowing the majority of the light to pass through and be captured by the next layer which may have a narrower E_g. There are clear advantages for having more than one bandgap; one need only review Figure 7.2 to see why—capturing and converting as much of the solar spectrum as possible will increase the efficiency of a PV device. For example, while crystalline silicon's E_g is 1.1 eV, gallium arsenide absorbs at 1.4 eV and aluminum gallium arsenide at 1.7 eV. By constructing a multiple junction

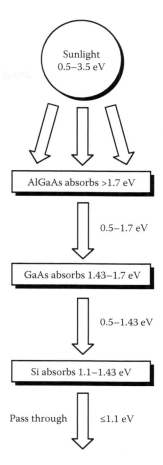

FIGURE 7.8   Light harvesting in a multijunction solar cell. (Adapted from http://www.eere. energy.gov/basics/renewable_energy/images/cspv_3.gif. Accessed January 25, 2013.)

solar cell consisting of all three of these materials, more of the incident photons can be harvested and converted (Figure 7.8). Obviously, the complexity of the cell architecture increases as well (U.S. Department of Energy 2011).

7.2.3   CURRENT–VOLTAGE CURVE AND EFFICIENCY

The current–voltage relationship for a typical solar cell is plotted in Figure 7.9.

> Note that sometimes current–voltage curves are shown representing current (I, units of mA) and sometimes current density (J, units of mA/cm^2). Current density simply takes into account the surface area of the PV device.

The maximum current density occurs when the cell is short-circuited (J_{sc}, zero load), and the maximum voltage occurs under open-circuit conditions when zero current

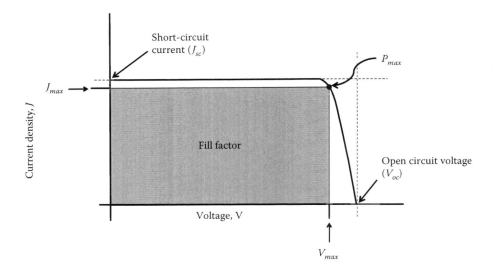

FIGURE 7.9   The current–voltage relationship for a typical PV device.

flows (the *open-circuit voltage*, V_{oc}). It is the product of the current density and the cell voltage that determines the power provided by the cell (Equation 7.2; a review of Figure 6.5 will reinforce this relationship).

$$P(\text{power}) = J(\text{current density}) \times V(\text{cell voltage}) \quad (7.2)$$

Recall that power equals the rate at which energy is expended over time and note that the product of the open-circuit voltage and the short-circuit current gives a *theoretical* maximum power, indicated by the dashed lines in Figure 7.9. The *actual* maximum power, indicated by the point P_{max} on the curve, reflects the balance between cell potential and current: there is a maximum current density (J_{max}) and a maximum cell voltage (V_{max}) that deliver the maximum power from the device. This maximum power is always less than the theoretical maximum and this deviation from ideal behavior is reflected in the *fill factor* (FF), the shaded rectangle in Figure 7.9 that links P_{max} to J_{max} and V_{max}. The fill factor, therefore, is always less than 1 and is related to P_{max} by

$$P_{max} = FF(J_{SC} \times V_{OC}) \quad (7.3)$$

Thus, the more rectangular the J–V curve, the more closely P_{max} approaches the theoretical maximum and the higher the efficiency of the PV device.

Solar cells that can absorb more of the solar spectrum (i.e., possessing a narrow E_g) will have a lower V_{oc}, while those with a large E_g will be unable to absorb the longer wavelengths in the solar spectrum. As a result, they will have a lower J_{sc}. Figure 7.10 illustrates the significant jump (as indicated by the asterisk) in current that a PV device exhibits under illuminated conditions. This photocurrent is

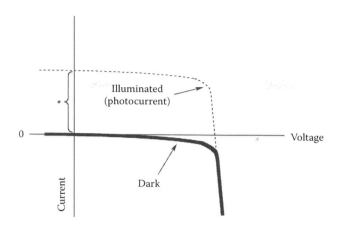

FIGURE 7.10 The dark current versus the photocurrent.

compared to the *dark current* that is measured without photons striking the device. This measured differential between the dark current and the photocurrent allows for the experimental determination of values that will allow calculation of the cell's efficiency.

Amazing improvement has been seen in the efficiency of PV devices over the past 60 years. It was in the mid-1950s when Bell Laboratories reported an efficiency of roughly 5% for a monocrystalline silicon solar cell (Alcatel · Lucent 2004) while as Figure 7.11 shows, steady improvement over the last 40 years has led to PV devices with efficiencies above 40% (National Renewable Energy Laboratory 2013). But ascertaining the efficiency of a solar cell can be more than a little confusing. As we have just seen, there is the theoretical efficiency and the actual efficiency associated with the photoevent. But cell efficiency is also affected by the cell materials, its fabrication, and the environmental conditions. For example, efficiency decreases with increasing temperature, since the band gap decreases with a resultant drop in the open-circuit voltage. In this section, we will try to explain some of the concepts associated with PV efficiencies and to clarify some important terms.

As for fuel cells, there are many factors responsible for the decrease in efficiency in a PV: recombination, photons with energies $<E_g$ or $>E_g$, the loss of part of the absorbed energy as heat, and so on. Another limit to a PV device's efficiency is, of course, the second law of thermodynamics. Returning to the Carnot efficiency, if we consider the Sun as a perfect black-body radiator ($T_{high} = 6000$ K) and take the ideal PV collection device to be at room temperature (300 K), a theoretical maximum of 95% efficiency could be attained, based on the thermal energy conversion (recall Equation 3.5, $\eta = 1 - T_{low}/T_{high} \times 100\%$ (Pagliaro et al. 2008)). Even the Sun cannot produce with 100% efficiency!

There is also the intuitive limit to a solar device's efficiency based on its bandgap. Shockley and Queisser calculated what has come to be known as the "Shockley–Queisser limit," a theoretical limit of 31% for conversion of (unconcentrated)

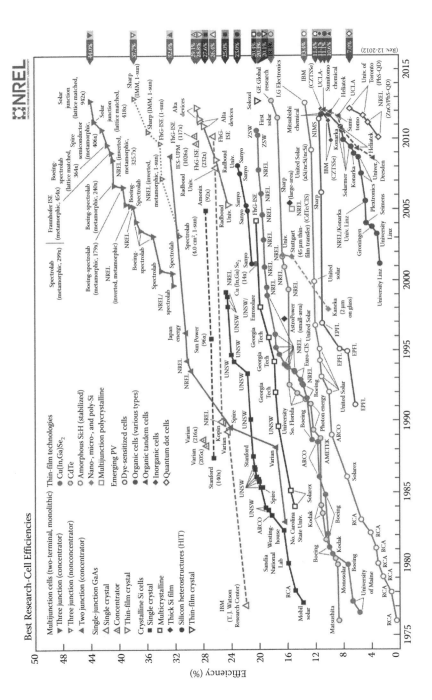

FIGURE 7.11 **(See color insert.)** Timeline of research cell efficiencies. (From National Renewable Energy Laboratory 2013a. *Best Research-Cell Efficiencies*. National Center for Photovoltaics, NREL.)

incident solar energy into electricity at a homojunction bandgap of 1.1 eV. This value reflects the reality of the unused solar spectrum (light energy $<E_g$ and wasted energy $>E_g$). This limit bumps up to about 33% for a bandgap of 1.4 eV (Nelson 2003) and increases to 66% when considering the conversion of photon energy for a hypothetical solar cell with an infinite number of bandgaps that *perfectly* match the solar spectrum (Nozik 2002).

In reporting the *actual* (measured) efficiency of a solar cell, the conditions under which the measurements are made must be standardized: what are the angle and intensity of the incident light? At what temperature is the measurement made, and through what medium does the light traverse? The AM1.5G (*air mass 1.5*) spectrum has been selected as the standard (Figure 7.12) for PV performance evaluation at 25°C. AM1.5G gives an incident power density of 1000 W/m² onto a flat PV device, defined as *one sun*. (While use of AM1.5G provides a standard test condition, it must be understood that this power density is wildly unrealistic given that the average sun power density on the Earth is a mere 170 W/m² (Pagliaro et al. 2008).)

Even when measured under standard conditions, there are several aspects to solar cell efficiency that can be puzzling. First there is the *quantum efficiency* (QE): how efficiently does each photon promote an electron, that is, for every 100 photons striking the PV, how many electron–hole pairs are generated? A perfect one-for-one match would be a quantum efficiency (QE) of one. But what about losses due to reflection of light? If these losses are not taken into consideration, the value is known as the *internal* QE (IQE); but if *every* photon hitting the cell surface is taken into consideration (even if reflected), this is referred to as the *external* QE (EQE).

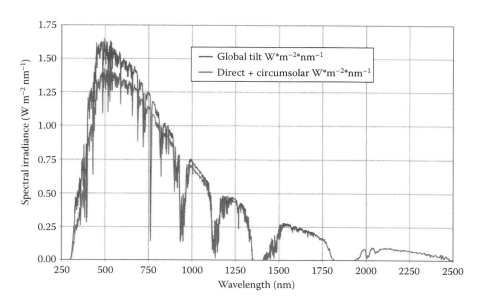

FIGURE 7.12 (**See color insert.**) The AM 1.5 Reference Spectra. (From http://rredc.nrel. gov/solar/spectra/am1.5/. Accessed May 23, 2013.)

The EQE is sometimes known as the IPCE ("incident photon to current conversion efficiency") and is determined from the equation

$$IPCE = \frac{electrons\ out}{incident\ photons\ in} = 1240 \cdot \frac{J_{sc}}{\lambda \cdot P_{in}} \qquad (7.4)$$

where J_{sc} is in units of A/cm^2 and P_{in} (W/cm^2) is the light intensity at wavelength λ (nm) (Hagfeldt et al. 2012). A good IPCE is directly related to a good J_{sc}.

The overall efficiency of a PV device in terms of power is known as the *power conversion efficiency* (PCE) and is measured and calculated by

$$\eta = \frac{J_{max}V_{max}}{P_{in}} \times 100\% \qquad (7.5)$$

where the product of the maximum current density and the maximum voltage output is P_{max}, and P_{in} is again the incident solar power striking the device. The maximum power output is (as noted above) related to the fill factor so that another way of quantifying the power conversion efficiency is shown in Equation 7.6:

$$\eta = \frac{V_{oc} \times J_{sc} \times FF}{P_{in}} \times 100\% \qquad (7.6)$$

7.3   INORGANIC SOLAR CELLS

7.3.1   SILICON

While they are expensive to manufacture and are limited in efficiency and application, the history and foundation of photovoltaics is built on silicon. The use of silicon-based photovoltaics is widespread: 84% of the PV cells shipped in the United States in 2011 were based on crystalline silicon (U.S. Energy Information Administration 2011). Given their prominence, we will take a detailed look at silicon solar cells before introducing other photovoltaics.

7.3.1.1   Architecture

A simple example of a single junction silicon solar cell is shown in Figure 7.13. The semiconductor materials are sandwiched between a back contact and a front contact to create the electrical circuit, with the p–n junction close to the light-receiving surface and a thicker p-layer behind to absorb the rest of the incident light. Any materials on the sunlight-facing side of the cell should be transparent or nearly so. In the case of a metallic contact such as silver, transparency and electrical conductivity is maximized by using thin strips as seen in Figure 7.14. However, metallic contacts (particularly on the light-receiving side of the cell) have been largely replaced by a very thin film of a *transparent conducting oxide* (TCO, e.g., SnO_2 or ZnO). An anti-reflective coating such as silicon nitride (Si_3N_4) is added to minimize reflectance, and

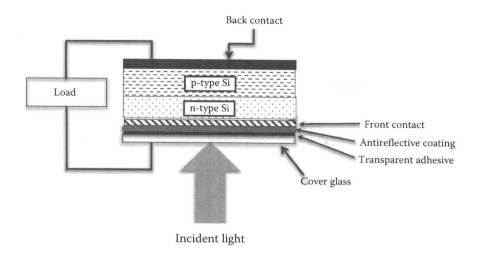

FIGURE 7.13   A schematic showing the architecture of a generic silicon solar cell.

the surface may be further passivated by etching it. In reality, the technology behind the fabrication of a highly efficient solar cell is quite detailed and complex.

7.3.1.2   Materials

The nature of the silicon in the PV device has a large impact on both its efficiency and cost. While the starting material, silicon dioxide (silica; SiO_2), is abundant and inexpensive, the process whereby pure solar grade Si is manufactured and converted into solar cells is not. Silica is first reduced to metallurgic grade silicon at very

FIGURE 7.14   (See color insert.) The thin "fingers" of solar cell metal contacts. (From Shutterstock 44807470.)

high temperature (>1414°C) using coke, the residue from heating coal in a limited amount of air (Equation 7.7). Coke consists of a mixture of porous carbon and mineral ash and is an inexpensive reducing agent. The reduced silicon is then further purified by conversion into trichlorosilane (Cl_3SiH; Equation 7.8), a low-boiling liquid that is easily purified by repeated distillation. Reduction of the highly pure trichlorosilane with hydrogen gas then yields semiconductor grade silicon (Equation 7.9) (Kitai 2011).

$$SiO_2 + 2C \rightarrow Si + 2CO \qquad (7.7)$$

$$Si\,(powder) + HCl \rightarrow Cl_3SiH + H_2 \qquad (7.8)$$

$$Cl_3SiH + H_2 \rightarrow Si + 3HCl \qquad (7.9)$$

Crystalline silicon, be it monocrystalline or multicrystalline, is abbreviated *c-Si*. High-purity monocrystalline silicon ingots (*sc-Si*, for single crystal) are formed by melting the element in a crucible made of pure quartz then "pulling an oriented seed" out of the molten mass (Ferrazza 2012). This energy-intensive method requires heating the Si above its melting point (>1410°C), adding to the manufacturing cost. These ingots are then sawed into thin wafers with as much as 50% kerf waste. Despite these disadvantages, monocrystalline Si cells are widely used because of their durability (lasting 20–30 years) and relatively high efficiency (as high as 27%; see Figure 7.11). Contributing to this efficiency is the fact that high-purity, highly ordered sc-Si PVs have the lowest rate of recombination of any of the silicon PV materials.

Preparation of multicrystalline silicon is simpler than that of single-crystal Si, as ribbon "wafers" can be pulled directly from the melt to a thickness of about 200–300 μm, avoiding the waste from sawing ingots. But while multicrystalline silicon is less expensive, the impurities and crystal defects known as *grain boundaries* make these Si solar cells more susceptible to recombination and therefore less efficient (Figure 7.11). In essence, grain boundaries are places in the solid matrix where adjacent crystals—*grains*—do not align properly. The current laboratory efficiencies for a single junction mc-Si solar cell are about 10% (Shah 2012).

Amorphous Si (*a-Si*) is the least-expensive material for a silicon PV device, but again this material suffers from low conversion efficiencies (4–8%) because it is a highly disordered material full of "dangling bonds"—essentially, atoms that are not bonded to another atom. As a result, a-Si shows a high rate of recombination. However, by treating a-Si with hydrogen, *hydrogenated amorphous silicon* (a-Si:H) results wherein the dangling bonds at the surface of the highly disordered a-Si are capped with hydrogen, improving durability and performance. The bandgap for a-Si:H is in the range of 1.6–1.85 eV (recall that the bandgap for crystalline Si is 1.1 eV) and efficiencies are as high as 13% (Figure 7.11) (Shah et al. 1999) (Shah 2012). Amorphous silicon is used extensively in thin-film inorganic solar cells.

7.3.2 Thin-Film Inorganic Solar Cells

Thin-film solar cells are significant for two important reasons: (1) by using a thin film of absorber material the cost of the cell is reduced, and (2) for some thin-film cells the option of flexibility is achieved. Several approaches to thin-film solar cells have been developed; this section focuses on thin-film *inorganic* PVs.

7.3.2.1 Thin-Film Silicon

Amorphous thin-film Si solar cells contain silicon in a film less than 50 μm thick, significantly decreasing the cost of the cell by reducing the amount of material needed (in contrast, the thickness of a c-Si wafer is around 300 μm) (Cheung 2010). However, with the decrease in cost comes a decrease in efficiency, resulting in the conundrum of balancing the cost of the a-Si layer with the efficiency of the cell. Thin-film silicon solar cells are typically manufactured by chemical vapor deposition (CVD) techniques, allowing several layers of thin films to be built up at relatively low temperatures (200–300°C) in a more complex configuration. If phosphine (PH_3) is mixed in with silane (SiH_4) during the deposition, a p-type material results. Similarly, borane (BH_3) can be used to prepare an n-type material (Shah et al. 1999).

Thin-film a-Si solar cells offer lower cost and low toxicity but suffer from low efficiency as noted above. One method of improving the efficiency of a-Si cells involves making a multijunction Si solar cell by building up layers of a-Si with microcrystalline silicon (μc-Si), improving the efficiency somewhat (Philibert 2011). For example, Sanyo's "HIT" cell (**h**eterojunction with **i**ntrinsic **t**hin layer) similarly uses both c-Si and a-Si in a complex cell architecture that boosts cell efficiency above 20% (Tsunomura et al. 2007). A simplified schematic of the HIT cell structure is shown in Figure 7.15 (Green 2012). The TCO layer is a highly textured surface of inverted pyramids that thoroughly scatter the incident light and improve light harvesting. The HIT cell illustrates the *p-i-n* junction, a common configuration

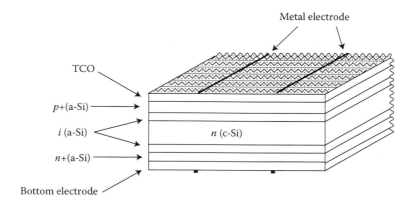

FIGURE 7.15 Schematic of the HIT cell. (Adapted with permission from Green, M.A. 2012. *Practical Handbook of Photovoltaics. Fundamentals and Applications*, edited by A. McEvoy, T. Markvart, L. Castañer. Waltham, MA/Oxford, UK: Academic Press.)

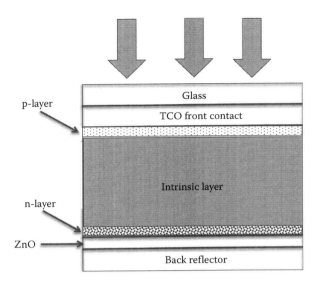

FIGURE 7.16   A schematic of the *p–i–n* architecture.

to improve movement of the charge carriers to the contacts. In a *p-i-n* junction, the doped layers are very thin and an undoped "intrinsic" region lies in-between (Figure 7.16). The free electrons and holes are generated in the intrinsic region, where the electric field separates them and they are more likely to make it to the electrodes prior to recombination.

Another approach to increasing the efficiency of a-Si PV devices is the incorporation of silver nanoparticles. By constructing a cell that includes silver "nucleated nanoparticles" (essentially, bumpy nanoparticles with enhanced surface roughness) in a zinc oxide layer, the incident light was scattered more strongly and over a larger range of angles. This leads to improved absorption of the light in the intrinsic absorbing layer. This cell also possesses a *p-i-n* junction and showed a 23% enhancement in efficiency that the researchers attributed to an improved fill factor, presumably due to lowered resistance as a result of the incorporation of the silver nanoparticles (Chen et al. 2012a). Thus, the performance of silicon solar cells continues to improve. In addition, silicon photovoltaics are fairly innocuous when it comes to sustainability as their use of resource-limited materials is relatively modest.

7.3.2.2   Copper Indium Selenide and Alloys

The chalcogens are those elements making up Group 16 (O, S, Se, Te, Po), and the chalcogenides are their (−2) ions: S^{2-}, Se^{2-}, and Te^{2-}, etc. Metal chalcongenides play an important role in photovoltaics. Copper indium diselenide ($CuInSe_2$ or CIS) and its alloys with sulfur and/or gallium have been combined in various ratios and configurations to make efficient and durable p-type materials for PVs. These materials possess the chalcopyrite crystal structure (Figure 7.17) and have a very high absorption coefficient in the visible to near-IR range, making them very well-suited for thin-film solar cells. However, the CIS bandgap, at ≈1 eV, is not optimum. Copper

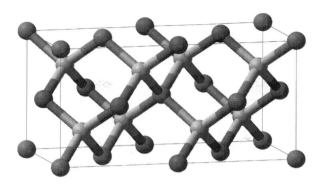

FIGURE 7.17 The chalcopyrite crystal structure. (Reprinted from http://en.wikipedia.org/wiki/File:Chalcopyrite-unit-cell-3D-balls.png.)

indium gallium (CIG) alloys are typically doped with either selenium or sulfur, making these absorber materials highly flexible. By partial substitution of indium atoms with gallium to obtain $Cu(In_{(1-x)}/Ga_x)Se_2$ (CIGSe), the bandgap can be tuned from ≈1.0 to 1.7 eV (Nakada 2012), making CIGS material a particularly good absorber for thin-film solar cells.

A standard CIGS/CIGSe cell configuration is shown in Figure 7.18; a cross-sectional scanning electron micrograph of an actual device is shown in Figure 7.19a for comparison. The molybdenum layer is the back contact upon which the CIGS layer lies as the p-type absorber. The thin cadmium sulfide (CdS) layer is the n-type material with a very large E_g of 2.4 eV (Bonnet 2012). This particular cell was doped

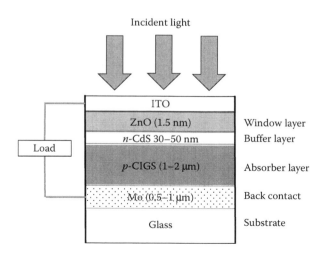

FIGURE 7.18 The architecture of a typical CIGS solar cell. (Reprinted from Saji, V.S., I.-H. Choi, and C.-W. Lee. Progress in electrodeposited absorber layer for $CuIn_{(1-x)}Ga_xSe_2$ (CIGS) solar cells. *Solar Energy* 85:2666–2678. Copyright 2011, with permission from Elsevier.)

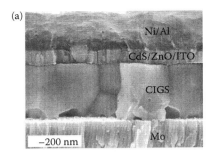 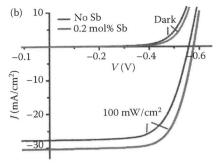

FIGURE 7.19 (a) A cross-sectional view of the layers in a CIGS device. (b) J–V curves illustrating the impact of antimony on cell performance. (Reprinted with permission from Yuan, M., D.B. Mitzi, W. Liu et al. 2009. Optimization of CIGS-Based PV device through Antimony Doping. *Chem. Mater.* 22 (2):285–287. Copyright 2010 American Chemical Society.)

with antimony and showed improved performance (Figure 7.19b) over the undoped cell, with an open-circuit voltage of 0.57 V, J_{sc} of 30.3 mA/cm^2, a fill factor of 71.2% and a moderate efficiency of 12.3% (Yuan et al. 2009). Even as long ago as 2008, a CIGS cell gave a record-setting 19.9% efficiency (Repins et al. 2008) and an efficiency of >20% for a CIGSe cell was recently reported (Green et al. 2012). A flexible variant prepared on a polymer foil by scientists in Switzerland set a record laboratory efficiency of 20.4%, matching the efficiencies of polycrystalline silicon cells (EMPA Materials Science & Technology 2013).

While CIGSe cells have clearly demonstrated some impressive efficiencies, the thin layers are often prepared by vacuum deposition techniques making fabrication of these devices particularly costly. Much research has been devoted to designing lower-cost, solution-based methods for preparation of these devices that do not use highly toxic and dangerous materials such as hydrazine (H_2NNH_2). To that end, researchers recently developed a method for the preparation of a CIG cell that contains both sulfur and selenium. A relatively benign butyldithiocarbamic acid solution was used to prepare the mixed $Cu(In,Ga)S_2$ absorber for fabrication into the CIGSSe cell as shown in Figure 7.20. Thus, an ethanolic solution containing carbon disulfide, *n*-butylamine, and indium hydroxide was heated to prepare the *n*-butyldithiocarbamic acid salt, to which gallium acetylacetonate and copper(I) oxide were added to make the $Cu(In,Ga)S_2$ precursor in various stoichiometric ratios. This solution was then spin-cast onto a molybdenum-coated glass substrate, annealed at 400°C, and a thin film of selenium deposited under vacuum at 540°C. Chemical bath deposition of the CdS layer (from cadmium sulfate and thiourea) was followed by deposition

$$CH_3CH_2CH_2CH_2NH_2 \xrightarrow[CH_3CH_2OH]{S=C=S} \text{[butyldithiocarbamic acid]} \xrightarrow{In(OH)_3} \text{[indium salt]}$$

FIGURE 7.20 Preparation of the precusor solution for a CIGSSe cell.

of a layer of zinc oxide, then ITO and an aluminum grid were applied to complete the cell. However, the CIGSSe cells prepared by this method showed an average efficiency of only 8.8% (Wang et al. 2012b).

CIGS solar cells show sufficiently high efficiencies to make them reasonable competition for the commercial market, but sustainability is a critical concern for these and other PV devices. Large-scale implementation is not sustainable due to the limited availability of the raw materials, particularly indium. An in-depth discussion of the issue of the sustainability of specific solar PV devices is presented in Section 7.7.

7.3.2.3 Cadmium Telluride

While both CIGS and cadmium thin-film solar cells have been studied for over two decades, it is the CdS/CdTe thin-film technology that has been the most successful commercially with a share of about 10% of the global PV market in 2008 (Kirchartz and Rau 2011). The main reasons for this are the relative ease of fabrication of CdS/CdTe solar cells, their material stability, cadmium telluride's excellent absorptivity coefficient (>5 × 10^5/cm; Armaroli and Balzani 2011), and nearly optimum E_g of 1.45 eV. However, these cells (like the CIGS/CdS cells described above) have the disadvantage of relying on highly toxic cadmium and concerns about its recycling and release into the environment persist. In addition, tellurium is a limited resource (recall Figure 1.8), making CdTe another technology of questionable sustainability.

Cadmium telluride can be prepared as either an n-type or a p-type material but it is most often used as p-type with CdS acting as the n-type donor. A typical configuration for a CdS/CdTe cell is shown in Figure 7.21. The electron–hole pairs are generated near the p–n junction that is activated by annealing at elevated temperature in the presence of a chlorine-containing salt (often CdCl$_2$) (Bonnet 2012). This activation is critical for the performance of the cell, as an unactivated CdTe/CdS junction leads to high rates of recombination. Such activation has been labeled "magic," but the improvement is somehow related to structural and electronic effects that take

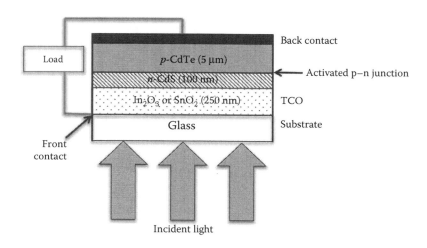

FIGURE 7.21   Architecture of the CdS/CdTe solar cell.

place by improving the crystalline quality of the grains directly at the p–n junction (Bonnet 2012).

Upon illumination, the electrons in a CdTe cell are swept into the CdS film and out the metal contact at the TCO layer, while the holes travel through the CdTe and complete the circuit out the back contact. The back contact, often a copper alloy, is a source of instability and loss of efficiency because the copper diffuses into the CdTe layer. An in-depth discussion of the physics associated with the back contact failings is beyond the scope of this chapter, but a *triple structure* has been developed to address the problem. This three-layer back junction consists of (1) a tellurium-enhanced surface followed by (2) a narrow bandgap p-type (chemically inert) semiconductor (the buffer layer), then (3) the metal contact. Cells fabricated using this triple structure approach are quite stable (Bonnet 2012). Given the initial success of CdTe cells, their use is likely to grow despite the questions of safety and sustainability.

7.4 ORGANIC PHOTOVOLTAICS

7.4.1 INTRODUCTION

Thin-film inorganic solar cells have already proven to be commercially viable and organic photovoltaic devices (OPV, also known as polymer solar cells) are rapidly attaining that distinction. While at first glance they may appear to be considerably more complex than a simple single junction silicon solar cell, we shall see that there are numerous parallels between OPVs and silicon-based cells. The major advantages to using polymer solar cells are lower cost, ease of processing, and flexibility both in terms of the wide variety of materials that can be synthesized for use in OPVs and in terms of their literal flexibility: OPVs can be printed on large rolls of polymer substrate for applications that demand light weight and plasticity (e.g., wearable electronics). In addition, organic materials generally have very high absorptivity coefficients ($>10^5$/cm). However, at this point, OPVs suffer from relatively low efficiencies, poor durability (the organic materials tend to degrade in light over time) and the charge transport of the electrons and holes through the material is inherently slow and inefficient. Furthermore, the materials used in these devices are, ultimately, carbon-based, meaning they currently derive from fossil fuels. Clearly, if these devices are to contribute significantly to a sustainable energy solution, the materials from which they are made must also be sustainable. While efficiencies for OPVs have lagged behind the standard Si cells (see Figure 7.11), improvement and growth in this area has been steady. Some believe OPVs are the future of renewable energy although it is generally agreed that a PCE of >10% for large area devices is needed in order for them to become commercially viable.

7.4.2 MECHANISM

In inorganic solid-state PV devices, as we have seen, the PV event is based on transfer of an electron from the valence band into the conduction band then separating the two with the internal electric field at the p–n junction and collecting the free

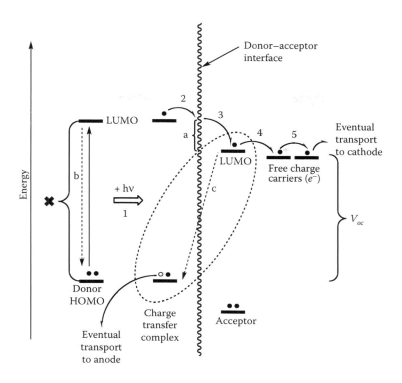

FIGURE 7.22 The photoelectric conduction process for OPV devices.

charge carriers at the appropriate contacts. For OPVs, the overall picture is similar, but instead of an electron and a hole, an *exciton* (an electron–hole pair) with a fairly tight binding energy is formed. Also, instead of promotion of an electron from a lower energy band to a higher energy band, the initial excitation is from the highest occupied molecular orbit (HOMO) of a *donor material* (HOMO$_D$) to the lowest unoccupied molecular orbital (LUMO) of that same material (LUMO$_D$). For this reason, the use of the term bandgap when discussing OPVs is, strictly speaking, incorrect. However, this misuse is widespread and *bandgap* will be used in this section to refer to the HOMO–LUMO gap of the light-absorbing donor material.

A simplified overview of the photoelectric conduction process for OPVs is shown in Figure 7.22.

- Upon illumination with sunlight of wavelength sufficient to meet or exceed the donor's HOMO$_D$–LUMO$_D$ gap (1), an exciton is generated which then must diffuse over a distance of a few nanometers to the donor–acceptor interface (2) (Carsten et al. 2011). Because of the tight binding energy of the exciton, it is crucial that the diffusion distance is very short or else recombination will compete and efficiency will plummet.
- At the donor–acceptor interface, the electron is transferred from the donor's LUMO (LUMO$_D$) to the acceptor's LUMO (LUMO$_A$) (3), generating a

charge transfer complex (also known as a "charge transfer state," that is, an electron and hole that are still attracted to one another by Coulombic forces, as indicated by the dashed oval in Figure 7.22).

- The electrons and holes become fully separated (4) and "hop" through the molecules of material (electrons through the acceptor and holes through the donor) and
- Conductivity results (5) as the free charge carriers are collected at the anode (holes) and cathode (electrons) to generate the electric current.

The mechanism for this process in OPVs is actually considerably more complex and continues to be studied. For example, excitons with energy $>>E_g$ ("hot excitons") can generate other species that allow for the extraction of the maximum possible energy from the incident light. For our purposes, the simple scheme of Figure 7.22 is sufficient. Interested readers are directed to Grancini et al. for further enlightenment (Grancini et al. 2013).

There must be a driving force at the D–A interface for overcoming the binding energy in the exciton and pushing the process forward to the charge transfer complex. It is estimated that this takes 0.1–0.5 eV for most systems, an amount of energy that must be provided by the ΔE_{LUMO} (Figure 7.22a; (He and Yu 2011)). If ΔE_{LUMO} is sufficient, the charge-transfer complex will fully dissociate into free charge carriers and back transfer and eventual decay is avoided. As with Si cells, reverse processes result in loss of efficiency. The exciton can simply decay (Figure 7.22b), or recombination of an electron with a hole can take place at either the charge transfer state or between free carriers (Figure 7.22c).

7.4.2.1 HOMO–LUMO Gap

The beauty of the OPV is that it is almost infinitely adjustable by synthesizing and matching different donor and acceptor materials. Indeed, the fine tuning of the HOMO–LUMO energy levels to improve the performance of the solar cell is the holy grail of OPV research. Two important energy differences—both relative and absolute—control the performance of the organic PV system: the $HOMO_D$–$LUMO_D$ gap (denoted as ✖, Figure 7.22) and the $HOMO_D$–$LUMO_A$ gap (labeled as V_{oc} in Figure 7.22)

- Since the V_{oc} is directly proportional to the difference in energy levels between $LUMO_A$ and $HOMO_D$, the larger this $HOMO_D$–$LUMO_A$ gap, the larger the V_{oc} and, accordingly, the maximum power conversion efficiency.
- If the donor $HOMO_D$–$LUMO_D$ gap is made smaller, the photon harvest can be maximized (a "low bandgap" polymer donor material) so that a greater number of photons with energy $\geq E_g$ are absorbed and the current density maximized.
- However, if this $HOMO_D$–$LUMO_D$ gap is reduced too much, the $LUMO_D$ energy level may drop below that of $LUMO_A$ with the result that the driving force for charge transport (ΔE_{LUMO}, (a) in Figure 7.21) will be lost.

Thus, a careful balance of the HOMO and LUMO energy levels of both the donor and acceptor molecular orbitals must be realized to optimize the power output of

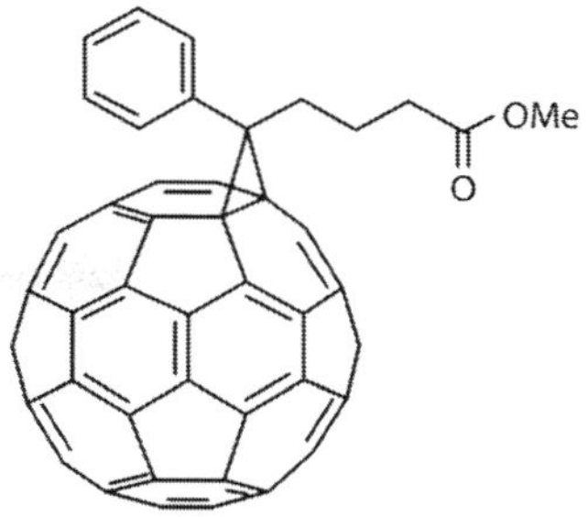

FIGURE 7.23 [6,6]-phenyl C_{61} butyric acid methyl ester.

the OPV. If we focus on the widely used acceptor material PCBM ([6,6]-phenyl C_{61} butyric acid methyl ester, Figure 7.23), for example, we can illustrate this concept. For an OPV made up of some hypothetical donor material and PCBM, the energy level of the donor molecule's LUMO ($LUMO_D$) should be about 0.2–0.3 eV above the $LUMO_{PCBM}$ energy level, that is, somewhat above −4.0 eV (the $LUMO_A$ of PCBM is −4.3 eV; see Figure 7.24). Based on the solar spectrum, we know that a reasonable

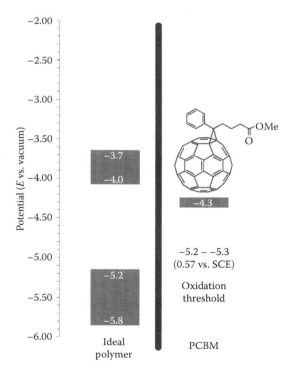

FIGURE 7.24 The ideal donor bandgap. (Adapted with permission from Blouin, N., A. Michaud, D. Gendron et al. 2007. Toward a rational design of poly(2,7-carbazole) derivatives for solar cells. *J. Am. Chem. Soc.* 130 (2):732–742. Copyright 2008 American Chemical Society.)

bandgap energy could be anywhere in the range of, roughly, 1–1.8 eV, with the narrower bandgaps resulting in increased capture of solar energy. Pulling this all together, the ideal range for a $HOMO_D$ partnered with PCBM would be somewhere in the range of −5.2 to −5.8 eV, as summarized in Figure 7.24 (Blouin et al. 2007). Narrower bandgaps can be achieved by either raising the energy of the HOMO or lowering the energy level of the LUMO (or both) (Winder et al. 2002).

7.4.2.2 Characterization of HOMO–LUMO Energy Levels

Before proceeding further, it is important to understand how the size of the HOMO–LUMO gaps and their relative energy are determined. Energy diagrams for comparison of relative energy levels can be constructed from calculated and/or experimentally determined data. For example, values for the bandgaps for a variety of donor–acceptor copolymers (Figure 7.25; *vide infra*) were recently calculated using density functional theory modified with periodic boundary conditions (PBC-DFT) and compared to experimentally determined values. In this particular series, the correlation of this calculated bandgap with the experimentally determined values was excellent, with $R = 0.9953$ (Pappenfus et al. 2011). The experimental determination of bandgaps can

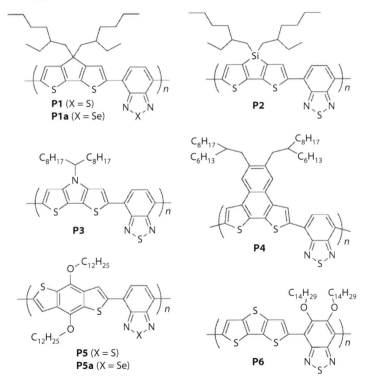

FIGURE 7.25 Copolymers used in the PBC-DFT study. (Reprinted with permission from Pappenfus, T.M., J.A. Schmidt, R.E. Koehn et al. 2011. PBC-DFT applied to donor–acceptor copolymers in organic solar cells: Comparisons between theoretical methods and experimental data. *Macromolecules* 44 (7):2354–2357. Copyright 2011 American Chemical Society.)

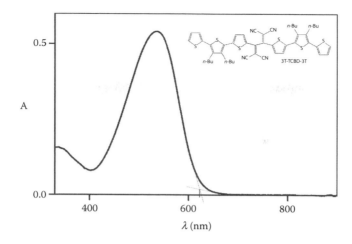

FIGURE 7.26 UV–visible spectrum of 3T-TCBD-3T. (From Pappenfus, T.M., D.K. Schneiderman, J. Casado et al. 2010. Oligothiophene tetracyanobutadienes: Alternative donor–acceptor architectures for molecular and polymeric materials. *Chem. Mater.* 23 (3):823–831. Copyright 2010 American Chemical Society.)

be easily determined with UV–visible absorption spectroscopy by using the simple conversion $\lambda(nm) = 1240/E(eV)$. To illustrate, the UV–visible spectrum of the tetracyanobutadiene-substituted oligothiophene 3T-TCBD-3T, a material of interest as a possible acceptor material in OPVs, is shown in Figure 7.26. The wavelength at which the *onset* of the low-energy absorption occurs represents the energy that most closely corresponds to the HOMO–LUMO gap. As shown in Figure 7.26, this wavelength is estimated by drawing tangents on the low-energy curve of the absorption spectrum. The obtained value of ≈624 nm results in a bandgap energy of 1.99 eV (Equation 7.10).

$$E_g = \frac{1240}{624} \text{nm} = 1.99 \, \text{eV} \qquad (7.10)$$

Cyclic voltammetry (CV), too, is widely used in the study of PV materials. By determination of the oxidation and reduction potentials of the donor and acceptor materials from the voltammogram, the frontier orbital energies can be estimated and from this the V_{oc} determined. While a more detailed discussion of CV is beyond the scope of this chapter, this experimental method is invaluable for understanding the electronic behavior of materials for use in solar cells. A caveat, however, is that these CV analyses are typically carried out in nonaqueous systems with the ferrocene/ferrocenium redox couple as a standard. This raises numerous issues with respect to consistency in measurements, making direct comparisons of various solar cell properties problematic (Cardona et al. 2011).

7.4.3 MATERIALS

Because the materials for OPV are products of organic synthesis there is large and ever-increasing number of different options, particularly for donor materials. In

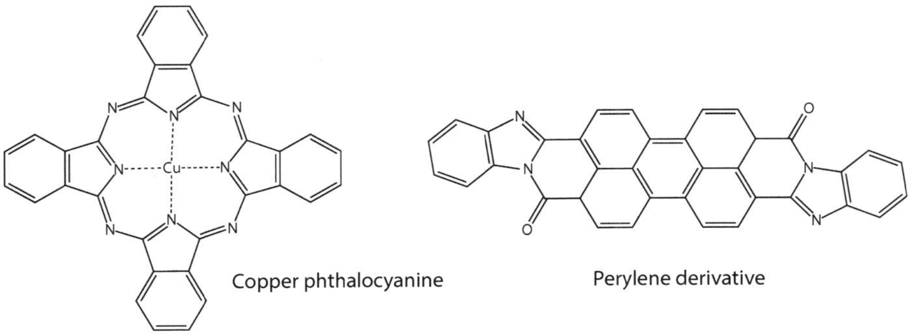

Copper phthalocyanine   Perylene derivative

FIGURE 7.27   Donor and acceptor materials for a novel two-layer OPV.

addition, novel approaches to the combination of these materials and their fabrication into PV devices continue to develop. For instance, a major advancement over "conventional" single-layer OPVs was the report of a two-layer organic PV cell in 1986 using copper phthalocyanine and a perylene tetracarboxylate derivative to create the donor–acceptor interface (Figure 7.27) (Tang 1986). In the next sections, the properties and preparations of some of the more commonly used materials will be described followed by the critically important aspect of their construction in an actual device (Section 7.4.4).

7.4.3.1   Donors

A donor material for an OPV must have excellent mechanical, electronic, and chemical properties. It should

- Be easy to process (this means it should be soluble, ideally in an environmentally friendly and inexpensive solvent)
- Form good films
- Be durable (including stability against air oxidation, which means the $HOMO_D$ energy level should be above -5.27 eV)
- Have a high absorption coefficient
- Possess good hole mobility
- Have well-matched HOMO–LUMO energy levels

The goal of the researcher in developing new materials for an OPV is to design a material that can meet as many of these demands as possible. Donor materials in particular are amenable to a vast array of structural modifications given the prowess of organic synthesis, making the potential assortment of OPVs almost unlimited. Some of the more common donor materials are triphenylamine-based benzothiadiazoles (**1**), polyalkylthiophenes (PAT) or polypyrroles (**2**), polyisothianaphthene (**3**), fluorene-containing materials (**4**), fused aromatics (**5**), or arylene vinylenes (**6**, Figure 7.28).

One of the most popular donor materials is poly-3-hexylthiophene (P3HT, **2**; Figure 7.28, R′ = *n*-hexyl and X = S). The bandgap of P3HT ($\approx$1.9 eV) is considered relatively large and, as a result, captures only a small portion of the solar spectrum.

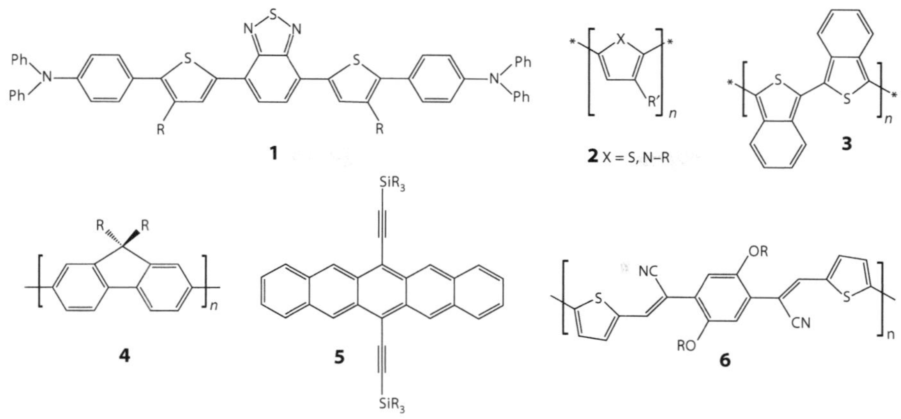

FIGURE 7.28 Commonly used donor materials for OPVs.

In addition, there is the issue of regioselectivity in the synthesis of polyalkylthio-phenes. Polymerization of 3-alkylthiophenes can lead to three possible couplings: head-to-tail (HT), head-to-head (HH), and tail-to-tail (TT) (Figure 7.29). Worse yet, there could be a variety of secondary couplings during polymerization, leading to HH-TT, HH-HT, TT-HT, among other linkages scattered throughout the polymer.

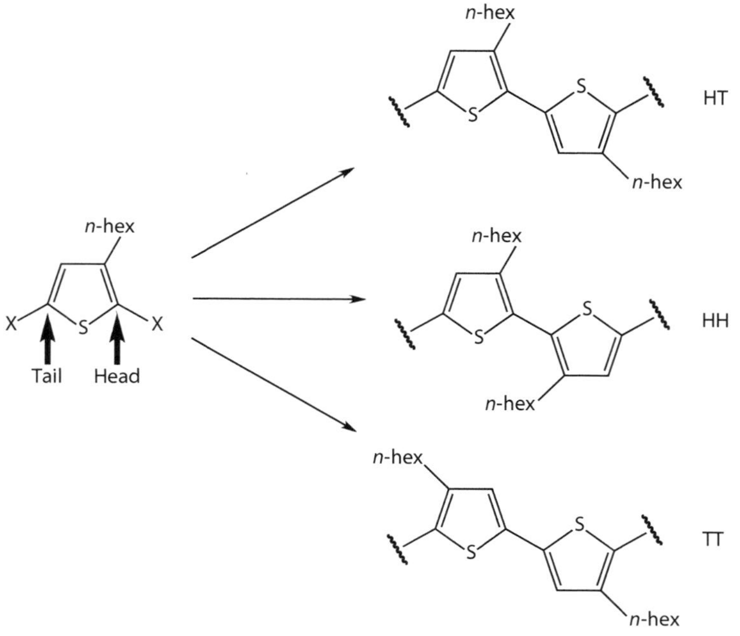

FIGURE 7.29 Possible coupling patterns for P3HT.

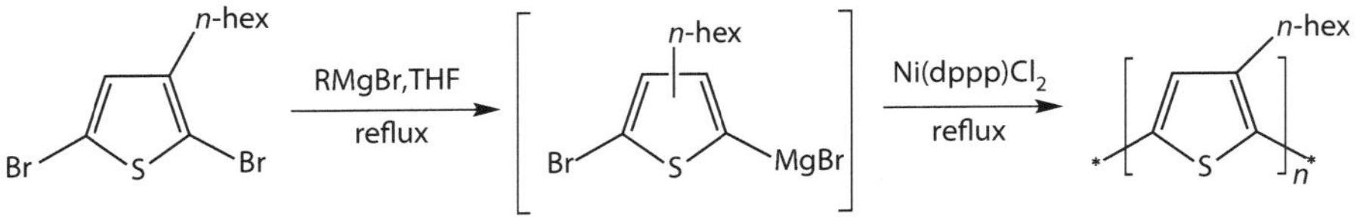

FIGURE 7.30   Synthesis of regioregular P3HT.

Fortunately, PATs can be synthesized with up to 99% regioselectivity with the desired HT coupling as shown in Figure 7.30 (Loewe et al. 1999).

Why does the regioselectivity matter? As we saw in Chapter 4, the structure of the polymer dramatically impacts its behavior in many ways, including its electronic behavior. For example, the extent of conjugation controls the HOMO–LUMO gap: if the coupling is such that the alkyl side chains sterically interact (as in head-to-head coupling), the extended pi system is twisted out of planarity to minimize these interactions. The HH isomer requires more than 20 kJ to force planarity, whereas regioregular P3HT with only the head-to-tail coupling avoids this torsional strain and therefore has a greater planarity and a lower bandgap. The very different electronic properties of these two isomers can be seen in their absorption spectra: the λ_{max} for a thin-film absorption spectrum of the regioregular P3HT is about 475 nm while that of regiorandom P3HT is shifted to $\approx$557 nm, directly related to the differing extent of conjugation.

Fortunately, regioregular P3HT is commercially available and can be easily amended: for example, modification of poly-3-hexylthiophene to the hydroxylated version allows its attachment to carbon nanotubes (Kuila et al. 2010) or graphene (Yu et al. 2010b) (Figures 7.31 and 7.32), opening an entirely new avenue to optimization of OPV device performance. Other synthetic modifications of poly-3-alkylthiophenes focus on fine tuning the bandgap, as discussed above. For example, structural modifications can be made that moderate the relative contributions of resonance contributors or increase or decrease the electron density of the donor.

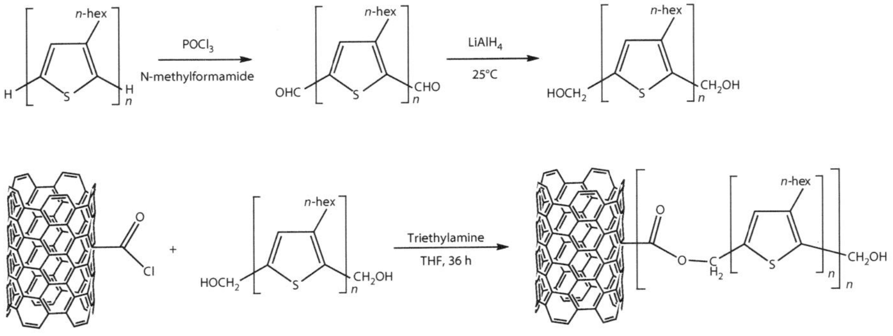

FIGURE 7.31   Synthesis of P3HT-grafted carbon nanotubes. (Reprinted with permission from Kuila, B.K., K. Park, and L. Dai. 2010. Soluble P3HT-grafted carbon nanotubes: Synthesis and photovoltaic application. *Macromolecules* 43 (16):6699–6705. Copyright 2010 American Chemical Society.)

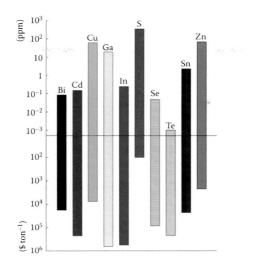

FIGURE 1.8 Cost versus availability of materials important in the solar photovoltaic industry. (Peter, L. M. 2011. Towards Sustainable Photovoltaics: The Search for New Materials. *Philos. Trans. R. Soc. Lond. A*, 369(1942): 1840–1856. By permission of the Royal Society.)

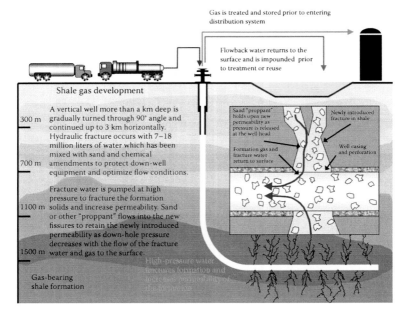

FIGURE 2.6 Hydraulic fracturing in the extraction of shale gas. (Reprinted with permission from Gregory, K.B., R.D. Vidic, and D.A. Dzombak. 2011. Water management challenges associated with the production of shale gas by hydraulic fracturing. *Elements* 7 (3):181–186. Copyright 2011, Mineralogical Society of America.)

FIGURE 2.8 An oil sands development in northern Alberta, Canada. (Shutterstock Image id 48011344.)

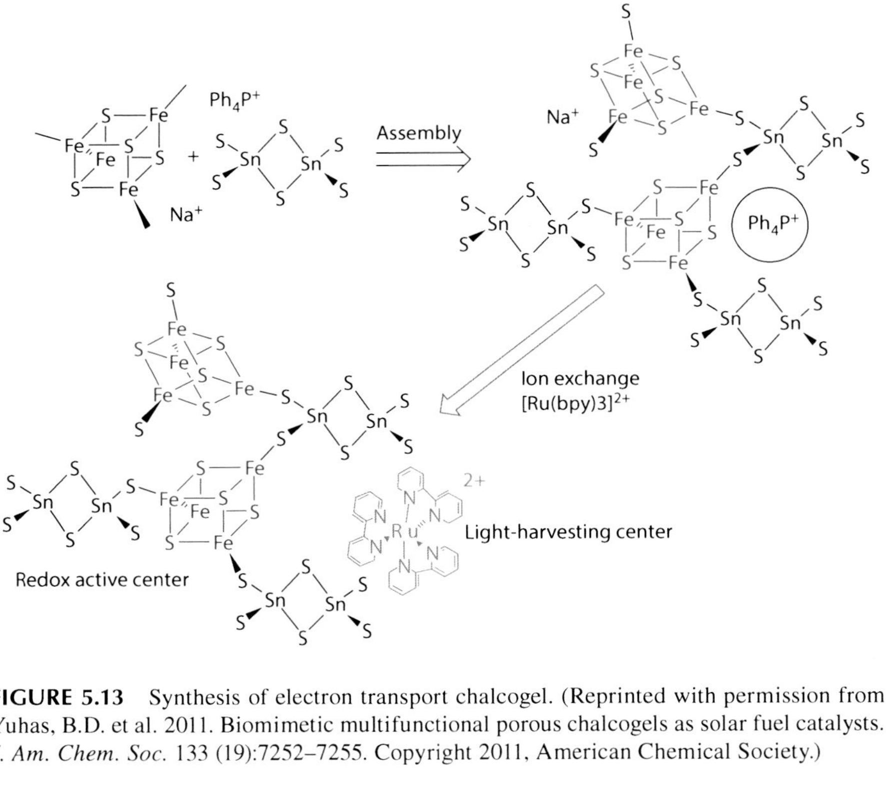

FIGURE 5.13 Synthesis of electron transport chalcogel. (Reprinted with permission from Yuhas, B.D. et al. 2011. Biomimetic multifunctional porous chalcogels as solar fuel catalysts. *J. Am. Chem. Soc.* 133 (19):7252–7255. Copyright 2011, American Chemical Society.)

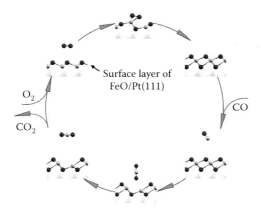

FIGURE 6.27 Proposed steps in the oxidation of CO on a FeO/Pt(111) surface. (Reprinted with permission from Giordano, L. and G. Pacchioni. 2011. Oxide films at the nanoscale: New structures, new functions, and new materials. *Acc. Chem. Res.* 44 (11):1244–1252. Copyright 2011, American Chemical Society.)

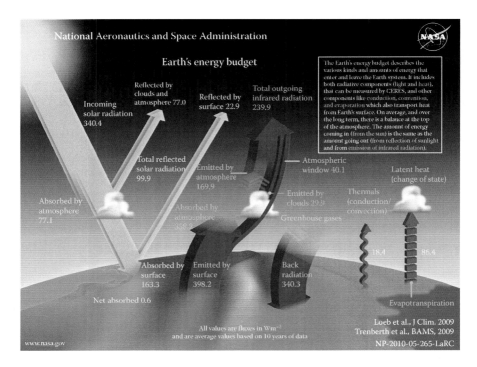

FIGURE 7.1 The Earth's energy budget. (From National Aeronautic and Space Administration. "Earth's Energy Budget." Retrieved May 23, 2013, from http://www.nasa.gov/audience/foreducators/topnav/materials/listbytype/Earths_Energy_Budget.html.)

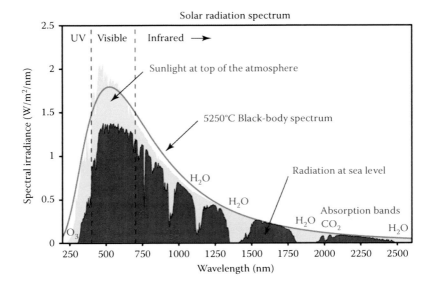

FIGURE 7.2 The solar radiation spectrum. (From Rohde, R. A. Solar radiation spectrum, Wikimedia Commons/Global Warming Art. Retrieved May 23, 2013, from http://commons. wikimedia.org/wiki/File:Solar_Spectrum.png.)

FIGURE 7.4 A solar thermal concentrator tower. (From Shutterstock 96405053.)

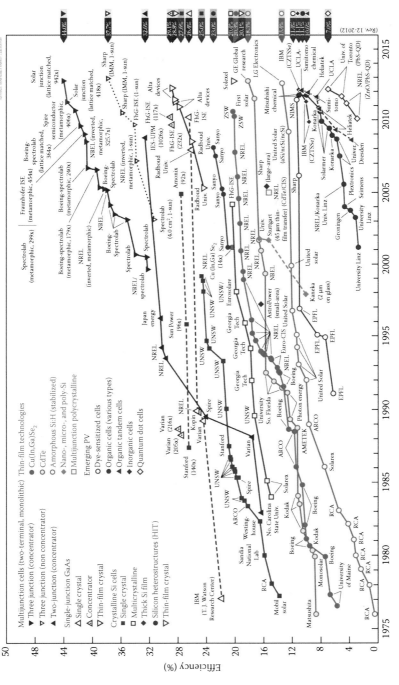

FIGURE 7.11 Timeline of research cell efficiencies. (From National Renewable Energy Laboratory 2013a. *Best Research-Cell Efficiencies.* National Center for Photovoltaics, NREL.)

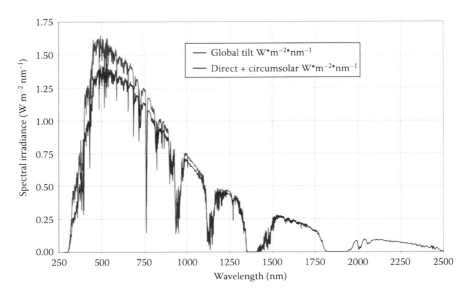

FIGURE 7.12 The AM 1.5 Reference Spectra. (From http://rredc.nrel.gov/solar/spectra/am1.5/. Accessed May 23, 2013.)

FIGURE 7.14 The thin "fingers" of solar cell metal contacts. (From Shutterstock 44807470.)

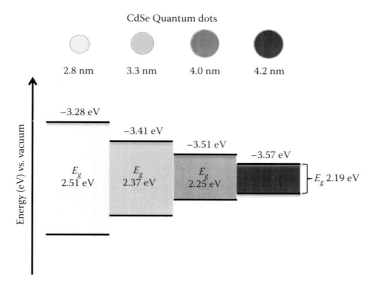

FIGURE 7.75 The relationship between QD size and bandgap. (Adapted with permission from Kamat, P. V. Boosting the efficiency of quantum dot sensitized solar cells through modulation of interfacial charge transfer. *Acc. Chem. Res.* 45 (11):1906–1915. Copyright 2012, American Chemical Society.)

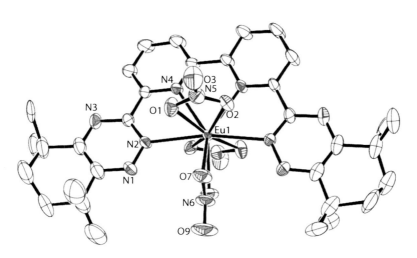

FIGURE 9.11 An ORTEP diagram of the europium(III) BTBPhen complex. (Reprinted with permission from Whittaker, D.M., T.L. Griffiths, M. Helliwell et al. 2013. Lanthanide speciation in potential SANEX and GANEX actinide/lanthanide separations using tetra-*N*-donor extractants. *Inorg. Chem.* 52 (7):3429–3444. Copyright 2013, American Chemical Society.)

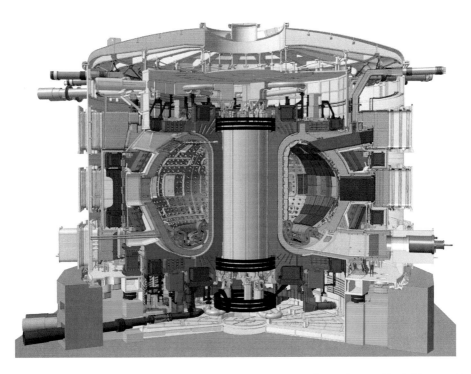

FIGURE 9.12   The ITER tokomak fusion reactor. (From http://www.iter.org/doc/all/content/com/gallery/Media/7%20-%20Technical/In-cryostat%20Overview%20130116.jpg)

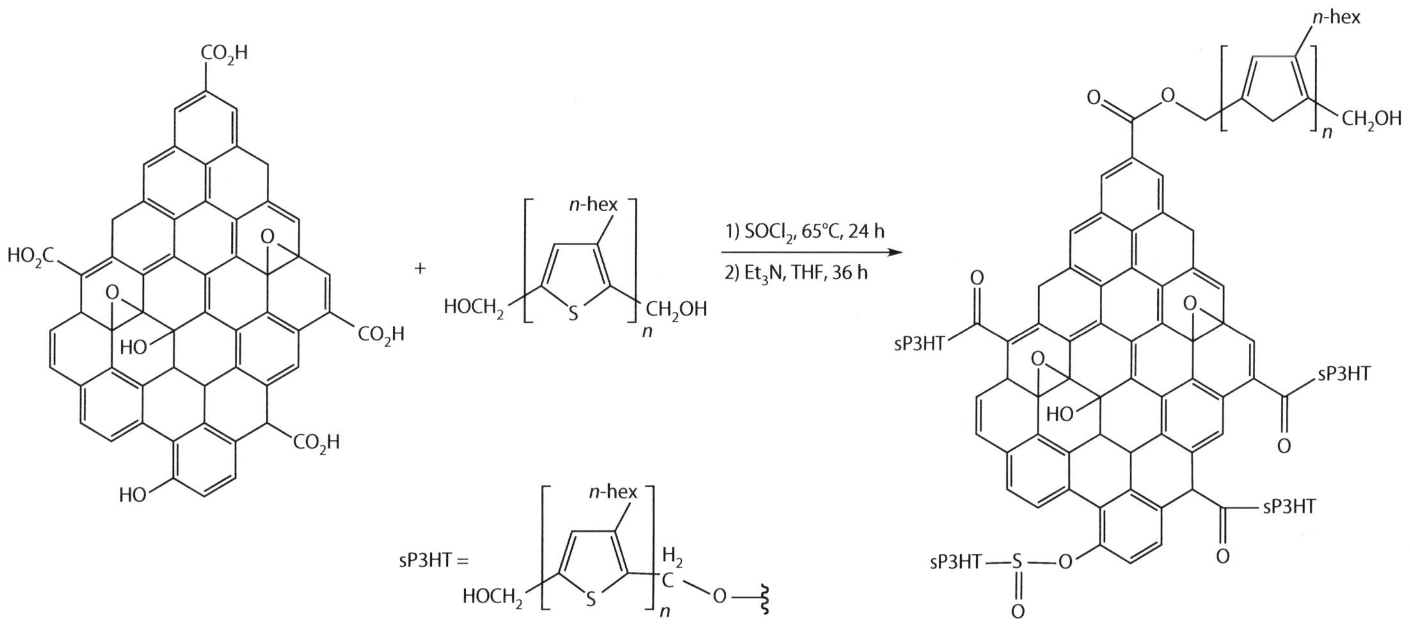

FIGURE 7.32 Synthesis of P3HT-grafted graphene. (Reprinted with permission from Yu, D., Y. Yang, M. Durstock et al. 2010b. Soluble P3HT-grafted graphene for efficient bilayer–heterojunction photovoltaic devices. *ACS Nano* 4 (10):5633–5640. Copyright 2010 American Chemical Society.)

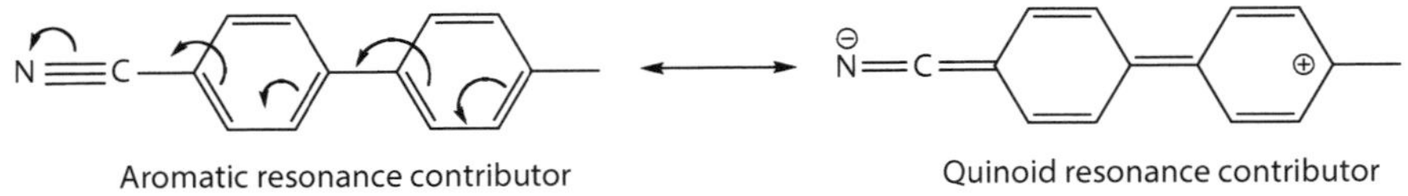

Aromatic resonance contributor   Quinoid resonance contributor

FIGURE 7.33 Comparison of the aromatic and quinoid resonance contributors for a cyano-substituted system.

With regard to resonance contributors, the extensive conjugation of the aromatic-annulated conducting polymer donor materials means they may exhibit more or less aromatic or quinoid character (Figure 7.33) that influences the HOMO–LUMO gap. Polythiophene retains complete aromaticity in representation **1** of Figure 7.34, but resonance contributor **2** shows the quinoid-like structure that disrupts the aromaticity of the thiophene ring. When a benzene ring is fused to the polythiophene backbone (as in polyisothianaphthene, structures **3/4** of Figure 7.34), the quinoid-like resonance contributor (**4**) contributes more strongly to the electronic distribution in the molecule due to the fact that the resonance stabilization energy for benzene is greater than that for thiophene. Thus, polyisothianaphthene has more quinoid character than does polythiophene and, as a result, the bandgap for polyisothianaphthene is smaller (1.0 eV vs. 2.0 eV; Cheng et al. 2009).

The electron density of the donor material (and hence the bandgap) can also be altered by adding electron-donating or electron-withdrawing substituents directly to the aromatic component of the polymer backbone. For example, the addition of electron-donating alkoxy groups to polythiophene (as in poly[3,4-(ethylenedioxy)

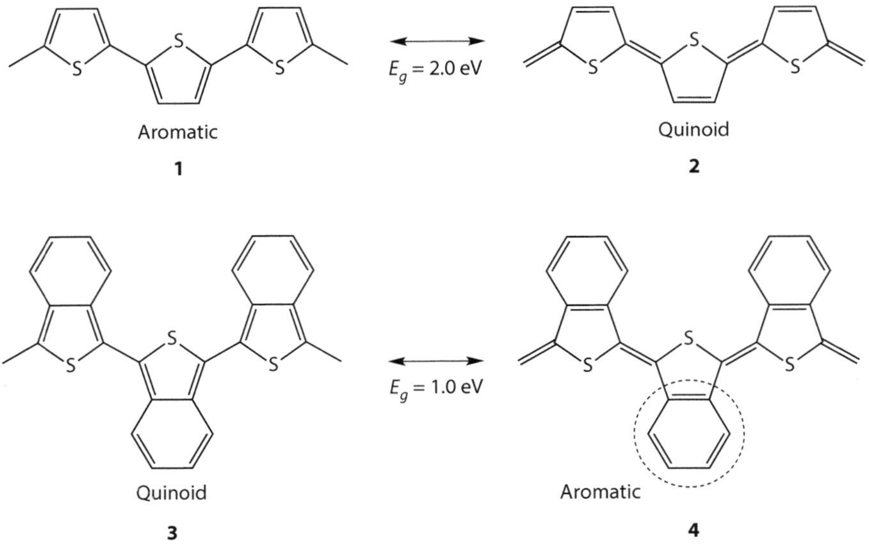

Aromatic $E_g = 2.0$ eV   Quinoid

1   **2**

Quinoid $E_g = 1.0$ eV   Aromatic

3   **4**

FIGURE 7.34 Quinoid versus aromatic thiophene structures and the impact on bandgap energies.

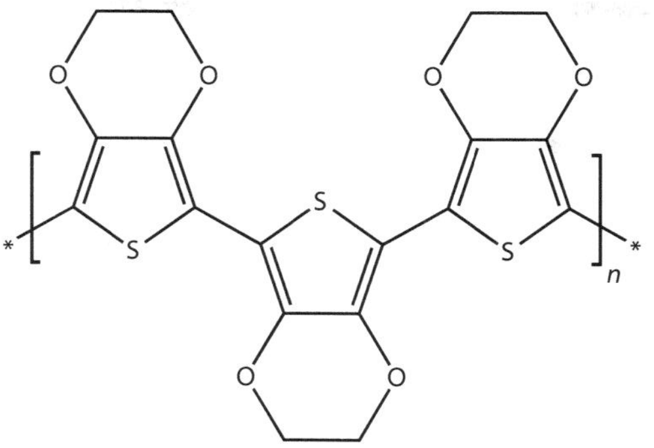

Poly[3,4-(ethylenedioxy)thiophene]

FIGURE 7.35 Electron-rich poly[3,4-(ethylenedioxy)thiophene].

thiophene], Figure 7.35) lowers the HOMO–LUMO gap by about 0.5 eV (Pei et al. 1994). However, one of the most prevalent approaches for decreasing the polymer bandgap and facilitating charge transfer is to design a "donor–acceptor" copolymer [-(donor–acceptor)$_n$...]. By copolymerizing the donor monomer (which has a higher HOMO) with a lower-LUMO acceptor monomer, the intramolecular charge transfer from donor to acceptor is facilitated and a lower bandgap polymer can result (Lee et al. 2011). In addition to the compounds shown in Figure 7.28, a few representative examples of other building blocks for these so-called *low bandgap*

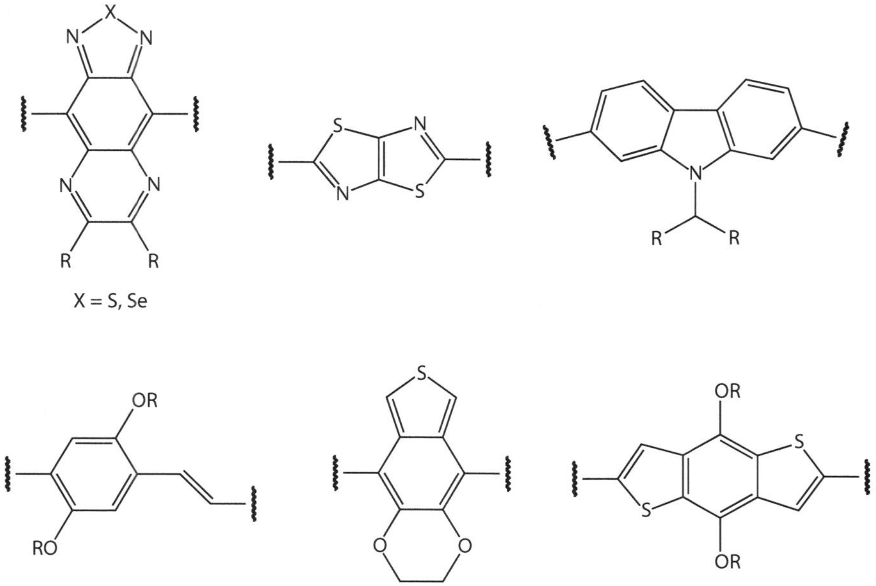

FIGURE 7.36 Some representative building blocks for low bandgap polymers.

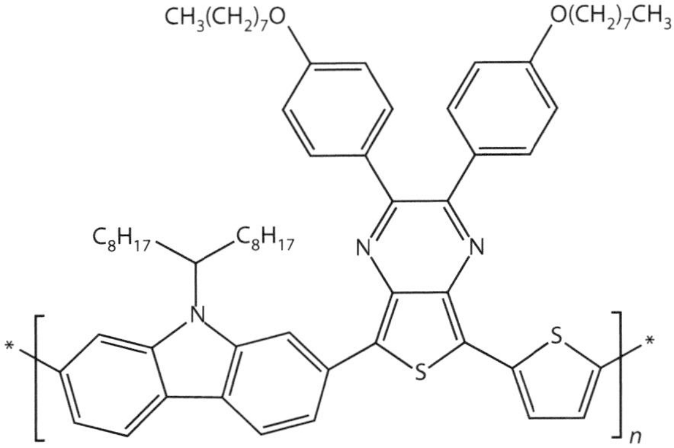

FIGURE 7.37 A quinoxaline-carbazole donor–acceptor copolymer.

donor–acceptor polymers are shown in Figure 7.36. A specific example of a promising low bandgap copolymer is shown in Figure 7.37. This material was synthesized using the Suzuki reaction (described below) in 60% yield and with a PDI of 2.8. The number-average molecular weight of 37 kg/mol was established by gel permeation chromatography and the glass transition temperature was determined to be 117°C by differential scanning calorimetry. In terms of its optical and electrochemical properties, the bandgap for this particular material was calculated from the onset of absorbance in its UV–vis spectrum and determined to be 1.97 eV; a solar cell fabricated with this material and a standard acceptor ($PC_{71}BM$; *vide infra*) provided a V_{oc} of 0.82 V, a J_{sc} of 9.96 mA/cm^2, a fill factor of 0.49, and a PCE of 4.0% (Lee et al. 2011).

These donor–acceptor copolymers can be prepared by taking advantage of high-yielding and regiospecific organometallic coupling reactions. This synthetic methodology links together an aryl halide (typically) with the appropriate coupling partner via a palladium-catalyzed oxidative addition/reductive elimination sequence (review Section 5.1 for information on these transition metal-catalyzed mechanistic steps). For example, the Stille reaction was used to prepare the alternating thieno–thiophene/benzodithiophene copolymer as shown in Figure 7.38 in 82% yield and

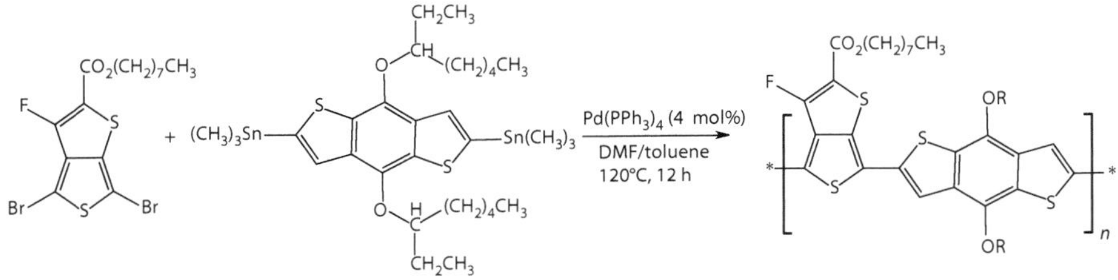

FIGURE 7.38 The Stille reaction used in the synthesis of a benzodithiophene copolymer.

with a M_w of 19.3×10^3 g/mol (as determined by GPC) and PDI of 1.32 (Liang et al. 2009). Use of aryl boronates as the coupling partner is the realm of the Suzuki reaction, another popular and effective coupling method. Suzuki coupling between a dibenzothiophene donor and a benzothiadiazole acceptor was used to prepare the D–A copolymer shown in Figure 7.39. The dark-red benzothiadiazole copolymer was formed in 42% yield and with $M_w = 6.4$ kg/mol, $M_n = 4.7$ kg/mol, and a PDI of 1.4. This material exhibited narrow bandgaps (<2 eV), and when paired with the acceptor material $PC_{71}BM$ (*vide infra*) in a solar cell, a PCE of $\approx 4.5\%$ for R = n-octyl was observed (Jin et al. 2012).

While these organometallic coupling reactions are high-yielding and well-established methods for coupling aromatics, they have some major drawbacks. Compound isolation and purification is difficult and the trialkyltin by-products formed in the Stille reaction are toxic. Furthermore, metal-catalyzed coupling reactions such as the Stille and Suzuki reactions often require additional steps in the synthesis of precursors to install the coupling functionality (i.e., the metal and halide) that allows for regiochemical control. Recent progress in the development of *direct arylation* (also known as dehydrohalogenative cross-coupling) has provided a promising alternative to synthesis of conducting polymers that is more benign. In direct arylation, a C–H bond is cleaved in the coupling reaction with regiocontrol based on electronic activation at the coupling site. Thus, for example, regioregular polyhexylthiophene was synthesized by the route shown in Figure 7.40 in quantitative yield and outstanding regioregularity (98%) (Wang et al. 2010). A terthiophene–thienopyrrolodione copolymer was synthesized using a similar method (Figure 7.41), using bulky alkyl side chains to help increase the solubility (and therefore ease of processing) of the polymer. The resultant polymer exhibited a "deep" HOMO energy level of -5.66 eV and bandgap of 1.8 eV, thus providing a good match for the other materials in the cell (recall Figure 7.24). Indeed, the solar cell fabricated from this material showed a PCE of >6% (Jo et al. 2012).

7.4.3.2 Acceptors

The acceptor materials in OPVs are largely fullerenes (popularly referred to as "buckyballs"), with [6,6]-phenyl-C_{61}-butyric acid methyl ester (PCBM, Figure 7.23) or [6,6]-phenyl-C_{71}-butyric acid methyl ester ($PC_{71}BM$, Figure 7.42) being the most commonly used. The main reason for their widespread use is that they are ultrafast (sub-picosecond) electron acceptors, with donation to the fullerene taking place much faster than back-transfer or decay processes (Kraabel et al. 1994; Sariciftci et al. 1992). The extensive delocalization of the π system makes these fullerenes quite polarizable, giving them a high dielectric constant and therefore helping to stabilize the charge transfer complex. The HOMO–LUMO levels are good (the LUMO of PCBM is a good match for many conducting polymers, although a higher LUMO would be advantageous, *vide infra*). Finally, the mobility of electrons within the fullerene framework is excellent.

Functionalization of fullerenes to form PCBM and $PC_{71}BM$ is fairly straightforward as the system is not aromatic and electrophiles can add to a double bond relatively easily (Figure 7.43) (Hummelen et al. 1995). Several other modifications

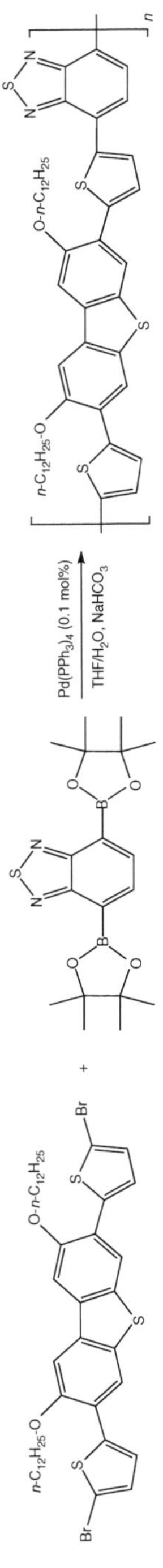

FIGURE 7.39 The Suzuki reaction used in the synthesis of a benzothiadiazole copolymer.

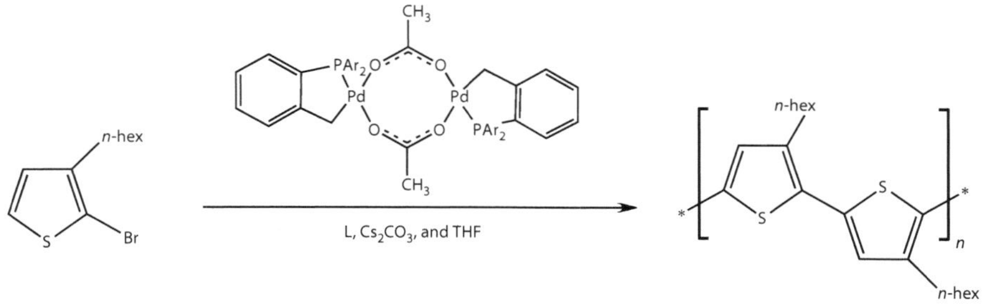

FIGURE 7.40   Synthesis of P3HT via the direct arylation method.

to the PCBM structure have been attempted to raise the level of the fullerene LUMO, as in the organocopper arylation protocol shown in Figure 7.44 (Xiao et al. 2012). The two arylated C_{70} fullerene isomers resulted in raised LUMO levels and, when incorporated into an OPV, PCEs of up to 2.87%. Attempts to raise the LUMO level by altering the aromatic substituent on the butyric acid arm (Figure 7.45) did, in general, improve the V_{oc} of fullerene OPVs (Kooistra et al. 2007). But these synthetic routes illustrate one of the disadvantages of attempting to modify the fullerene framework: isomer formation leads to tedious and yield-lowering separation steps. Isolation of the two isomers shown in Figure 7.44 required silica gel column chromatography followed by GPC. Even after these two separation steps, further heating was required to convert isomeric contaminants in one of the fractions into the isolated compound.

An additional issue associated with fullerenes is that they are notoriously insoluble: the solubility of PCBM is only 25 mg/mL in chlorobenzene and 30 mg/mL in orthodichlorobenzene (Kronholm and Hummelen 2009). In addition, fullerenes have a tendency to aggregate which can have a detrimental effect on the solar cell's performance. Researchers have also found that the solvent from which the acceptor material is processed can make a significant difference with respect to the overall efficiency of the OPV. Finally, the absorption coefficient for fullerenes—and match with the solar spectrum—is relatively poor. The absorption spectrum of P3HT, PCBM, and a 1:2 PCBM:P3HT film is shown in Figure 7.46 showing that absorption by PCBM is largely in the UV region below 400 nm where spectra irradiance tapers off significantly (Cook et al. 2009).

7.4.4   ARCHITECTURE AND MORPHOLOGY

7.4.4.1   Architecture

The typical OPV is assembled as shown in Figure 7.47. Some features warrant particular explanation. First, there are additional layers that are fundamental to improving the efficiency of the device, one of which is the PEDOT:PSS layer between the transparent conducting oxide (TCO) anode and the donor:acceptor layer. PEDOT is **p**olymerized 3,4-**e**thylene**d**ioxy**t**hiophene and PSS is **p**oly-4-**s**tryene**s**ulfonic acid. The PEDOT:PSS mix is prepared by the aqueous oxidative polymerization of EDOT

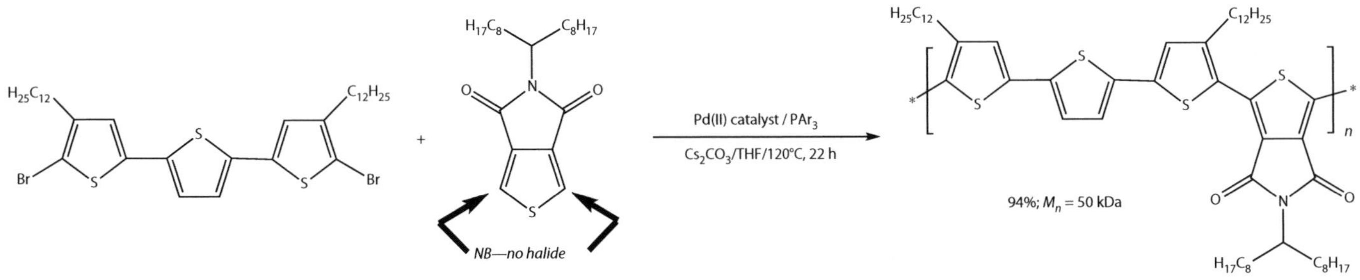

FIGURE 7.41 Direct arylation method for the synthesis of terthiophene–thienopyrrolodione copolymer.

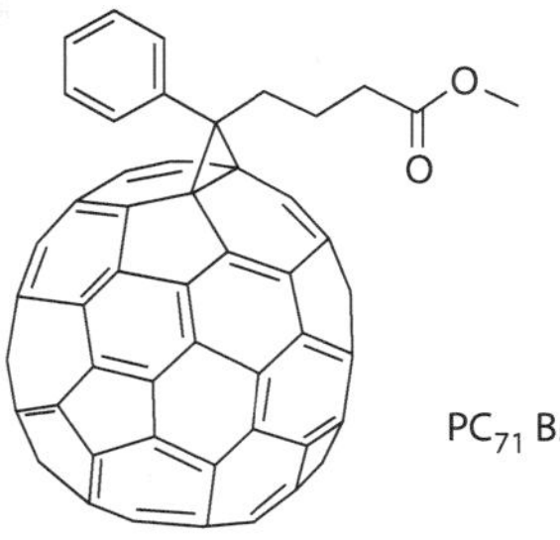

PC$_{71}$BM

FIGURE 7.42 [6,6]-Phenyl C$_{71}$ butyric acid methyl ester.

(3,4-ethylenedioxythiophene) to give the charged polymer pair shown in Figure 7.48. The presence of the PEDOT:PSS layer provides extensive π-stacking to improve charge transport and it smooths the transparent conducting oxide layer (which is typically indium tin oxide) for better contact. It is believed that the presence of the PEDOT:PSS layer also provides a better match to the HOMO level of the donor, but much remains to be understood with respect to the mechanism. In addition to the PEDOT:PSS layer, buffer layers may be added to improve the contact to the electrodes (e.g., a thin film of LiF below an aluminum cathode).

7.4.4.2 Morphology

One of the most important features of the OPV in terms of overall performance is the morphology of the donor–acceptor layer. The morphology of the materials involved is strongly impacted by structure, fabrication, and processing methods. Unlike c-Si cells, OPV materials are not constrained to a layer design. Since the photoconduction process depends entirely upon the interface between the donor and acceptor materials, increasing the contact area between these surfaces should improve efficiency, otherwise excitons formed far from the interface recombine before they diffuse to the D–A interface for separation and charge collection. In a seminal paper by Yu and coworkers, interspersing the donor material (in this case, a phenylvinylene polymer) with PCBM created a bicontinuous heterojunction, now more commonly referred to as a *bulk heterojunction* (BHJ), a cartoon of which is shown in Figure 7.49. The device with the BHJ showed an increase in quantum efficiency by more

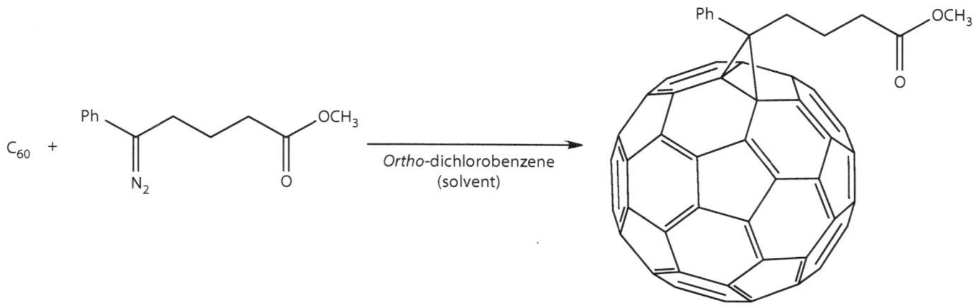

FIGURE 7.43 Preparation of PCMB from the C60 fullerene.

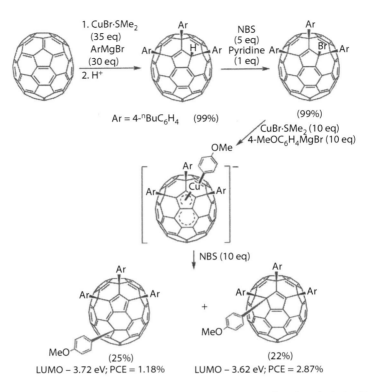

FIGURE 7.44 Modification of PCBM by organometallic-mediated arylations. (Reprinted with permission from Xiao, Z., Y. Matsuo, I. Soga et al. 2012. Structurally defined high-LUMO-level 66π-[70]fullerene derivatives: Synthesis and application in organic photovoltaic cells. *Chem. Mater.* 24 (13):2572–2582. Copyright 2012 American Chemical Society.)

than two orders of magnitude (Yu et al. 1995). Figure 7.50a shows a TEM of a 60 nm slice of regioregular P3HT mixed with PCBM in a 1:0.7 ratio, clearly illustrating the interpenetrating nature of the bulk heterojunction. Figure 7.50b is a modified image that better illustrates the morphology of the slice (Moon et al. 2008). The BHJ approach is the standard for OPV devices and the importance of optimizing the

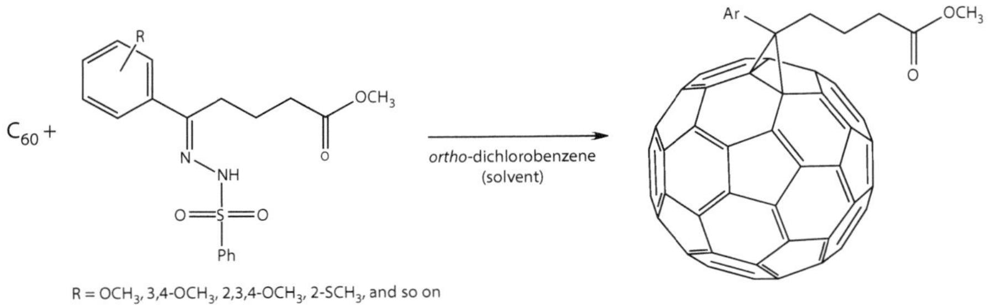

FIGURE 7.45 Altering the butyric acid substituent of PCBM.

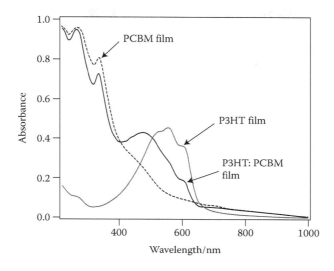

FIGURE 7.46 Comparative UV–vis absorbance spectra of PCBM, P3HT and the donor–acceptor film. (Reprinted from Cook, S., R. Katoh, and A. Furube. 2009. Ultrafast studies of charge generation in PCBM:P3HT blend films following excitation of the fullerene PCBM. *J. Phys. Chem. C* 113 (6):2547–2552. Copyright 2009 American Chemical Society.)

BHJ morphology on device performance has been confirmed by theoretical studies (Lyons et al. 2012). Interestingly enough, the BHJ is thermodynamically disfavored and phase separation will eventually result at higher temperatures. The BHJ is another reason that fullerenes are exceptional as acceptor materials: the symmetry of fullerenes makes them especially well-suited to accept electrons from *all* directions, as required in a BHJ device.

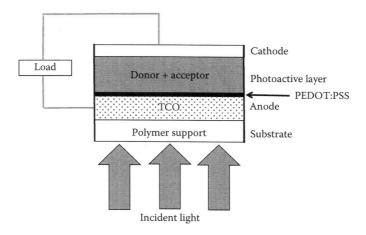

FIGURE 7.47 Schematic of a typical organic PV cell.

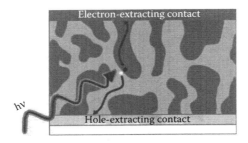

FIGURE 7.48   Structures and preparation of PEDOT:PSS.

The processing of the BHJ materials has been found to significantly impact the morphology and, hence, the performance of the OPV device. The choice of solvent, method of drying, annealing—all play a large role in the orientation of the molecular materials and overall efficiency of the heterojunction. Figure 7.51 outlines the many aspects of processing that impact morphology (Dang et al. 2013). First, casting

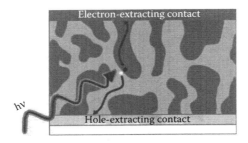

FIGURE 7.49   Cartoon depicting the interpenetrating layers of the bulk heterojunction. (Adapted with permission from Giridharagopal, R., and D.S. Ginger. 2010. Characterizing morphology in bulk heterojunction organic photovoltaic systems. *J. Phys. Chem. Lett.* 1 (7):1160–1169. Copyright 2010 American Chemical Society.)

100 nm

FIGURE 7.50   (a) TEM results showing the P3HT:PCBM BHJ. (b) Modified binary image of same. (Adapted with permission from Moon, J.S., J.K. Lee, S. Cho et al. 2008. 'Columnlike' structure of the cross-sectional morphology of bulk heterojunction materials. *Nano Lett.* 9 (1):230–234. Copyright 2009 American Chemical Society.)

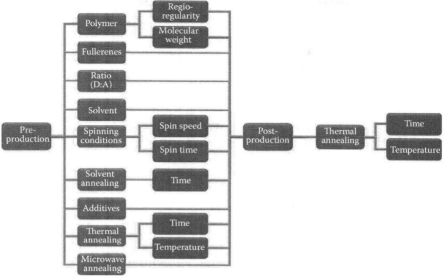

FIGURE 7.51 The many variables that impact OPV morphology. (Reprinted with permission from Dang, M.T., L. Hirsch, G. Wantz et al. 2013. Controlling the morphology and performance of bulk heterojunctions in solar cells. Lessons learned from the benchmark poly(3-hexylthiophene):[6,6]-phenyl-C61-butyric acid methyl ester system. *Chem. Rev.* 113 (5):3734–3765. Copyright 2013 American Chemical Society.)

the thin film from solvent and allowing it to dry slowly allows for equilibration of molecular organization and self-assembly into better pi-stacking as the concentration of the material increases with decreasing solvent. Well-defined nanostructures and regions of crystalline material can also form from careful film casting (Ewbank et al. 2009). Annealing leads to even better crystallinity, which leads to better hole transport and improves the external quantum efficiency of P3HT:PCBM OPVs. However, there is an optimum annealing time and temperature, as segregation of the D–A materials can occur with extended annealing times.

Recent research results by Zhao and coworkers nicely demonstrate the impact of the solvent from which the BHJ film is cast on the organization of the BHJ. The top image in Figure 7.52 is a TEM of a P3HT:PCBM film cast from neat chlorobenzene while the bottom image is the same material cast from 5:95 acetone:chlorobenzene. The addition of the acetone results in the formation of nanofibrils which were found to improve both the absorption and the hole conduction. The *J–V* curves for cells fabricated from these materials are shown in Figure 7.53a and b, where (a) plots the results of the P3HT:PCBM film without annealing and (b) is after annealing at 140°C for 3 min. Comparison of these two plots clearly shows the positive impact of both the nanofibril formation and annealing on the cell performance (Zhao et al. 2009).

The structures of the individual donor and acceptor molecules, too, play a role in the quality of the BHJ because they impact the overall properties of the polymer. For

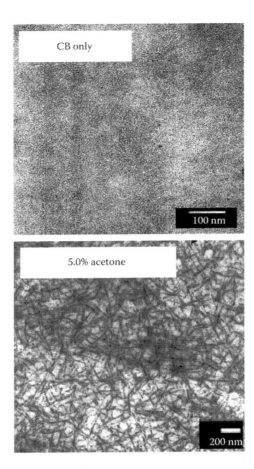

FIGURE 7.52 The impact of the addition of acetone on the morphology of the P3HT:PCBM film. (Adapted with permission from Zhao, Y., S. Shao, Z. Xie et al. 2009. Effect of poly (3-hexylthiophene) nanofibrils on charge separation and transport in polymer bulk heterojunction photovoltaic cells. *J. Phys. Chem. C* 113 (39):17235–17239. Copyright 2009 American Chemical Society.)

example, it is generally believed that the higher the molecular weight, the better the device performance. The degree of interpenetration between the donor and acceptor materials is also impacted by the molecular properties of the materials involved. For example, the T_g of the polymeric donor material has been found to influence the ability of PCMB to diffuse through the BHJ (Watts et al. 2009). The steric and electronic nature of substituents on the polymer backbone can impede molecular organization or lead to better pi-stacking or crystallinity. More recent studies suggest that the *ordered bulk heterojunction* consisting of columns of conjugated polymer donor material (as nanorods) surrounded by the electron-accepting material and set perpendicular to the device electrodes is an improved morphology (Po et al. 2009). A recent example of the fabrication of such a cell is representative. An indium tin

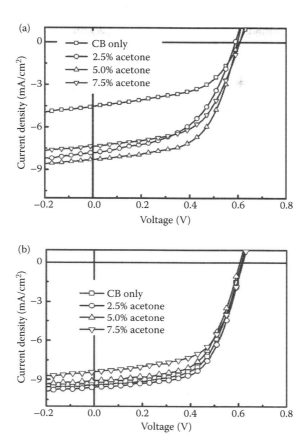

FIGURE 7.53 Cell performance as a function of solvent (a) without annealing and (b) with annealing. (Adapted with permission from Zhao, Y., S. Shao, Z. Xie et al. 2009. Effect of poly (3-hexylthiophene) nanofibrils on charge separation and transport in polymer bulk heterojunction photovoltaic cells. *J. Phys. Chem. C* 113 (39):17235–17239. Copyright 2009 American Chemical Society.)

oxide glass substrate was coated with PEDOT:PSS then the P3HT nanopillars were constructed using nanoprint lithography—essentially growing the nanopillars on an aluminum oxide template, then dissolving the Al_2O_3 away (see Figure 7.54 for a scanning electron micrograph of the free-standing nanopillars). The PCBM was spin-coated directly onto the P3HT nanopillars and an aluminum cathode was evaporated onto the surface to complete the cell. The enhancement in performance from a regular bilayer P3HT:PCBM cell is substantial, as shown in the *J–V* curves in Figure 7.55 (Chen et al. 2012b).

As can be gathered, improvements and innovation in OPVs are continuing at a frenetic pace, but efficiency and durability have yet to reach a level where commercial viability is promising. OPVs have a great deal of catching up to do to unseat the dominance of silicon solar cells.

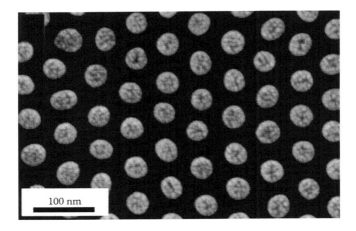

FIGURE 7.54   A bird's eye view of P3HT nanopillars. (Adapted with permission from Chen, D., W. Zhao, and T.P. Russell. 2012b. P3HT Nanopillars for organic photovoltaic devices nanoimprinted by AAO templates. *ACS Nano* 6 (2):1479–1485. Copyright 2012 American Chemical Society.)

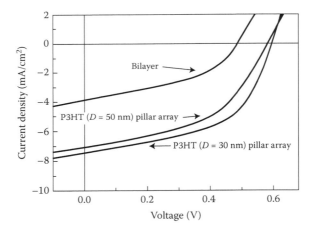

FIGURE 7.55   Cell performance of nanopillars versus a standard bilayer P3HT:PCBM cell. (Adapted with permission from Chen, D., W. Zhao, and T.P. Russell. 2012b. P3HT nanopillars for organic photovoltaic devices nanoimprinted by AAO templates. *ACS Nano* 6 (2):1479–1485. Copyright 2012 American Chemical Society.)

7.5   DYE-SENSITIZED SOLAR CELLS

7.5.1   INTRODUCTION

In 1991, a seminal finding presented a new approach in PV architecture: by using a dye as the light harvester in a solar cell, a phenomenal improvement in harvesting incident light was obtained (O'Regan and Grätzel 1991). This finding paved the way for the development of *dye-sensitized solar cells* (DSSC), an area of research that has been growing ever since (Meyer 2010).

How are DSSCs different from the solar cells we have examined thus far? Here's a quick recap:

- In crystalline silicon solar cells, a bilayer architecture between a p-type material and an n-type material sets up the photocurrent via a bandgap promotion of an electron from the valence band to the conduction band. Both materials are essentially the same (a homojunction).
- In inorganic thin-film solar cells, the layer architecture and mechanism are similar to silicon cells, but heterojunctions between different p- and n-type materials are the norm.
- In organic solar cells, the p- and n-type inorganic semiconducting materials are replaced with organic materials: a conducting polymer donor (or electron-rich small molecule) and a fullerene acceptor. Band theory no longer applies; the pertinent energy differential is between HOMO and LUMO energy levels. The bilayer architecture is replaced with the interspersed bulk heterojunction.
- In DSSCs, the number of working materials increases and both inorganic and organic materials are part of the cell's key components. An organic or organometallic *dye sensitizer* is used to absorb the light and supply an excited state electron to a *metal-oxide semiconductor*. A *redox couple* (commonly referred to as the *electrolyte*) is required to regenerate the dye. The electrons and holes are conducted to the appropriate electrodes as usual.

DSSCs are basically our attempt to mimic photosynthesis: instead of chlorophyll in a leaf absorbing light energy and injecting an electron into an electron transport chain, a synthetic dye is used in a solar cell with the result being generation of electricity instead of synthesis of carbohydrates. In conventional PV processes, the semiconductor is both the light absorber *and* the charge carrier. In DSSC, the absorber (the dye) and the charge carrier (the metal oxide) are *separate species*. In this section, we will focus on n-type DSSCs, where the semiconductor involved is an n-type semiconductor. DSSCs in which the sensitized semiconductor is a p-type material have also been developed, but we will leave those to the interested reader to pursue independently.

7.5.2 ARCHITECTURE

The DSSC is built up in layers upon a glass substrate just as for other solar cells, but the addition of the dye sensitizer makes the architecture slightly more complex, as shown in Figure 7.56. The glass substrate is coated with a transparent conducting oxide (TCO) [indium tin oxide (ITO) or fluorine-doped tin oxide (FTO)]; the glass/TCO combination makes up the anode. The metal-oxide semiconductor is deposited on the anode to a thickness of about 4–10 μm (Meyer 2010). It will come as no surprise to learn that the nanoscale morphology of the metal-oxide layer is important; a *mesoporous* titanium oxide is the most commonly used semiconductor. A mesoporous solid is intermediate in porosity between macroporous (>50 nm) and

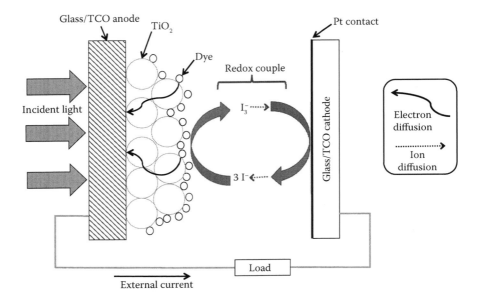

FIGURE 7.56   Schematic of a typical DSSC.

microporous (<2 nm) materials (Rouquerol et al. 1994). Applied on top of the metal oxide is a thin (ca. 10 μm) layer of the light-absorbing dye, of which there are many variants. By using a mesoporous semiconductor material, the surface area of the dye is increased enormously, markedly improving the light-harvesting ability of the device. However, the thickness of the dye must be optimal for a good balance of light harvesting and electron injection to the semiconductor. If the layer is too thick, the electron injection is impeded. An atomic monolayer of dye sensitizer is ideal for good diffusion of the exciton to the metal oxide. Some kind of redox electrolyte—typically the I_3^-/I^- couple—is applied next, followed by the cathode. In the case of DSSC, a "stronger" working electrode is required, thus platinum-coated glass is often used to collect the electrons. More detail on the TCO, dye, and redox mediator is presented in Section 7.5.4.

7.5.3   MECHANISM

How the various materials work together in a DSSC to create a photocurrent is graphically illustrated in Figure 7.57.

- Absorption. First and foremost, the incident light excites the dye molecules to an excited state (1), indicated by S* (S because the dye molecules are the *sensitizers* in a DSSC).
- Injection. The excited dye molecule can then inject an electron [on the pico-second time scale, (2)] into the conduction band of the metal-oxide semi-conductor (in this example, TiO_2) (Listorti et al. 2011).

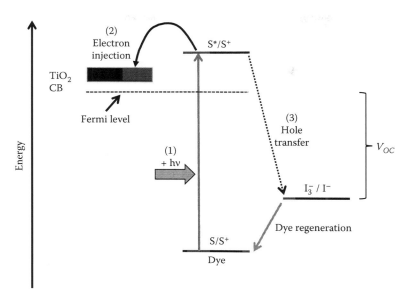

FIGURE 7.57 Diagram of the conduction process in a DSSC.

- Collection. The circuitous path through the metal-oxide semiconductor to the anode must be fast. The mesoporous structure means that the diffusion path is relatively long; failure to diffuse quickly to the electrode leads to recombination and back-electron transfer mechanisms.
- Regeneration. Upon injection of an electron into the mesoporous TiO_2, the dye becomes oxidized (as indicated by S^+) and the hole is transferred to the redox couple (3). Something must be present in the cell to reduce S^+ back to the ground, unoxidized state. In this figure, I^- plays that role, rapidly ($\approx 10~\mu s$) reducing the dye (Boschloo and Hagfeldt 2009). Reduction of the oxidized dye must take place fairly quickly lest a buildup of S^+ results increasing the likelihood of back-electron transfer to the oxidized dye. More detail regarding this step of the mechanism is presented in Section 7.5.4.3.
- The I_3^- anion is reduced back to I^- by electrons collected at the platinum electrode to complete the electrochemical circuit.

Key to the success of the DSSC is appropriate matching of the energy levels of the metal-oxide semiconductor conduction band, the HOMO–LUMO of the dye sensitizer and the redox potential of the mediator. The maximum voltage that can be obtained is based on the Fermi level of the metal oxide and the redox potential of the I_3^-/I^- couple. Given the complexity of the DSSC, the potential for inefficiencies are legion. For example, the excited state dye can decay and the injected electron can recombine with the oxidized dye or the I_3^-. Despite these issues, the efficiencies of the DSSC are above 10% and increasing.

7.5.4 Materials

7.5.4.1 Metal Oxide

Tin, zinc, and titanium oxides are three wide bandgap semiconductor metal oxides that have found widespread use in DSSC and Figure 7.58 gives their HOMO and LUMO energies. It is titania (TiO_2), however, that is far and away the most commonly used n-type semiconductor. As noted above, the morphology of the TiO_2 layer is a crucial aspect, since the high surface area of the metal-oxide layer—and, consequently, the dye—is required for efficient light harvesting. However, there is much more to the morphology of TiO_2 than meets the eye and the progression from the empirical study of TiO_2 performance in DSSCs to development of a mechanistic understanding is taking place both theoretically and empirically.

Titanium dioxide exists in nature in one of three forms: *rutile* is the most common and the most stable form, where the oxygen atoms are octahedral about the Ti atom. The other two forms—*anatase* and *brookite*—exhibit distorted octahedra about Ti, with the most stable form of anatase existing as a tetragonal bipyramid (Lazzeri et al. 2001). Evidence suggests that anatase is the most photocatalytically active form of TiO_2 and, therefore, the best choice for use in DSSC and a commercial product that is primarily anatase is available. A method for the controlled synthesis of anatase TiO_2 in either microflowers or microspheres is shown in Figure 7.59 (Li et al. 2013). Other syntheses of anatase nanocrystals often make use of the *sol–gel method*. In the sol–gel method, a precursor material (often Ti(IV) isopropoxide) is hydrolyzed and heated in the appropriate solvent(s) to generate a colloid which is then refluxed to

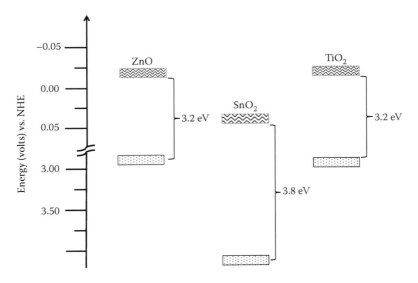

FIGURE 7.58 The wide bandgaps of metal-oxide semiconductors used in DSSC. (Adapted from *Renew. Sustain. Energy Rev.*, 16, Gong, J., J. Liang, and K. Sumathy, Review on dye-sensitized solar cells (DSSCs): Fundamental concepts and novel materials. 5848–5860. Copyright 2012 with permission from Elsevier.)

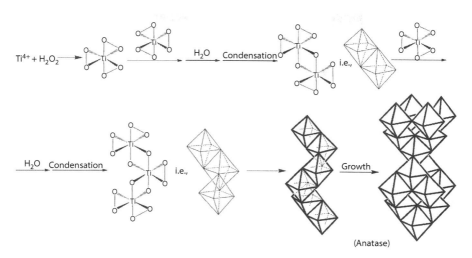

FIGURE 7.59 An overview of controlled anatase synthesis. (Reprinted with permission from Li, T., B. Tian, J. Zhang et al. 2013. Facile tailoring of anatase TiO_2 morphology by use of H_2O_2: From microflowers with dominant {101} facets to microspheres with exposed {001} facets. *Ind. Eng. Chem. Res.* 52 (20):6704–6712. Copyright 2013 American Chemical Society.)

generate the TiO_2 nanoparticles as a gel (Isley and Penn 2006). Coating the gel onto the TCO/glass substrate requires subsequent heating to sinter the particles together slightly for electrical contact between the particles. Small changes in the synthetic parameters—solvent, temperature, pH, and so on—can result in substantial changes in the shape, surface, size, and, therefore, properties of the resultant TiO_2 nanoparticles and the PV device.

Much more elaborate methods continue to be developed for preparation of more strictly controlled TiO_2 morphologies. Templating, that is, growing the TiO_2 film in the presence of something that will cause it to grow with a defined porosity, can lead to mesoporous structures of well-defined and -controlled pore sizes. For example, by using a nonionic block copolymer made up of ethylene oxide and propylene oxide repeat units (Figure 7.60), miscelles form in the mixture upon which the mesoporous structure can build. A three-layer film of this templated mesoporous TiO_2 was found to give an enhancement of about 50% in the photon conversion efficiency due to a significant increase in the short-circuit photocurrent, likely a result of increased light-harvesting attributable to the improved porosity of the titania (Zukalová et al. 2005).

New methods of synthesis have been developed that allow the controlled growth of metal-oxide semiconductors in various shapes, sizes, and with varying facets to maximize photocatalytic activity, and XRD and electron microscopy have been used extensively to characterize these materials and morphologies. It was known that rod-like anatase gave better performance in a DSSC than ball-shaped anatase (6.2% photoelectric conversion efficiency vs. 5.4%) (Wu et al. 2007) and two facets of the anatase crystal—{001} (flat) and {101} (somewhat corrugated)—have been of

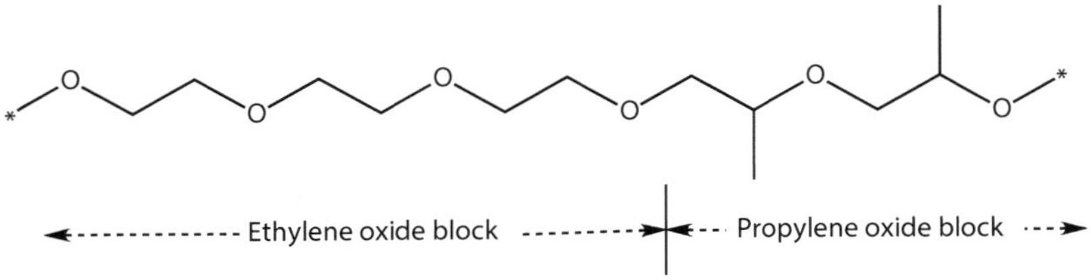

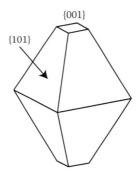

FIGURE 7.60   A representative ethylene oxide–propylene oxide block copolymer.

particular interest in unraveling the source of the photocatalytic activity (Figure 7.61) (Lazzeri et al. 2001). The synthesis of nanocrystals of anatase from $TiCl_4$ or TiF_4 (Figure 7.62) can be carried out to give a preponderance of either {001} or {101} facets depending upon the surfactant used as seen in the TEM images shown in Figure 7.63. Research is ongoing to determine which shape of nanocrystal gives the best results for a DSSC (Gordon et al. 2012).

Even more elaborate high-surface-area structures for the metal-oxide semiconductor dye support have been investigated. "Bamboo-like" ridged TiO_2 nanotube arrays were prepared by the relatively mild anodic oxidation of titanium foil in the

FIGURE 7.61   Descriptors for facets of the anatase crystal structure.

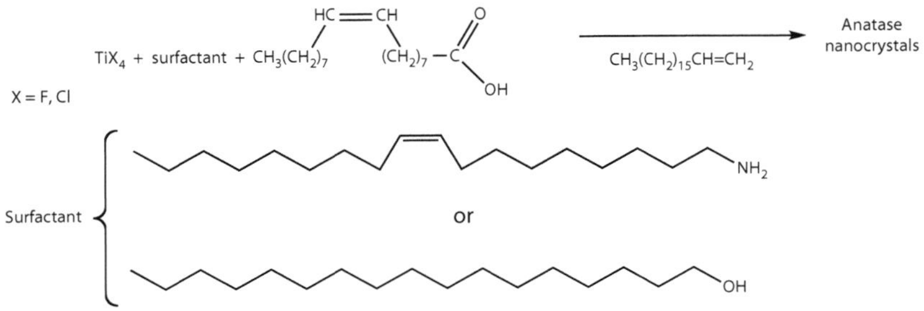

FIGURE 7.62   Synthetic preparation of anatase nanocrystals.

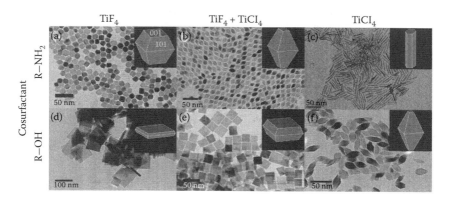

FIGURE 7.63 Impact of surfactant and halide on TiO_2 crystal shape; a–f represents different combinations of cosurfactant and titanium halide precursor. (Reprinted with permission from Gordon, T.R., M. Cargnello, T. Paik et al. 2012. Nonaqueous synthesis of TiO_2 nanocrystals using TiF_4 to engineer morphology, oxygen vacancy concentration, and photocatalytic activity. *J. Am. Chem. Soc.* 134 (15):6751–6761. Copyright 2012 American Chemical Society.)

presence of an ethylene glycol-based electrolyte. The relevant chemical transformations are shown in Equations 7.11 through 7.13 (Luan et al. 2012).

$$H_2O \rightarrow 2H^+ + O^{2-} \text{ (decomposition of water)} \tag{7.11}$$

$$Ti^\circ + 2O^{2-} \rightarrow TiO_2 + 4e^- \text{ (metal oxidation)} \tag{7.12}$$

$$TiO_2 + 6F^- + 4H^+ \rightarrow TiF_6^{2-} + 2H_2O \text{ (oxide dissolution)} \tag{7.13}$$

As indicated in Equation 7.13, the growing TiO_2 nanotube is dissolved in the presence of fluoride ions; by alternating the voltage, pH, and electrolyte composition, the TiO_2 nanotube growth can be controlled to make the sectioned bamboo-like structure with pronounced ridges between the sections. The nanotubes were converted into anatase TiO_2 by annealing in air at 450°C as confirmed by the XRD results shown in Figure 7.64; an SEM image of an array is shown in Figure 7.65. Apparently, the rough edges of the nanotubes increased their ability to adsorb the dye sensitizer. Smooth-walled TiO_2 nanotubes gave a PCE of 3.90%, while the same length bamboo-like nanotubes showed a PCE of 5.64% (Luan et al. 2012).

A final example of titania morphology that has shown considerable promise for DSSC is that known as the *inverse opal*. The inverse opal morphology can be described as the fused spherical cavities that are left behind after templating a material on close-packed spheres. Figure 7.66 graphically illustrates the process, and Figure 7.67 shows a scanning electron micrograph of such a structure prepared by dipping a substrate templated with close-packed polymethylmethacrylate spheres into a TiO_2 precursor solution. The inverse opal voids were then filled with meso-structured titania. A DSSC prepared with this TiO_2 showed a 10-fold improvement in PCE over the regular mesoporous titania (Mandlmeier et al. 2011).

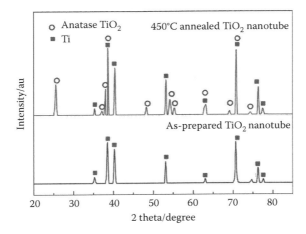

FIGURE 7.64 XRD results for bamboo-like TiO_2 showing transformation to anatase upon annealing. (Reprinted with permission from Luan, X., D. Guan, and Y. Wang. 2012. Facile synthesis and morphology control of bamboo-type TiO_2 nanotube arrays for high-efficiency dye-sensitized solar cells. *J. Phys. Chem. C* 116 (27):14257–14263. Copyright 2012 American Chemical Society.)

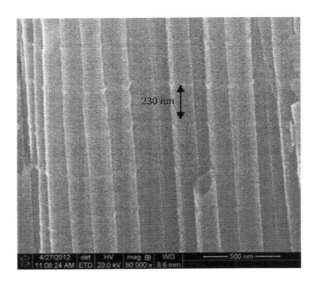

FIGURE 7.65 Scanning electron micrograph showing cluster of bamboo-like TiO_2. (Reprinted with permission from Luan, X., D. Guan, and Y. Wang. 2012. Facile synthesis and morphology control of bamboo-type TiO_2 nanotube arrays for high-efficiency dye-sensitized solar cells. *J. Phys. Chem. C* 116 (27):14257–14263. Copyright 2012 American Chemical Society.)

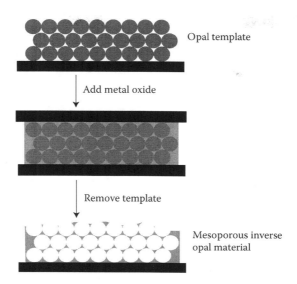

FIGURE 7.66 Process for synthesis of templated inverse opal TiO$_2$. (Adapted with permission from Wu, G., Y. Jiang, D. Xu et al. 2010. Thermoresponsive inverse opal films fabricated with liquid-crystal elastomers and nematic liquid crystals. *Langmuir* 27(4): 1505–1509. Copyright 2011 American Chemical Society.)

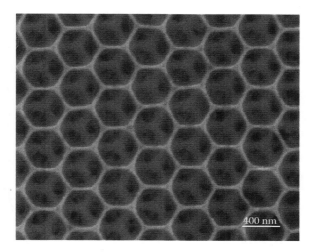

FIGURE 7.67 An SEM image of the inverse opal TiO$_2$ scaffold. (Adapted with permission from Mandlmeier, B., J. M. Szeifert et al. 2011. Formation of interpenetrating hierarchical titania structures by confined synthesis in inverse opal. *J. Am. Chem. Soc.* 133(43): 17274–17282. Copyright 2011 American Chemical Society.)

7.5.4.2 Dye Sensitizer

As with any solar cell, maximizing the capture of photon flux from the solar spectrum is an important goal and in the DSSC, this is the role of the dye. The ideal dye should absorb as much sunlight as possible, ideally capturing all wavelengths below 1000 nm. The LUMO level of the dye must be high enough for thermodynamically favorable electron transfer to the metal oxide, and for regeneration of the dye the reduction potential of the oxidized sensitizer must be more positive than that of the redox mediator. Of course, the dye needs to be stable in sunlight and at elevated temperatures, but with DSSCs comes a new requirement: the dye should be functionalized with a group that allows it to anchor strongly to the metal-oxide surface in a tight monolayer to prevent access of the redox mediator to the metal oxide. The most common covalent linker is the carboxyl group, but phosphonates, sulfonates, and even borates and silyl linkers (Figure 7.68) have been used (Hagfeldt et al. 2010).

Octahedral ruthenium pyridyl systems are the historical dye sensitizers, with the "gold standard" for DSSC being the so-called N3 dye, shown in Figure 7.69; the

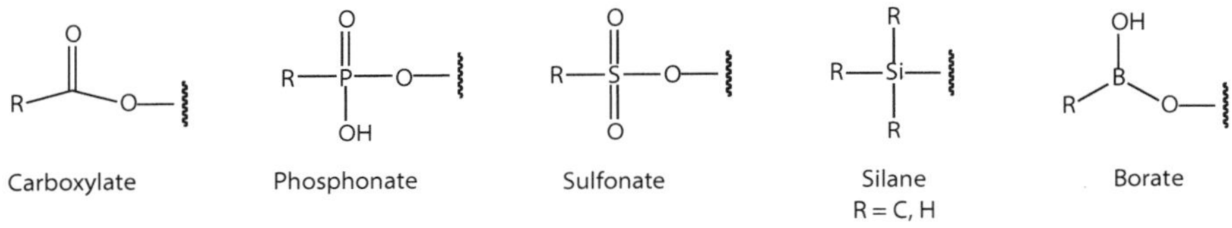

FIGURE 7.68   Functional group classes used as linkers for DSSC.

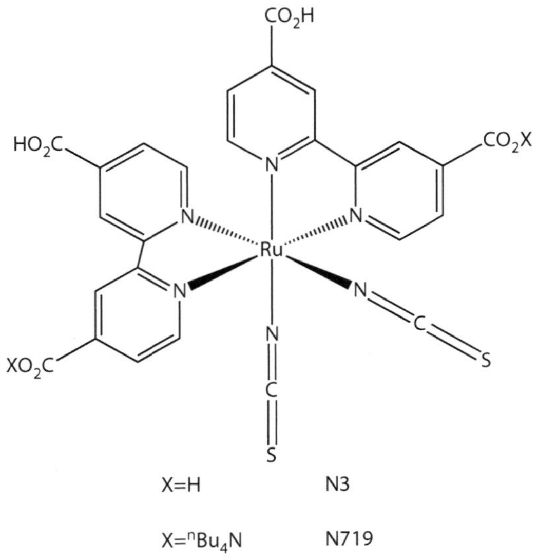

| X=H | N3 |
| X=^{n}Bu$_4$N | N719 |

FIGURE 7.69   Ruthenium dyes for DSSCs; N3 = *cis*-bis(isothiocyanato)bis(2,2′-bipyridyl-4,4′-dicarboxylato)ruthenium(II) dye, a.k.a. N3.

analogous tetra-*n*-butylammonium salt is the widely used N719 dye. The carboxyl groups on the bipyridyl ligands react with hydroxy functionality on the metal-oxide surface to anchor the dye. The isothiocyanate ligands (–N=C=S) on ruthenium increase the energy level of the HOMO so that the complex has a red-shifted absorption to a longer wavelength (Hagfeldt et al. 2010). The primary reason these dyes work so well for DSSC is because of the fact that, upon absorption of a photon and promotion of an electron to the dye LUMO, the metal complex can shift electron density between the metal center and the ligands to help stabilize the excited complex. This is known as a *low-energy metal-to-ligand charge transfer state*. Because of this ability to stabilize the excited state, many other successful Ru dye sensitizers have been designed, including the broadly successful "black dye" (Figure 7.70). As Figure 7.71 illustrates, these dyes are much more effective at absorbing light through the visible region into the near IR than the wide bandgap material TiO_2, leading to the high efficiencies characteristic of the DSSC.

Ruthenium is one of the Earth's rare metals (Section 1.2.3), so use of the Ru dyes adds to the cost of the solar cell and its extensive use is not sustainable. Other metal complexes have been explored as alternatives (osmium, rhenium, iron, platinum, and copper) and research into alternatives proceeds apace (Hagfeldt et al. 2010). Porphyrins and phthalocyanines are good candidates because of their overall stability and excellent absorption in the near-IR region (>800 nm). More inexpensive and abundant metals like zinc have also yielded promising results as sensitizers. The zinc phthalocyanine in Figure 7.72, for example, showed an IPCE of 45% in the near-IR (Nazeeruddin et al. 1999).

Another alternative to rare, toxic, or expensive metals is to avoid them altogether, and the use of metal-free dyes is increasing. These organic dyes offer immense structural diversity subject to design and synthesis, leading to:

- Tunable HOMO–LUMO levels
- Very high molar absorptivity
- Broad absorption spectrum into the far red and near-IR portion of the electromagnetic spectrum
- Low cost and low toxicity
- Sustainability

There are several classes of metal-free dyes and, not suprisingly, several share structural features with the conducting polymer donor materials used in OPVs (triarylamines, phenylvinylenes, fluorenes, thiophenes, and pyrrole derivatives). A representative example from a few of the more common classes of metal-free dyes is given in Table 7.2. Most possess a general donor–π system–acceptor framework, injecting the electron into the semiconductor through the acceptor group. Natural dyes (e.g., porphyrins and anthocyanin-based flavonoids (Figure 7.73) have also been explored. For example, a solar cell capable of achieving a J_{sc} of 1–2 mA/cm² (in bright sunlight) can be constructed from the natural anthocyanin dyes found in a variety of berries (Smestad and Grätzel 1998; Wang and Kitao 2012). Extraction of natural dyes sidesteps the potential agony of synthesis, but also

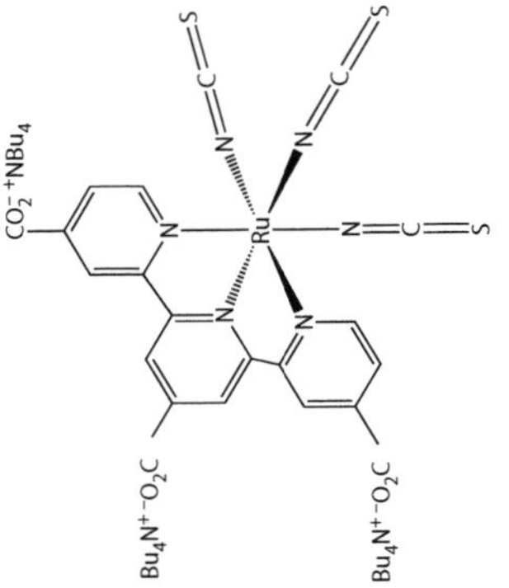

FIGURE 7.70 Other ruthenium dye sensitizers.

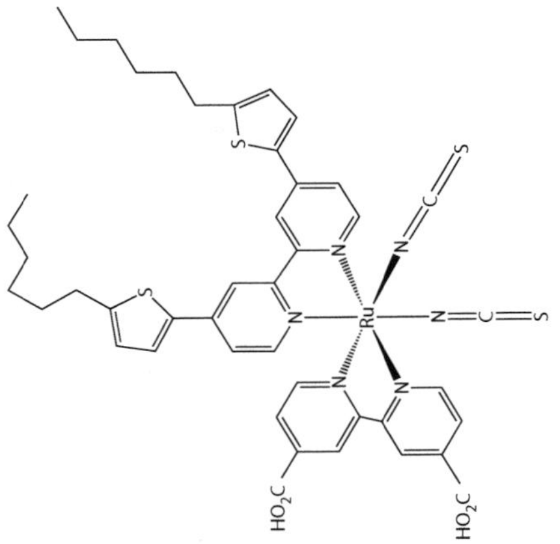

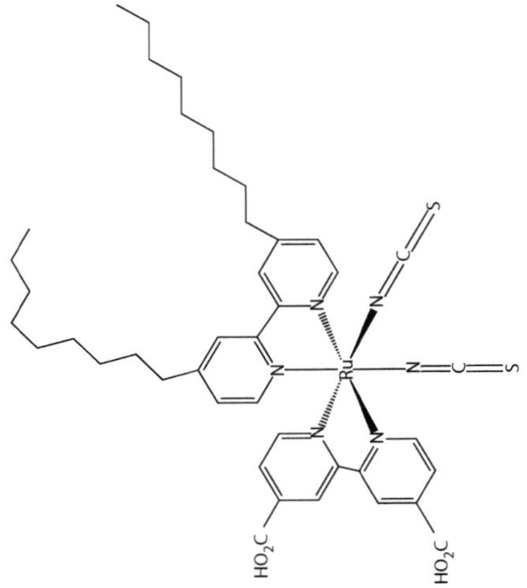

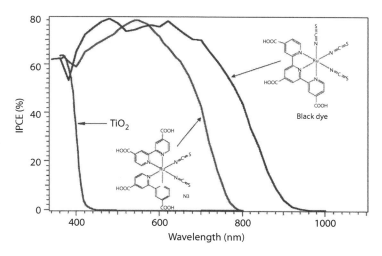

FIGURE 7.71 The incident photon-to-current efficiencies for titania, N3, and black dye. (Reprinted with permission from Grätzel, M. 2009. Recent advances in sensitized mesoscopic solar cells. *Acc. Chem. Res.* 42 (11):1788–1798. Copyright 2009 American Chemical Society.)

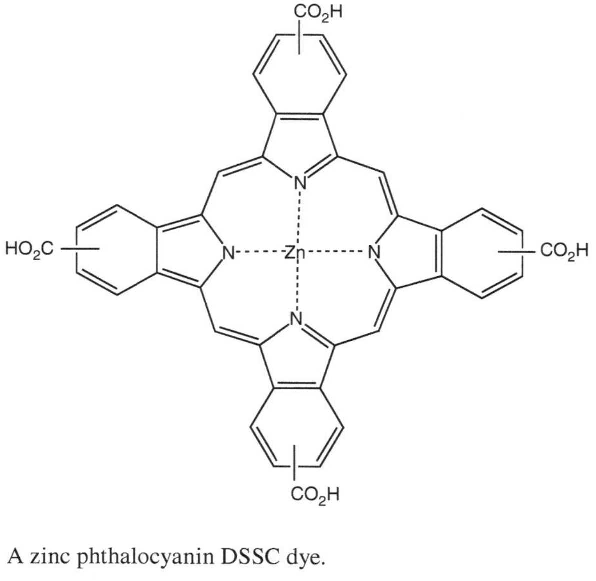

FIGURE 7.72 A zinc phthalocyanin DSSC dye.

removes the option of structural design to fine-tune the HOMO–LUMO gap. The key strengths of these highly conjugated metal-free dyes is their increased molar extinction coefficient and the broadened absorbance into the near IR. As a result, the short-circuit currents for metal-free dyes are usually very good. However, the V_{oc}s are generally quite low, with overall as yet uninspiring PCEs.

TABLE 7.2
Organic Dye Sensitizers

Classification	Structure	Comments	Reference
Triphenylamine	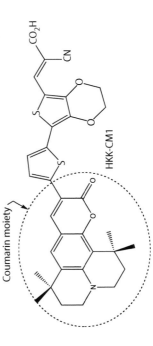 SD-6	Triphenylamines are resistant to aggregation, show good stability, and their propeller-shaped structure minimizes recombination. A DSSC with this dye typically gives an IPCE of 73% and PCE of ≈7%. SD-6 contains the R_3N and a fluorene donor with a cyanoacrylate acceptor and thiophene π-linker (a donor–donor–π-acceptor architecture). By having two donors, the electron density is increased along with an improved molar extinction coefficient.	Shen et al. (2011)
Coumarin	Coumarin moiety HKK-CM1	Coumarin donor with cyanoacrylate acceptor. Excellent activity in the visible region, good stability, and LUMO level matches well with CB of TiO_2, however, HKK-M1 showed only a modest IPCE of ≈6%.	Seo et al. (2011)

Squarine

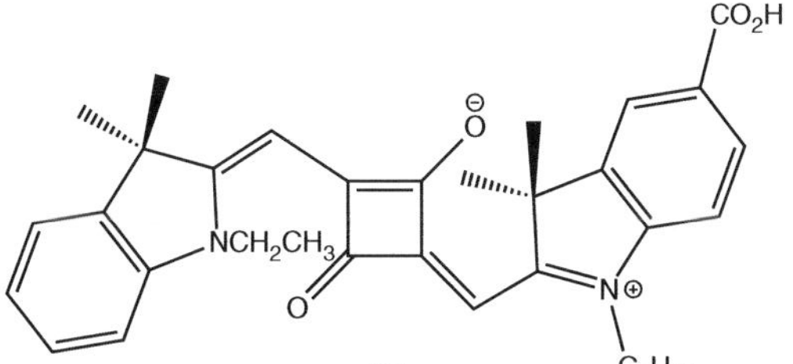

Extremely high molar absorption coefficients from green to near-IR region of solar spectrum. Susceptible to aggregation and poor stability. S7's dissymmetry is a "major breakthrough" that led to a PCE of ≤4.5% and an IPCE of 85%.

Yum et al. (2007)

S7

Carbazole

Carbazoles possess a wide bandgap, good emission, and adsorption properties. MK1 was among the first donor–π–linker–acceptor organic dyes. Addition of the alkyl chains suppresses aggregation and increases the electron lifetime, leading to higher V_{oc}s.

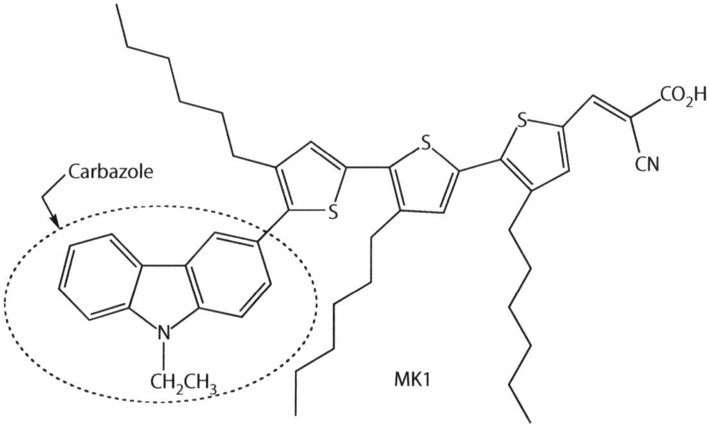

Koumura et al. (2006)

MK1

Carbazole

continued

TABLE 7.2 (continued)
Organic Dye Sensitizers

Classification	Structure	Comments	Reference
Perylene	PK-0002	Superior stability and excellent molar extinction coefficient, but PCEs are low (<4%). PK-0002 has HOMO–LUMO energy levels of −5.31 and −2.88 eV, respectively, making it a good match for a TiO_2 DSSC and it showed absorption into the near-IR (≈800 nm), but PCE was 3.1%.	Otsuki et al. (2011)
Anthracene		By addition of phenylethylene units to the anthracene core, the molar extinction coefficient of this dye was increased. The dimethyloctyl groups prevent dye aggregation and a DSSC based on this dye gave a PCE of 4.1% with a fill factor of 0.71.	Yang et al. (2011)

Source: Adapted from Kanaparthi, R. K., J. Kandhadi, and L. Giribabu. 2012. *Tetrahedron* 68 (40):8383–8393; Koumura, N., Z.-S. Wang, S. Mori et al. 2006. *J. Am. Chem. Soc.* 128 (44):14256–14257; Otsuki, J., Y. Takaguchi, D. Takahashi et al. 2011. *Adv. Optoelectron.* 2011: 7pp. Article I.D. 860486; Seo, K.D., H.M. Song, M. J. Lee et al. 2011. *Dyes Pigm.* 90 (3):304–310; Shen, P., Y. Tang, S. Jiang et al. 2011.*Org. Electron.* 12 (1):125–135; Yang, X., J-K. Fang, Y. Suzuma et al. 2011. *Chem. Lett.* 40 (6):620–622; Yum, J-H., P. Walter, S. Huber et al. 2007.*J. Am. Chem. Soc.* 129 (34):10320–10321.

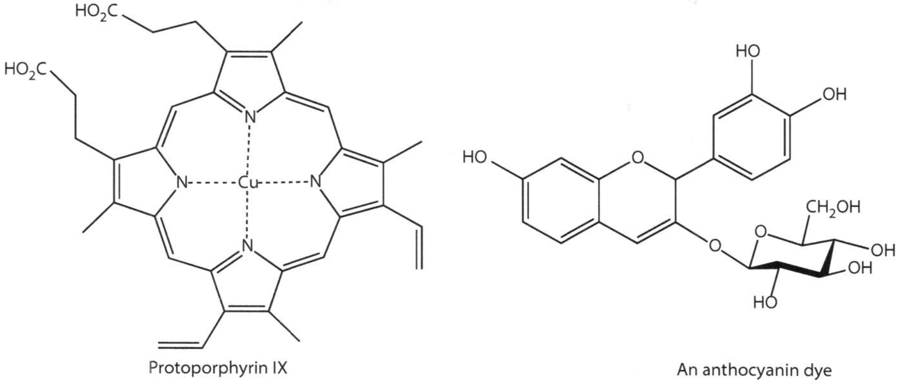

Protoporphyrin IX   An anthocyanin dye

FIGURE 7.73   Some natural dyes used as sensitizers in DSSCs.

7.5.4.3   Redox Mediator

If you have read this far, the complexity of the DSSC should be quite evident. The demands on the materials involved are electronic, chemical, and mechanical. The requirements for the redox mediator (frequently used synonymously with *electrolyte*) are equally strict. Of course, it must reduce the oxidized dye meaning the redox potentials must be appropriately balanced. It, too, must be photochemically and thermally stable. If liquid, it must be of fairly low viscosity to minimize transport problems and the solvent in which it is dissolved must not dissolve (or react with) the dye or, for that matter, the metal oxide! In addition, these materials must be compatible with whatever compound is used to seal the cell, lest the solvent evaporate away or leak out. Given these stringent requirements, there is little variation in the redox mediator: the iodide/triiodide couple is virtually unsurpassed for the DSSC. An average redox potential of I_3^-/I^- in acetonitrile has been reported to be about 0.3 V (vs. NHE), so that in a DSSC with N3 or N719, the potential difference between the dye and I_3^-/I^- is approximately 0.75 V, driving the redox reactions in the cell (Boschloo and Hagfeldt 2009).

Preparation of the redox mediator for a DSSC is carried out by following a protocol that includes iodine-containing compounds, a polar organic solvent, and various additives (see, e.g., Figure 7.74). The iodide species are typically at a fairly high

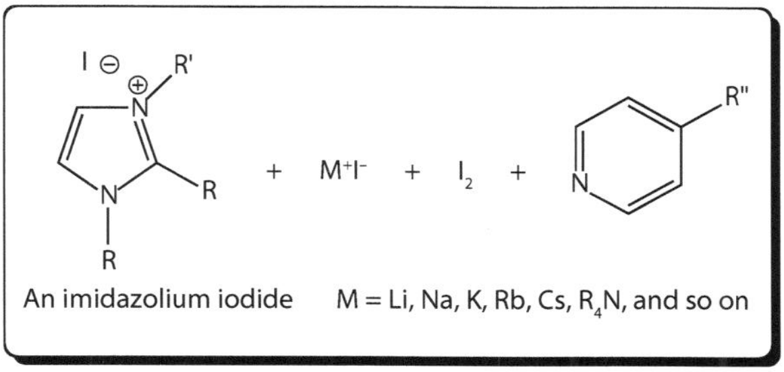

An imidazolium iodide   M = Li, Na, K, Rb, Cs, R_4N, and so on

FIGURE 7.74   A DSSC redox "recipe."

concentration (ca. 0.1–0.7 M in iodide and 10–200 mM iodine) (Listorti et al. 2011), with the pertinent overall reaction being

$$I_2 + I^- = I_3^-$$ (7.14)

Solubility of the I_3^-/I^- pair is an obvious concern and a wide variety of organic solvents have been tried including tetrahydrofuran, various alcohols, dimethylformamide, and dimethylsulfoxide. One of the most commonly used solvents is 3-methoxyproprionitrile ($CH_3OCH_2CH_2CN$) (Hagfeldt et al. 2010), but alternatives to the organic solvent are being increasingly explored. Room-temperature ionic liquids, mostly imidazolium-based, have been tested, but the increased viscosity leads to poor mass transport of the triiodide species and overall poorer performance of the cell (Yu et al. 2010a). A "quasi solid-state" electrolyte approach adds an inorganic additive to the electrolyte to increase the viscosity without sacrificing the good diffusion and transport properties of a liquid. For example, a Mg–Al nanoclay infused with nitrate ions was used with liquid I_3^-/I^- electrolyte in a TiO_2/N719 DSSC with the resultant cell showing a 9.6% PCE, a 10% improvement over the corresponding liquid electrolyte (Wang et al. 2012a). The use of polyvinyl acetate to create a polymer gel matrix for the I_3^-/I^- electrolyte was recently explored. A DSSC cell fabricated with this gel electrolyte achieved a PCE of 8.27% (comparable to that of a liquid electrolyte) and showed much improved long-term stability compared to a liquid electrolyte (Wang et al. 2012c).

While the I_3^-/I^- couple is nearly ubiquitous, iodine is corrosive and absorbs in the visible spectrum thus reducing the IPCE. Replacement of the I_3^-/I^- electrolyte continues to be a goal and alternative redox couples have been studied including Co^{2+}/Co^{3+}, SCN^-/$(SCN)^{3-}$ and ferrocene/ferrocenium Fe^{2+}/Fe^{3+} (Tian and Sun 2011). Solid-state DSSCs containing organic or inorganic hole conductors are another attractive option, where the diffusion of redox molecules is replaced with the hopping mechanism of holes. Conducting polymers are examples of organic hole conductors; inorganic semiconductors such as copper(I) iodide or copper thiocyanate (CuSCN) have also been studied as hole conductors in DSSC. A DSSC using cesium tin iodide ($CsSnI_3$) doped with fluoride (TiO_2/N719) gave a PCE of up to 10.2% (Chung et al. 2012). The problem with solid hole conductors is that the dye must not only coat the mesoporous TiO_2, it must also effectively infiltrate the pores to leave few gaps in the hole-conducting path (Hagfeldt et al. 2010). Overall, the progress in optimizing redox mediators to address the limitations of the I_3^-/I^- in a liquid electrolyte has been encouraging but it is in its initial stages.

7.5.4.3.1 Mechanism

The redox chemistry of the iodide/iodine system is quite complex, consisting of at least seven individual redox reactions with corresponding electrochemical potentials, thus a detailed discussion is beyond the scope of this chapter. A simplified mechanistic pathway for the iodide/triiodide-mediated redox chemistry in the DSSC is shown in Equations 7.15 through 7.18, where S stands for the dye sensitizer.

$$S^* \rightarrow S^+ + e^- \text{ (electron injection to } TiO_2)$$ (7.15)

$$S^+ + I^- \rightarrow (S \cdots I) \text{ (one electron transfer to oxidized dye)}$$ (7.16)

$$(S \cdots I) + I^- \rightarrow S + I_2^{-\bullet} \tag{7.17}$$

$$2I_2^{-\bullet} \rightarrow I_3^- + I^- \tag{7.18}$$

The overall reduction of the dye proceeds via production of an oxidized dye–iodide complex, $S \cdots I$ (Equation 7.16) which breaks down in the presence of iodide to form an iodine radical anion (Equation 7.17). The radical anions undergo a disproportionation reaction to produce triiodide and iodide to complete the cycle (Equation 7.18). By making several assumptions, it can be derived that the concentration of the radical anion is about two orders of magnitude higher than that of iodine in a typical acetonitrile-based electrolyte system (Boschloo and Hagfeldt 2009).

High-efficiency dye-sensitized solar cells burst into the PV scene in 1991 with a combination of materials that has been hard to improve upon in terms of performance. As usual, issues of sustainability—ruthenium for the dye sensitizer or palladium for the counter electrode—continue to provide room for improvement, and the number of research publications on DSSCs has grown at an exponential pace. Despite their current limitations in terms of commercial viability, the results have been encouraging and the future looks very bright for DSSC.

7.6   QUANTUM DOT SOLAR CELLS

7.6.1   INTRODUCTION

As we have seen, a big shortfall in being able to harvest all of the energy provided by the Sun is the failure to match energies: solar cell materials are limited by their bandgaps and cannot take advantage of the full solar spectrum. Dye-sensitized solar cells were developed, in part, to address this shortcoming by inserting a tunable absorber with a high molar absorptivity into the device. In this section, we introduce another sensitizer that has been shown to improve the efficiency of light harvesting: the *quantum dot* (QD). In essence, QDs replace the dye in a DSSC and have an even higher absorptivity than conventional dyes (Toyoda and Shen 2012). QDs used in PV devices are inorganic semiconductor nanocrystals with diameters in the range of 2–10 nm, meaning that each QD contains only hundreds to thousands of atoms. QDs have been used in PV devices since the early 1990s and there are many variations, for example, solid-state heterojunction solar cells, depleted heterojunction cells, and hybrid polymer solar cells. Our focus is strictly on quantum-dot *sensitized* solar cells (QDSSC).

Like dyes, QDs are tunable. An electron confined within a QD must have a wavelength equal to the QD circumference (or a whole number fraction thereof). By changing the size of the QD, one can tune the wavelength of the absorption and emission spectra of the QD and therefore the bandgap for the solar cell, just as different dyes are synthesized to match the solar spectrum (see Figure 7.75) (National Renewable Energy Laboratory June 2010). Compared to organic or inorganic coordination compound sensitizers, however, inorganic QDs are very stable and hold the potential to address the durability issues that dog dye-sensitized solar cells.

Another especially attractive feature of QDSSC is that they are able to take advantage of surplus energy—energy that is otherwise wasted in those photons with

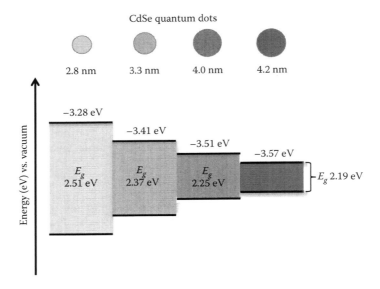

FIGURE 7.75 (**See color insert.**) The relationship between QD size and bandgap. (Adapted with permission from Kamat, P. V. Boosting the efficiency of quantum dot sensitized solar cells through modulation of interfacial charge transfer. *Acc. Chem. Res.* 45 (11):1906–1915. Copyright 2012, American Chemical Society.)

energy $>E_g$. As noted above, the excess kinetic energy for these photons is usually lost nearly instantaneously as heat—temperatures as high at 3000 K can exist in a lattice at 300 K (Nozik 2002)! But what if that excess energy could be captured and used? In that case, the theoretical solar-to-electric power efficiency could reach as high as 66% (Ross and Nozik 1982)! Capture of this excess energy generates *hot carriers* (hot electrons and hot holes), and transfer of this energy to electrons and holes nearby can produce additional electron–hole pairs, aka *electron–hole pair multiplication* (EHPM) [synonymous with *carrier multiplication* (CM) or *multiple exciton generation* (MEG)]. Figure 7.76 depicts this process graphically, where the hot carrier is designated as n_1^*. Transfer of the excess energy by EHPM generates an additional hot carrier (n_2^*) which can in turn generate yet another hot carrier, and so on. Electron–hole pair multiplication must take place quickly—within a picosecond or two—before cooling occurs as indicated by k_{cool} in Figure 7.76. This is where QDs enter in. Because QDs are unfathomably small, confined, and three-dimensional spaces, the hot electrons and hot holes generated therein cool at reduced rates giving enough time for the EHPM process. The *quantum yield* (QY) is defined as the number of electron–hole pairs (excitons) that are generated per absorbed photon. The use of QDs and their propensity for EHPM means that the quantum yields can be much greater than 1.

Given the possibility of enhanced electron–hole pair multiplication, QD PVs hold great promise for improved efficiencies, but they have yet to best DSSC performance. In a recent example, a colloidal QD device prepared with lead sulfide QDs as the p-type absorber atop n-TiO$_2$ donor material on a fluorine-doped tin oxide (FTO)

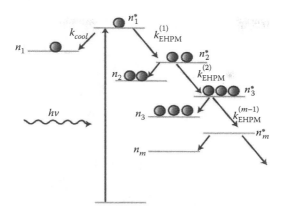

FIGURE 7.76   The electron–hole pair multiplication process. (Reprinted with permission from Beard, M.C., A.G. Midgett, M.C. Hanna et al. 2010. Comparing multiple exciton generation in quantum dots to impact ionization in bulk semiconductors: Implications for enhancement of solar energy conversion. *Nano Lett.* 10(8): 3019–3027. Copyright 2010 American Chemical Society.)

transparent conducting electrode gave a PCE of 8.5%. This much improved efficiency (for QD solar cells, see Figure 7.11) was attributed to a new design concept: a *donor supply electrode* prepared by layering a very thin film of TiO_2 onto a relatively thick layer of the FTO. With this configuration the depletion zone in the PV was increased and as a result more photocarriers could be extracted leading to improved efficiency (Maraghechi et al. 2013). Despite this finding, the conversion efficiencies for QD solar cells currently lag behind that of DSSC, although advancements will undoubtedly continue to be seen.

7.6.2   ARCHITECTURE AND MATERIALS

As one might expect, the architecture of the QD-sensitized solar cell is not unlike that of a dye-sensitized solar cell. The anode is typically a TCO-coated glass substrate upon which a mesoporous layer of a metal oxide is affixed. The key difference is, of course, that it is now QDs—not dye molecules—that are layered atop of the semiconductor metal-oxide particles, as illustrated in Figure 7.77. Owing to the requirements of the QD materials, the redox mediator is no longer I^-/I_3^- and the electrode is not typically platinum. More detail regarding each of these materials follows.

7.6.2.1   Semiconductor

Just as for the DSSC (and for the same reasons), the morphology of the semiconductor material is critical to the performance of the QDSSC. Zinc oxide and titanium oxide are the most often used, and, like DSSC, various morphologies have been explored from the standard mesoporous film to nanotubes, nanowires, and the inverse opal structure. It is the makeup and synthesis of the QD sensitizers, however, that we will examine in more detail.

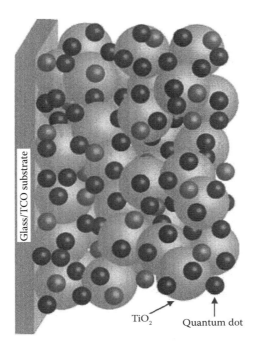

Glass/TCO substrate

TiO$_2$ Quantum dot

FIGURE 7.77   Depiction of the anode–semiconductor–QD layers in a QDSSC. (Adapted with permission from Santra, P.K. and P.V. Kamat. 2012. Tandem-layered quantum dot solar cells: Tuning the photovoltaic response with luminescent ternary cadmium chalcogenides. *J. Am. Chem. Soc.* 135 (2):877–885. Copyright 2012 American Chemical Society.)

7.6.2.2   Quantum Dots

The materials typically used for QDs in QDSSC are the metal chalcogenides, just as for thin-film inorganic solar cells (Section 7.3.2). While cadmium sulfide (CdS) is probably the most widely studied QD for PV applications, cadmium selenide (CdSe), antimony sulfide (Sb$_2$S$_3$), indium phosphide (InP), indium arsenide (InAs), lead sulfide (PbS), and lead selenide (PbSe) are among the others that have been used. There are two primary methods for connecting the dots to the metal oxide: (a) by direct growth of the QDs onto the oxide surface or (b) by synthesis of the QDs first, then linking them to the oxide surface. As with dye sensitization, the amount of coverage, quality of the attachment, and overall morphology strongly impact the cell performance since it directly effects recombination and charge separation and transport. Good coverage with good interfacial connectivity results in a high J_{sc} by reducing recombination of the injected electrons from TiO$_2$ to the electrolyte. However, a thick layer of QDs makes electron transport more difficult and the porosity of the QD layer is diminished, limiting the contact between the QD sensitizers and the redox electrolyte and therefore diminishing the solar cell performance (Zhu et al. 2011).

Direct growth. The first method for preparing QDs—direct growth on the metal-oxide electrode—can take place by *either chemical bath deposition* (CBD) or by *successive ionic layer adsorption and reaction* (SILAR). In the CBD method, the appropriate precursors are dissolved together and the metal-oxide electrode immersed into this solution. For example, CdS QDs were grown on a TiO_2 electrode by mixing together cadmium nitrate $[Cd(NO_3)_2]$ and thiourea $[NH_2C(S)NH_2]$ in water, placing the electrode into the bath, then treating the assembly to microwave irradiation whereupon the individual reagents presumably reacted on the TiO_2 surface to make the CdS QDs (Zhu et al. 2011). X-ray diffraction was used to confirm the desired outcome; the resultant XRD pattern (Figure 7.78) shows the expected peaks for the cubic phase of cadmium and sulfur. The optimum concentration of the bath components was found to be 0.5 M as the cell fabricated from this concentration gave the best conversion efficiency (albeit only 1.8%).

The SILAR method of QD deposition takes place by cycling the electrode repeatedly into separate solutions of the cationic and anionic precursors. SILAR is considered the better method for deposition of QDs onto mesoporous metal oxides (Emin et al. 2011) and can be illustrated by the work of Guijarro et al. Cadmium acetate (0.5 M, aqueous) was used as the Cd^{2+} source and sodium selenosulfate (Na_2SeSO_3, 1 M aqueous) was the anionic precursor. As the transmission electron microscopy images in Figure 7.79 illustrate, after twenty successive immersions of the TiO_2 electrode in the cadmium and selenium solutions discrete CdS particles ($\approx$10 nm in size) are forming on the TiO_2 surface (Guijarro et al. 2010).

The linker method. Attaching presynthesized nanocrystalline QDs to the metal-oxide surface has also been used extensively, allowing more control of the diameter and shape of the QDs—an important consideration given that the size tunability of QDs is what makes their use in PVs attractive in the first place. However, the performance of these QDSSCs is generally lower, likely due to a lower loading of QDs on the electrode surface (Margraf et al. 2013). One of the most commonly used class of linkers is a mercapto acid, for example, thioglycolic acid ($HSCH_2CO_2H$) or

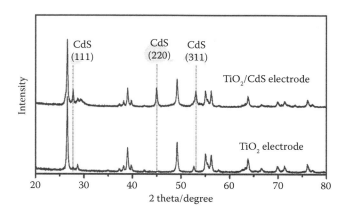

FIGURE 7.78 X-ray diffraction pattern confirming the presence of the desired phases of CdS quantum dots. (Reprinted with permission from Zhu, G., L. Pan, T. Xu et al. 2011. One-step synthesis of CdS sensitized TiO_2 photoanodes for quantum dot-sensitized solar cells by microwave assisted chemical bath deposition method. *ACS Appl. Mater. Interfaces* 3 (5):1472–1478. Copyright 2011 American Chemical Society.)

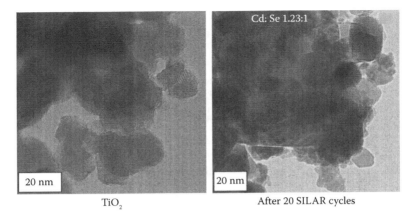

FIGURE 7.79 TEM images for TiO$_2$/CdSe QDs prepared by the SILAR method. (Adapted with permission from Guijarro, N., T. Lana-Villarreal, Q. Shen et al. 2010. Sensitization of titanium dioxide photoanodes with cadmium selenide quantum dots prepared by Silar: Photoelectrochemical and carrier dynamics studies. *J. Phys. Chem. C* 114 (50):21928–21937. Copyright 2010 American Chemical Society.)

mercaptopropionic acid (HSCH$_2$CH$_2$CO$_2$H). Cysteine [HSCH$_2$CH(NH$_2$)CO$_2$H], too, has been found to be an effective linker, particularly with respect to efficiency of charge injection to the TiO$_2$ semiconductor that must proceed indirectly through the linker. A cartoon illustrating the connection of CdSe QDs linked to a TiO$_2$ surface is shown in Figure 7.80. The TiO$_2$ electrode surface is first modified by immersion deposition of the linkers followed by immersion of the QD solution. With any of these methods, careful optimization of the experimental variables (concentration of solutions, time and temperature of immersion, even pH) is important to obtain the best thickness of QD layer, connection between the QD and TiO$_2$, and minimization of aggregation. Needless to say, it is a work in progress.

7.6.2.3 Redox Mediator and Electrode Materials

While the iodide/triiodide couple is just about ideal for the DSSC, the reactivity of this pair with metal chalcogenides (particularly CdS) make it an unacceptable choice for the QDSSC. Although cobalt(II/III) and ferrocene/ferrocenium (Fe^{2+}/Fe^{3+}) couples have been used in QDSSC, the polysulfide electrolyte S^{2-}/S$_x^{2-}$ is the most commonly chosen. In aqueous media, sodium sulfide exists in equilibrium with the thiolate anion (HS$^-$; Equation 7.19)

$$S^{2-} + H_2O \rightleftarrows HS^- + HO^- \tag{7.19}$$

Although the equilibrium lies to the right for this acid–base reaction, the sulfide ion is the active reducing agent (Chakrapani et al. 2011).

Further, because of the use of sulfide in the electrolyte, platinum is an incompatible material for the counter electrode in a QDSSC. Carbon, gold, copper sulfide,

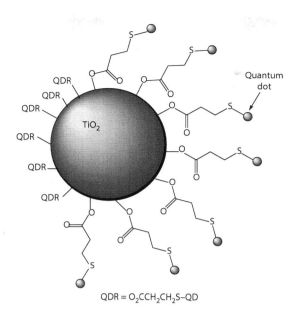

QDR = O₂CCH₂CH₂S–QD

FIGURE 7.80 A depiction of the linker method for attaching QDs to TiO₂ particles; relative sizes are *not* to scale.

and even conducting polymers have been used in this role. For example, researchers found that a porous matrix of PEDOT (see Figure 7.48) was "very suitable" for use with the polysulfide electrolyte and performed slightly better than a gold-based electrode (Yeh et al. 2011). Similarly, 2-μm microspheres made up of copper, zinc, tin, and sulfur (Cu_2ZnSnS_4) were prepared and used effectively as the counter electrode in a QDSSC with the polysulfide electrolyte, showing improvement in both J_{sc} and FF (Xu et al. 2012).

7.6.3 MECHANISM

The mechanism for a QDSSC is, as expected, quite similar to that for a DSSC. Figure 7.81 graphically summarizes both the architecture and the conduction process. Several postulated steps in the mechanism are given in Equations 7.20 through 7.22, where "h" stands for hole and with CdS as the QD sensitizer, TiO₂ as the metal-oxide semiconductor, and polysulfide as the electrolyte (Emin et al. 2011).

$$CdS + h\nu \rightarrow CdS(h + e)(\text{exciton generation}) \qquad (7.20)$$

$$CdS(h + e) + TiO_2 \rightarrow CdS(h) + TiO_2(e)(\text{charge separation}) \qquad (7.21)$$

$$CdS(h) + S^{2-} \rightarrow CdS + S^{-\bullet} \text{ (hole scavenging by redox mediator)} \qquad (7.22)$$

Incident photons striking the device generate an exciton that undergoes dissociation at the QD/metal oxide (MO_x) interface (Equation 7.20). Electron injection into the conduction band of MO_x results in charge separation and the electron forging its way to the anode (Equation 7.21). The QD is oxidized and the holes are scavenged by the redox mediator, reducing it (Equation 7.22) to give a radical anion, $S^{-\bullet}$, which in turn complexes with S^{2-} to make a polysulfide radical anion. Reduction of the QD to its ground state takes place by the sulfide anion (S^-), and the oxidized species S_x^{2-} is reduced at the cathode to complete the electrochemical cycle, as indicated by the dashed double-headed arrow in Figure 7.81.

As noted above, QDSSCs are not as yet very efficient PV devices. This poor performance is attributed to a multitude of recombination processes including some based on "surface states" of the QD (for our purposes, a surface state is where the material ends with dangling bonds at a surface, a defect much like a grain boundary in multicrystalline silicon (Section 7.3.1.2). The various routes of recombination are shown in Figure 7.82. Pathways (1) and (2) illustrate recombination of an electron with a hole via the surface state, where (1) is recombination of the already injected electron in the TiO_2 with a trapped hole in the QD and (2) is recombination within the QD. It is a short hop for these surface-trapped electrons to the electrolyte (3). Alternatively, prior to injection into the TiO_2, the separated electron in the QD (4) or the TiO_2 (5) can be captured by the electrolyte (Shalom et al. 2009). Back electron transfer from the injected electron in TiO_2 (pathway 5) has been shown to occur to $S^{-\bullet}$ much more rapidly than the analogous process in DSSC with the I^-/I^{3-} couple, making this a major hurdle to overcome for improved QDSSC performance (Chakrapani et al. 2011). Some improvement can be obtained in limiting these paths of recombination by passivation of the surface states. We have already seen one example of improving solar cell performance by passivating these dangling bonds, that being

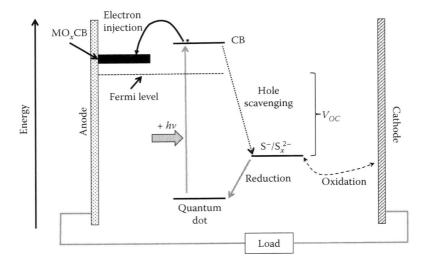

FIGURE 7.81 The photoconduction process in a QD-sensitized solar cell.

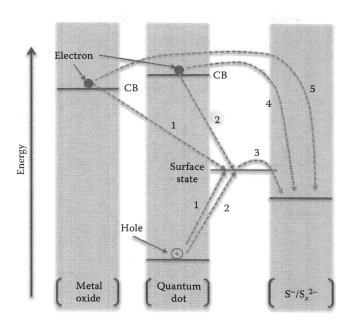

FIGURE 7.82 Decay processes in a QD-sensitized solar cell. (Adapted with permission from Shalom, M., S. Dor, S. Rühle et al. 2009. Core/CdS quantum dot/shell mesoporous solar cells with improved stability and efficiency using an amorphous TiO_2 coating. *J. Phys. Chem. C* 113 (9):3895–3898. Copyright 2009 American Chemical Society.)

a-Si:H (Section 7.3.1.2). In QDs, surface states can be modified by coating the QD with another material such as ZnS, TiO_2, or even organic molecules (Emin et al. 2011). In so doing, this coating serves as a block between the QD and the electrolyte, minimizing pathways (4) and (5) and thus improving the performance of the cell.

7.7 SUSTAINABILITY, PHOTOVOLTAICS, AND THE CZTS CELL

Solar is one of the fastest growing areas of renewable energy, but PV technology is very much limited by *resource availability*. While silicon is one of the most abundant materials on the Earth, silver—used for the electrodes in c-Si cells—will eventually restrict their implementation (Feltrin and Freundlich 2008). In addition, several rare elements are critical components in photovoltaics (recall Figure 1.8 in Chapter 1). Gallium and selenium limit the installation of CIGS/Se solar cells—gallium and selenium are both past their production peak, and the demand for gallium is projected to outpace production within years (Knowledge Transfer Network 2010). Indium, too, is a scarce element and indium tin oxide is used in a very wide variety of PV devices. Germanium, cadmium, and tellurium also pose resource limitation issues. A hypothetical situation in which CdTe cells supply 25 TW would require over 100 times more cadmium than available in the current world reserves (Peter 2011). These constraints in material availability are one of the major driving forces behind thinner layer technology and increasing efficiencies, but there are limits to each (efficiency is *literally* limited by the Shockley–Queisser limit). The alternative is to (a) use less, (b) use material science and engineering concepts to utilize these materials more effectively (e.g., nanotechnology), or (c) find replacements for resource-scarce materials. It is the latter approach that has led to the development of the "earth abundant" photovoltaics based on $Cu_2Zn(Sn_{1-x}Ge_x)S,Se$ (CZTS or CZTSe).

CZTS and derivatives are members of the *kesterite* crystal family where kesterite refers to a relatively abundant mineral of formula $Cu_2Zn_{0.75}Fe^{2+}_{0.25}Sn_{1.3}S_4$. In effect, the CZTS/Se cells replace the rare elements of indium and gallium with their neighbors zinc and tin on the periodic table. Like the chalcogenides covered in Section 7.3.2.2, the kesterites have the same basic copper/sulfur structure, but the M(III) ions have been replaced with an equal number of M(II) and M(IV) atoms (Figure 7.83). The CZTS/Se materials are p-type absorbers in which modification of the ratio of S to Se allows for tuning the band gap, as can be gleaned from comparing the values for CZTS (1.45 eV) and CZTSe (0.94 eV; Table 7.1). Band gaps in the range of 1–1.5 eV can be attained (Mitzi et al. 2011). Recently, CZTS or CZTSe cells substituted with germanium in varying ratios (replacing tin) have been investigated to further tune the band gap and address device limitations. A 40% Ge-doped CZTSe ($Cu_{1.5}ZnSn_{0.5}Ge_{0.4}Se_4$) shows a shift to a slightly larger band gap (1.15 eV vs. 1.08 eV for pure CZTSe) but with essentially no improvement in PCE (the Ge-substituted cell gives a PCE of 9.1% compared to the pure CZTSe cell's 9.07%) (Bag et al. 2012).

Research effort into the use of CZTS/Se materials in solar cells is very active. A comparison of a variety of CZTSSe cells with a CIGSSe standard (with a PCE of 13.8%) was carried out to begin to understand the limitations of CZTSSe cell efficiencies. The devices were fabricated on a molybdenum-coated glass substrate followed by the absorber material (either CZTSSe or CIGSSe), then CdS/ZnO/ITO

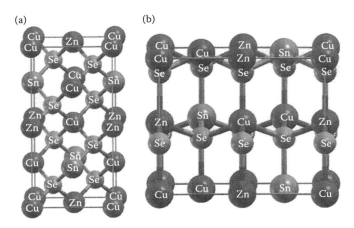

FIGURE 7.83 The unit cells of (a) kesterite and (b) CZTSSe. (Reprinted with permission from Fan, F.-J., L. Wu, M. Gong et al. 2013. Composition- and band-gap-tunable synthesis of Wurtzite-derived $Cu_2ZnSn(S_{1-x}Se_x)_4$ nanocrystals: Theoretical and experimental insights. *ACS Nano* 7 (2):1454–1463. Copyright 2013 American Chemical Society.)

with a Ni-Al contact grid and MgF_2 antireflective coating. Three major findings were reported:

- V_{OC}. The CZTS/Se cells showed low open-circuit voltages relative to their band gap. This was attributed to significant recombination at the buffer/ absorber interface.
- *Fill factor*. Compared to the CIGSSe standard, the fill factors of the CZTSSe cells were low, presumably a result of high series resistance. This may be due to a thick $MoSe_2$ layer at the back contact in the hydrazine-processed cells.
- *External quantum efficiency*. The CZTSSe cells showed a poor response at long wavelengths, possibly because of a very low carrier lifetime (which would also result in the low V_{OC}s and could be related to the high recombination at the interface) (Mitzi et al. 2011; Todorov et al. 2010).

Because the research and development are in the early stages, these modest results are actually very promising. One major hurdle with respect to their further improvement, however, lies in the development of better methods of preparing the absorber material as a pure, single-phase thin film. The complexity of the kesterite material leads to challenges in phase segregation and control of the elemental composition which naturally impacts device performance. Both vacuum and nonvacuum techniques for film preparation have been explored and the nonvacuum technique of wet chemical deposition using hydrazine appears to be the most successful (although use of hydrazine presents its own issues in terms of toxicity) (Bag et al. 2012). Naturally, various other methods are under development, including synthesis of CZTS nanocrystals followed by sintering with selenium vapor (Ford et al. 2011) or electrodeposition (Peter 2011). Ultimately an ink-based approach—where the precursor material could be rolled onto the substrate—is a goal for large-scale production of these and many other PV devices (Mitzi et al. 2011).

7.8 CONCLUSIONS

Beyond resource availability, the sustainability of photovoltaics requires careful assessment of the cradle-to-grave cycle at the end of the PV life cycle: what do we do with expended PVs? Recycling is a highly desirable option because of the limited supply of many PV materials. In the case of CdTe solar cells, both cadmium and tellurium are toxic so that reclaiming these elements from CdTe solar cells is imperative. But recovery is complicated and costly. Land use, too, is a piece of the PV sustainability puzzle: given the generally low efficiency of photovoltaics, a huge amount of land area is required to generate large amounts of usable solar energy. Obviously, a candid focus on sustainability in all of its aspects is required as we contemplate the use of photovoltaics to help meet our growing energy appetite.

Innovations to increase the efficiency of solar PV devices continue at a rapid pace; this chapter has only introduced a few of the more common configurations. Tandem cells, intermediate band solar cells, inorganic–organic hybrid cells, organometallic photovoltaics, dye-sensitized QD devices—every day encouraging new findings are reported. The key concerns associated with solar energy persist, however: persistently low efficiency, high cost, and questionable sustainability. In summary,

- Given the current levels of efficiency, the contributions of solar photovoltaics to our huge energy needs are likely to be small.
- The cost of materials and processing is, of course, an important concern. The switch from crystalline silicon cells to amorphous silicon meant lower manufacturing costs, but also decreased efficiency. And cost is also related to...
- Sustainability. As noted above, limited resources, toxicity of some materials, and concerns about land use cloud the picture for solar PV technology. Nevertheless, PVs are much more environmentally friendly with respect to lifecycle air emissions per GWh than fossil fuel-based electricity (Fthenakis et al. 2008). With the current astonishing growth rate in PV production and installations, plus the astonishing amount of research effort being devoted to solar PV technology, it is certain that this area of energy generation will continue to grow.

OTHER RESOURCES

BOOKS

Brabec, C., V. Dyakonov, and U. Scherf. 2008. *Organic Photovoltaics*. Weinheim, FRG: Wiley-VCH.
Brendel, R. 2003. *Thin-Film Crystalline Silicon Solar Cells. Physics & Technology*. Weinheim, FRG: Wiley-VCH.
Fonash, S.J. 2010. *Solar Cell Device Physics*, 2nd ed. Burlington, MA: Academic Press/ Elsevier.
Kitai, A. 2011. *Principles of Solar Cells, LEDs and Diodes*. West Sussex, UK: John Wiley & Sons.
Nelson, J. 2003. *The Physics of Solar Cells*. London, UK: Imperial College Press.
Pagliaro, M., G. Palmisano, and R. Ciriminna. 2008. *Flexible Solar Cells*. Weinheim, FRG: Wiley-VCH.
Shah, A. 2012. Thin-film silicon solar cells. In *Practical Handbook of Photovoltaics. Fundamentals and Applications*, edited by A. McEvoy, T. Markvart and L. Castañer. Waltham, MA/Oxford, UK: Academic Press.

ONLINE RESOURCES

U.S. Department of Energy/Energy Efficiency and Renewable Energy/Solar: http://www.eere.energy.gov/topics/solar.html
International Energy Agency/Solar PV: http://www.iea.org/topics/solarpvandcsp/
European Photovoltaic Industry Association: http://www.epia.org/home/

REFERENCES

Alcatel·Lucent. 2013. *Bell Labs Celebrates 50th Anniversary of the Solar Cell*. Alcatel·Lucent 2004 [cited May 23, 2013]. Available from http://www3.alcatel-lucent.com/wps/portal/!ut/p/kcxml/04_Sj9SPykssy0xPLMnMz0vM0Y_QjzKLd4w3MfQFSYGYRq6mpEoYgbxjgiRIH1vfV-P_NxU_QD9gtzQiHJHR0UAAD_zXg!!/delta/base64xml/L0lJayEvUUd3QndJQSEvNElVRkNBISEvNl9BXzdNVC9lbl93dw!!?LMSG_CABINET=Bell_Labs&LMSG_CONTENT_FILE=News_Features/News_Feature_Detail_000134.xml.

Armaroli, N. and V. Balzani. 2011. *Energy for a Sustainable World*. Weinheim, FRG: Wiley-VCH.

Bag, S., O. Gunawan, T. Gokmen et al. 2012. Hydrazine-processed Ge-substituted CZTSe solar cells. *Chem. Mater.* 24 (23):4588–4593.

Beard, M.C., A.G. Midgett, M.C. Hanna et al. 2010. Comparing multiple exciton generation in quantum dots to impact ionization in bulk semiconductors: Implications for enhancement of solar energy conversion. *Nano Lett.* 10(8): 3019–3027.

Blouin, N., A. Michaud, D. Gendron et al. 2007. Toward a rational design of poly(2,7-carbazole) derivatives for solar cells. *J. Am. Chem. Soc.* 130 (2):732–742.

Bonnet, D. 2012. CdTe thin-film PV modules. In *Practical Handbook of Photovoltaics*, edited by A. McEvoy, T. Markvart, and L. Castañer. Waltham, MA/Oxford, UK: Academic Press. P. 284.

Boschloo, G. and A. Hagfeldt. 2009. Characteristics of the iodide/triiodide redox mediator in dye-sensitized solar cells. *Acc. Chem. Res.* 42 (11):1819–1826.

Cardona, C.M., W. Li, A.E. Kaifer et al. 2011. Electrochemical considerations for determining absolute frontier orbital energy levels of conjugated polymers for solar cell applications. *Adv. Mater.* 23:2367–2371.

Carsten, B., J.M. Szarko, H.J. Son et al. 2011. Examining the effect of the dipole moment on charge separation in donor–acceptor polymers for organic photovoltaic applications. *J. Am. Chem. Soc.* 133 (50):20468–20475.

Chakrapani, V., D. Baker, and P.V. Kamat. 2011. Understanding the role of the sulfide redox couple (S^{2-}/Sn^{2-}) in quantum dot-sensitized solar cells. *J. Am. Chem. Soc.* 133 (24):9607–9615.

Chen, X., B. Jia, J.K. Saha et al. 2012a. Broadband enhancement in thin-film amorphous silicon solar cells enabled by nucleated silver nanoparticles. *Nano Lett.* 12 (5):2187–2192.

Chen, D., W. Zhao, and T.P. Russell. 2012b. P3HT nanopillars for organic photovoltaic devices nanoimprinted by AAO templates. *ACS Nano* 6 (2):1479–1485.

Cheng, Y.-J., S.-H. Yang, and C.-S. Hsu. 2009. Synthesis of conjugated polymers for organic solar cell applications. *Chem. Rev.* 109 (11):5868–5923.

Cheung, N. 2013. *Solar Cells Fabrication Technologies*. U.C. Berkeley, 2010 [cited 9 Jan 2013].

Chung, I., B. Lee, J. He et al. 2012. All-solid-state dye-sensitized solar cells with high efficiency. *Nature* 485 (7399):486–489.

Cook, S., R. Katoh, and A. Furube. 2009. Ultrafast studies of charge generation in PCBM:P3HT blend films following excitation of the fullerene PCBM. *J. Phys. Chem. C* 113 (6):2547–2552.

Dang, M.T., L. Hirsch, G. Wantz et al. 2013. Controlling the morphology and performance of bulk heterojunctions in solar cells. Lessons learned from the benchmark poly(3-hexylthiophene):[6,6]-phenyl-C61-butyric acid methyl ester system. *Chem. Rev.* 113 (5):3734–3765.

Eisberg, R.M. 1961. *Fundamentals of Modern Physics*. New York: John Wiley & Sons.

Emin, S., S.P. Singh, L. Han et al. 2011. Colloidal quantum dot solar cells. *Solar Energy* 85 (6):1264–1282.

EMPA Materials Science & Technology. *A New World Record for Solar Cell Efficiency*. EMPA Materials Science and Technology, January 18, 2013. Available from http://www.empa.ch/plugin/template/empa/3/131438/—/l=2.

Ewbank, P.C., D. Laird, and R.D. McCullough. 2009. Regioregular polythiophene solar cells: Material properties and performance. In *Organic Photovoltaics*, 1–55. Weinheim, FRG: Wiley-VCH, Verlag GmbH & Co KGaA.

Fan, F.-J., L. Wu, M. Gong et al. 2013. Composition- and band-gap-tunable synthesis of Wurtzite-derived $Cu_2ZnSn(S_{1-x}Se_x)_4$ nanocrystals: Theoretical and experimental insights. *ACS Nano* 7 (2):1454–1463.

Feltrin, A. and A. Freundlich. 2008. Material considerations for terawatt level deployment of photovoltaics. *Renew. Energy* 33 (2):180–185.

Ferrazza, F. 2012. Crystalline silicon: Manufacture and properties. In *Practical Handbook of Photovoltaics. Fundamentals and Applications*, edited by A. McEvoy, T. Markvart, and L. Castañer. Waltham, MA/Oxford, UK: Academic Press.

Ford, G.M., Q. Guo, R. Agrawal et al. 2011. Earth abundant element $Cu_2Zn(Sn_{1-x}Ge_x)S_4$ nanocrystals for tunable band gap solar cells: 6.8% efficient device fabrication. *Chem. Mater.* 23 (10):2626–2629.

Fthenakis, V.M., H.C. Kim, and E. Alsema. 2008. Emissions from photovoltaic life cycles. *Environ. Sci. Technol.* 42:2168–2174.

Giridharagopal, R., and D.S. Ginger. 2010. Characterizing morphology in bulk heterojunction organic photovoltaic systems. *J. Phys. Chem. Lett.* 1 (7):1160–1169.

Gong, J., J. Liang, and K. Sumathy, Review on dye-sensitized solar cells (DSSCs): Fundamental concepts and novel materials. *Renew. Sustain. Energy Rev.*, 16, 5848–5860.

Gordon, T.R., M. Cargnello, T. Paik et al. 2012. Nonaqueous synthesis of TiO_2 nanocrystals using TiF_4 to engineer morphology, oxygen vacancy concentration, and photocatalytic activity. *J. Am. Chem. Soc.* 134 (15):6751–6761.

Grancini, G., M. Maiuri, D. Fazzi et al. 2013. Hot exciton dissociation in polymer solar cells. *Nat. Mater.* 12 (1):29–33.

Grätzel, M. 2009. Recent advances in sensitized mesoscopic solar cells. *Acc. Chem. Res.* 42 (11):1788–1798.

Green, M.A. 2012. High-efficiency silicon solar cell concepts. In *Practical Handbook of Photovoltaics. Fundamentals and Applications*, edited by A. McEvoy, T. Markvart, and L. Castañer. Waltham, MA/Oxford, UK: Academic Press.

Green, M.A., K. Emery, Y. Hishikawa et al. 2012. Solar cell efficiency tables (Version 39). *Prog. Photovolt. Res. Appl.* 20 (1):12–20.

Guijarro, N., T. Lana-Villarreal, Q. Shen et al. 2010. Sensitization of titanium dioxide photoanodes with cadmium selenide quantum dots prepared by SILAR: Photoelectrochemical and carrier dynamics studies. *J. Phys. Chem. C* 114 (50):21928–21937.

Hagfeldt, A., G. Boschloo, L. Sun et al. 2010. Dye-sensitized solar cells. *Chem. Rev.* 110 (11):6595–6663.

Hagfeldt, A., U.B. Cappel, G. Boschloo et al. 2012. Dye-sensitized photoelectrochemical cells. In *Practical Handbook of Photovoltaics*, edited by A. McEvoy, T. Markvart, and L. Castañer. Waltham, MA/Oxford, UK: Academic Press/Elsevier. P. 284.

Hanna, M.C., M.C. Beard, and A.J. Nozik. 2012. Effect of solar concentration on the thermodynamic power conversion efficiency of quantum-dot solar cells exhibiting multiple exciton generation. *J. Phys. Chem. Lett.* 3 (19):2857–2862.

He, F. and L. Yu. 2011. How far can polymer solar cells go? In need of a synergistic approach. *J. Phys. Chem. Lett.* 2 (24):3102–3113.

Hummelen, J.C., B.W. Knight, F. LePeq et al. 1995. Preparation and characterization of fulleroid and methanofullerene derivatives. *J. Org. Chem.* 60:532–538.

IEA/International Energy Agency. 2013. *2012 Key World Energy Statistics*. Paris, France.

Isley, S.L. and R.L. Penn. 2006. Relative brookite and anatase content in sol–gel-synthesized titanium dioxide nanoparticles. *J. Phys. Chem. B* 110 (31):15134–15139.

Jäger-Waldau, A. PV Status Report 2012. 2012. Ispra, Italy, European Union.

Jin, E., C. Du, M. Wang et al. 2012. Dibenzothiophene-based planar conjugated polymers for high efficiency polymer solar cells. *Macromolecules* 45 (19):7843–7854.

Jo, J., A. Pron, P. Berrouard et al. 2012. A new terthiophene-thienopyrrolodione copolymer-based bulk heterojunction solar cell with high open-circuit voltage. *Adv. Energy Mater.* 2 (11):1397–1403.

Kamat, P. V. Boosting the efficiency of quantum dot sensitized solar cells through modulation of interfacial charge transfer. *Acc. Chem. Res.* 45 (11):1906–1915.

Kanaparthi, R.K., J. Kandhadi, and L. Giribabu. 2012. Metal-free organic dyes for dye-sensitized solar cells: Recent advances. *Tetrahedron* 68 (40):8383–8393.

Katagiri, H., N. Sasaguchi, S. Hando et al. 1997. Preparation and evaluation of Cu_2ZnSnS_4 thin films by sulfurization of E-B evaporated precursors. *Sol. Energy Mater. Sol. Cells* 49:407–414.

Kirchartz, T. and U. Rau. 2011. Introduction to thin-film photovoltaics. In *Advanced Characterization Techniques for Thin Film Solar Cells*, edited by D. Abou-Ras, T. Kirchartz and U. Rau. Weinheim, FRG: Wiley-VCH Verlag GmbH & Co. KGaA.

Knowledge Transfer Network. 2013. *Minerals and Elements Review*. Chemistry Innovation Ltd. 2010 [cited 8 May 2013]. Available from http://www.chemistryinnovation.co.uk/stroadmap/files/dox/MineralsandElementspages.pdf.

Kooistra, F.B., J. Knol, F. Kastenberg et al. 2007. Increasing the open circuit voltage of bulk-heterojunction solar cells by raising the LUMO level of the acceptor. *Org. Lett.* 9 (4):551–554.

Koster, L.J.A., V.D. Mihailetchi, M. Lenes et al. 2009. Performance improvement of polymer: Fullerene solar cells due to balanced charge transport. In *Organic Photovoltaics*, 281–297. Weinheim, FRG: Wiley-VCH Verlag GmbH & Co. KGaA.

Koumura, N., Z.-S. Wang, S. Mori et al. 2006. Alkyl-functionalized organic dyes for efficient molecular photovoltaics. *J. Am. Chem. Soc.* 128 (44):14256–14257.

Kraabel, B., D. McBranch, N.S. Sariciftci et al. 1994. Ultrafast spectroscopic studies of photoinduced electron transfer from semiconducting polymers to C60. *Phys. Rev. B. Condens. Matter* 50:18543–18552.

Kronholm, D.F. and J.C. Hummelen. 2009. Fullerene-based acceptor materials. In *Organic Photovoltaics*, 153–178. Weinheim, FRG: Wiley-VCH Verlag GmbH & Co. KGaA.

Kuila, B.K., K. Park, and L. Dai. 2010. Soluble P3HT-grafted carbon nanotubes: Synthesis and photovoltaic application. *Macromolecules* 43 (16):6699–6705.

Lazzeri, M., A. Vittadini, and A. Selloni. 2001. Structure and energetics of stoichiometric TiO_2 anatase surfaces. *Phys. Rev. B* 63 (15):155409.

Lee, S.K., W.-H. Lee, J.M. Cho et al. 2011. Synthesis and photovoltaic properties of quinoxaline-based alternating copolymers for high-efficiency bulk-heterojunction polymer solar cells. *Macromolecules* 44 (15):5994–6001.

Li, T., B. Tian, J. Zhang et al. 2013. Facile tailoring of anatase TiO_2 morphology by use of H_2O_2: From microflowers with dominant {101} facets to microspheres with exposed {001} facets. *Ind. Eng. Chem. Res.* 52 (20):6704–6712.

Liang, Y., D. Feng, Y. Wu et al. 2009. Highly efficient solar cell polymers developed via fine-tuning of structural and electronic properties. *J. Am. Chem. Soc.* 131 (22):7792–7799.

Listorti, A., B. O'Regan, and J.R. Durrant. 2011. Electron transfer dynamics in dye-sensitized solar cells. *Chem. Mater.* 23 (15):3381–3399.

Loewe, R.S., S.M. Khersonsky, and R.D. McCullough. 1999. A simple method to prepare head-to-tail coupled, regioregular poly(3-alkylthiophenes) using Grignard metathesis. *Adv. Mater.* 11 (3):250–253.

Luan, X., D. Guan, and Y. Wang. 2012. Facile synthesis and morphology control of bamboo-type TiO_2 nanotube arrays for high-efficiency dye-sensitized solar cells. *J. Phys. Chem. C* 116 (27):14257–14263.

Lyons, B.P., N. Clarke, and C. Groves. 2012. The relative importance of domain size, domain purity and domain interfaces to the performance of bulk-heterojunction organic photovoltaics. *Energy Environ. Sci.* 5:7657–7663.

Mandlmeier, B., J.M. Szeifert, D. Fattakhova-Rohlfing et al. 2011. Formation of interpenetrating hierarchical titania structures by confined synthesis in inverse opal. *J. Am. Chem. Soc.* 133 (43):17274–17282.

Maraghechi, P., A.J. Labelle, A.R. Kirmani et al. 2013. The donor–supply electrode enhances performance in colloidal quantum dot solar cells. *ACS Nano* 7(7):6111–6116.

Margraf, J.T., A. Ruland, V. Sgobba et al. 2013. Quantum dot sensitized solar cells: Understanding linker molecules through theory & experiment. *Langmuir* 29:2434–2438.

McElroy, M.B. 2010. *Energy: Perspectives, Problems & Prospects.* New York: Oxford University Press.

Mehos, M. 2008. Another pathway to large-scale power generation: Concentrating solar power. *MRS Bull.* 33 (4):364–366.

Meyer, G.J. 2010. The 2010 millennium technology grand prize: Dye-sensitized solar cells. *ACS Nano* 4 (8):4337–4343.

Mitzi, D.B., O. Gunawan, T.K. Todorov et al. 2011. The path towards a high-performance solution-processed kesterite solar cell. *Solar Energy Mater. Solar Cells* 95 (6):1421–1436.

Moon, J.S., J.K. Lee, S. Cho et al. 2008. 'Columnlike' structure of the cross-sectional morphology of bulk heterojunction materials. *Nano Lett.* 9 (1):230–234.

Nakada, T. 2012. CIGS-based thin film solar cells and modules: Unique material properties. *Electron. Mater. Lett.* 8 (2):179–185.

National Aeronautic and Space Administration. "Earth's Energy Budget." Retrieved 23 May, 2013, from http://www.nasa.gov/audience/foreducators/topnav/materials/listbytype/Earths_Energy_Budget.html.

National Renewable Energy Laboratory 2013a. *Best Research-Cell Efficiencies.* National Center for Photovoltaics, NREL. Golden, CO.

National Renewable Energy Laboratory 2013b. *Quantum Dots Promise to Significantly Boost Photovoltaic Efficiencies (NREL/FS-6a4-47571).* June 2010. Golden, CO.

Nazeeruddin, M.K., R. Humphry-Baker, M. Grätzel et al. 1999. Efficient near-IR sensitization of anocrystalline TiO_2 films by zinc and aluminum phthalocyanines. *J. Porphyr. Phthalocyanines* 3 (3):230–237.

Nozik, A.J. 2002. Quantum dot solar cells. *Physica E Low Dimens. Syst. Nanostruct.* 14 (1–2):115–120.

O'Regan, B. and M. Grätzel. 1991. A low-cost, high-efficiency solar cell based on dye-sensitized colloidal TiO_2 films. *Nature* 353:737–740.

Otsuki, J., Y. Takaguchi, D. Takahashi et al. 2011. Piperidine-substituted perylene sensitizer for dye-sensitized solar cells. *Adv. Optoelectron.* 2011, 7pp, Article I.D. 860486, doi:10.1155/2011/860486.

Pappenfus, T.M., D.K. Schneiderman, J. Casado et al. 2010. Oligothiophene tetracyanobutadienes: Alternative donor—Acceptor architectures for molecular and polymeric materials. *Chem. Mater.* 23 (3):823–831.

Pappenfus, T.M., J.A. Schmidt, R.E. Koehn et al. 2011. PBC-DFT applied to donor–acceptor copolymers in organic solar cells: Comparisons between theoretical methods and experimental data. *Macromolecules* 44 (7):2354–2357.

Pei, Q., G. Zuccarello, M. Ahlskog et al. 1994. Electrochromic and highly stable poly(3,4-ethylenedioxythiophene) switches between opaque blue-black and transparent sky blue. *Polymer* 35:1347–1351.

Peter, L.M. 2011. Towards sustainable photovoltaics: The search for new materials. *Philos. Trans. R. Soc. Lond. A* 369 (1942):1840–1856.

Philibert, C. 2011. *Renewable Energy Technologies. Solar Energy Perspectives*. Paris: International Energy Agency.

Po, R., M. Maggini, and N. Camaioni. 2009. Polymer solar cells: Recent approaches and achievements. *J. Phys. Chem. C*. 114:695–706.

Repins, I., M.A. Contreras, B. Egaas et al. 2008. 19.9%-efficient $ZnO/CdS/CuInGaSe_2$ solar cell with 81.2% fill factor. *Prog. Photovolt. Res. Appl.* 16:235–239.

Rohde, R. A. Solar radiation spectrum, Wikimedia Commons/Global Warming Art. Retrieved 23 May, 2013, from http://commons.wikimedia.org/wiki/File:Solar_Spectrum.png.

Ross, R.T. and J. Nozik. 1982. Efficiency of hot-carrier solar energy converters. *J. Appl. Phys.* 53:3813–3818.

Rouquerol, J., D. Avnir, C.W. Fairbridge et al. 1994. Recommendations for the characterization of porous solids. *Pure Appl. Chem.* 66 (8):1739–1758.

Saji, V.S., I.-H. Choi, and C.-W. Lee. 2011. Progress in electrodeposited absorber layer for $CuIn_{(1-x)}Ga_xSe_2$ (CIGS) solar cells. *Solar Energy* 85:2666–2678.

Santra, P.K. and P.V. Kamat. 2012. Tandem-layered quantum dot solar cells: Tuning the photovoltaic response with luminescent ternary cadmium chalcogenides. *J. Am. Chem. Soc.* 135 (2):877–885.

Sariciftci, N.S., L. Smilowitz, A. J. Heeger et al. 1992. Photoinduced electron transfer from a conducting polymer to buckminsterfullerene. *Science* 258:1474–1476.

Seo, K.D., H.M. Song, M.J. Lee et al. 2011. Coumarin dyes containing low-band-gap chromophores for dye-sensitized solar cells. *Dyes Pigm.* 90 (3):304–310.

Shah, A., P. Torres, R. Tscharner et al. 1999. Photovoltaic technology: The case for thin-film solar cells. *Science* 285 (30 July):6.

Shalom, M., S. Dor, S. Rühle et al. 2009. Core/CdS quantum dot/shell mesoporous solar cells with improved stability and efficiency using an amorphous TiO_2 coating. *J. Phys. Chem. C* 113 (9):3895–3898.

Shen, P., Y. Tang, S. Jiang et al. 2011. Efficient triphenylamine-based dyes featuring dual-role carbazole, fluorene and spirobifluorene moieties. *Org. Electron.* 12 (1):125–135.

Siemens. 2012. *High Concentrated Photovoltaics*. Germany: Siemens.

Smestad, G.P. and M. Grätzel. 1998. Demonstrating electron transfer and nanotechnology: A natural dye-sensitized nanocrystalline energy converter. *J. Chem. Ed.* 75 (6):752.

Tang, C.S. 1986. Two-layer organic photovoltaic cell. *Appl. Phys. Lett.* 48 (2):183–185.

Tian, H. and L. Sun. 2011. Iodine-free redox couples for dye-sensitized solar cells. *J. Mater. Chem.* 21 (29):10592–10601.

Todorov, T. K., Reuter, K.M. and Mitzi, D.B. 2010. High-efficiency solar cell with earth-abundant liquid-processed absorber. *Adv. Mat.* 22 (20):E156-E159.

Toyoda, T. and Q. Shen. 2012. Quantum-dot-sensitized solar cells: Effect of nanostructured TiO_2 morphologies on photovoltaic properties. *J. Phys. Chem. Lett.* 3 (14):1885–1893.

Tsunomura, Y., Y. Yoshimine, M. Taguchi et al. 22%-Efficiency HIT solar cell. Sanyo Electric Co. 2007. Hyogo, Japan. us.sanyo.com/Dynamic/customPages/docs/solarPower_22_3_Cell_Efficiency_White_Paper_Dec_07.pdf.

U.S. Department of Energy. 2013. *Crystalline Silicon Photovoltaic Cells*. U.S. Department of Energy, 8/12/2011 2011 [cited Jan. 9, 2013]. Available from http://www.eere.energy.gov/basics/renewable_energy/crystalline_silicon.html

U.S. Energy Information Administration. *Annual Photovoltaic Cell/Module Shipments Report* 2011 [cited EIA-63B]. Available from http://www.eia.gov/renewable/annual/solar_photo/pdf/table5.pdf

Wang, X.-F. and O. Kitao. 2012. Natural chlorophyll-related porphyrins and chlorins for dye-sensitized solar cells. *Molecules* 17:4484–4497.

Wang, Q., R. Takita, Y. Kikuzaki et al. 2010. Palladium-catalyzed dehydrohalogenative polycondensation of 2-bromo-3-hexylthiophene: An efficient approach to head-to-tail poly(3-hexylthiophene). *J. Am. Chem. Soc.* 132 (33):11420–11421.

Wang, X., S.A. Kulkarni, B.I. Ito et al. 2012a. Nanoclay gelation approach toward improved dye-sensitized solar cell efficiencies: An investigation of charge transport and shift in the TiO_2 conduction band. *ACS Appl. Mater. Interfaces* 5 (2):444–450.

Wang, G., S. Wang, Y. Cui et al. 2012b. A novel and versatile strategy to prepare metal–organic molecular precursor solutions and its application in $Cu(In,Ga)(S,Se)_2$ solar cells. *Chem. Mater.* 24 (20):3993–3997.

Wang, L., H. Zhang, C. Wang et al. 2012c. Highly stable gel-state dye-sensitized solar cells based on high soluble polyvinyl acetate. *ACS Sust. Chem. Eng.* 1 (2):205–208.

Watts, B., W.J. Belcher, L. Thomsen et al. 2009. A quantitative study of PCBM diffusion during annealing of P3HT:PCBM blend films. *Macromolecules* 42 (21):8392–8397.

Winder, C., G. Matt, J.C. Hummelen et al. 2002. Sensitization of low bandgap polymer bulk heterojunction solar cells. *Thin Solid Films* 403–404:373–379.

Wu, J., S. Hao, J. Lin et al. 2007. Crystal morphology of anatase titania nanocrystals used in dye-sensitized solar cells. *Cryst. Growth Des.* 8 (1):247–252.

Wu, G., Y. Jiang, D. Xu et al. 2010. Thermoresponsive inverse opal films fabricated with liquid-crystal elastomers and nematic liquid crystals. *Langmuir* 27(4): 1505–1509.

Xiao, Z., Y. Matsuo, I. Soga et al. 2012. Structurally defined high-LUMO-level 66π-[70]fullerene derivatives: Synthesis and application in organic photovoltaic cells. *Chem. Mater.* 24 (13):2572–2582.

Xu, J., X. Yang, Q.-D. Yang et al. 2012. Cu_2ZnSnS_4 hierarchical microspheres as an effective counter electrode material for quantum dot sensitized solar cells. *J. Phys. Chem. C* 116 (37):19718–19723.

Yang, X., J.-K. Fang, Y. Suzuma et al. 2011. Synthesis and properties of 9,10-anthrylene-substituted phenyleneethynylene dyes for dye-sensitized solar cell. *Chem. Lett.* 40 (6):620–622.

Yeh, M.-H., C.-P. Lee, C.-Y. Chou et al. 2011. Conducting polymer-based counter electrode for a quantum-dot-sensitized solar cell (QDSSC) with a polysulfide electrolyte. *Electrochim. Acta* 57:277–284.

Yu, G., J. Gau, J.C. Hummelen et al. 1995. Polymer photovoltaic cells: Enhanced efficiencies via a network of internal donor–acceptor heterojunctions. *Science* 270:1789–1791.

Yu, Z., M. Gorlov, J. Nissfolk et al. 2010a. Investigation of iodine concentration effects in electrolytes for dye-sensitized solar cells. *J. Phys. Chem. C* 114 (23):10612–10620.

Yu, D., Y. Yang, M. Durstock et al. 2010b. Soluble P3HT-grafted graphene for efficient bilayer–heterojunction photovoltaic devices. *ACS Nano* 4 (10):5633–5640.

Yuan, M., D.B. Mitzi, W. Liu et al. 2009. Optimization of CIGS-based PV device through antimony doping. *Chem. Mater.* 22 (2):285–287.

Yum, J.-H., P. Walter, S. Huber et al. 2007. Efficient far red sensitization of nanocrystalline TiO_2 films by an unsymmetrical squaraine dye. *J. Am. Chem. Soc.* 129 (34):10320–10321.

Zhao, Y., S. Shao, Z. Xie et al. 2009. Effect of poly (3-hexylthiophene) nanofibrils on charge separation and transport in polymer bulk heterojunction photovoltaic cells. *J. Phys Chem. C* 113 (39):17235–17239.

Zhu, G., L. Pan, T. Xu et al. 2011. One-step synthesis of Cd-S sensitized TiO_2 photoanodes for quantum dot-sensitized solar cells by microwave assisted chemical bath deposition method. *ACS Appl. Mater. Interfaces* 3 (5):1472–1478.

Zoppi, G., I. Forbes, R.W. Miles et al. 2009. $Cu_2ZnSnSe_4$ thin film solar cells produced by selenisation of magnetron sputtered precursors. *Prog. Photovolt.: Res. Appl.* 17:315–319.

Zukalová, M., A. Zukal, L. Kavan et al. 2005. Organized mesoporous TiO_2 films exhibiting greatly enhanced performance in dye-sensitized solar cells. *Nano Lett.* 5 (9):1789–1792.

8 Biomass

8.1 INTRODUCTION

When people think "sustainable energy," the energy source that often springs to mind is, naturally, *biomass*. Humans have been using biomass for energy since the dawn of our time. Not only is it our personal source of energy (as in food), we have also used it to cook our food and heat our homes for millennia. As a potential energy source, biomass is relatively abundant (ranked third, after oil and coal) and has supplied more than 90% of the fuel and energy needs of the United States until the mid-nineteenth century (Champagne 2008). In several European countries, biomass makes a considerable contribution to energy supply and consumption and it is still a significant source of energy in developing countries (Pereira et al. 2012). Among the various energy solutions presented in this book, biomass-to-energy conversions are arguably the most sustainable and are considerably cleaner than coal, for example, in that biomass-derived energy generates far fewer NO_x or SO_x emissions. Another advantage of the use of biomass is its amenability to small-scale installations with the concomitant promise of energy availability and economic development in rural and developing areas. Biomass led the way in renewable energy consumption in 2011 in the United States, with 4.4 quadrillion BTU being consumed (hydroelectric power was a distant second with 3.2 quad) (U.S. Energy Information Administration 2012). Given the availability of biomass, its very low levels of pollutants and the wide variety of conversion options for deriving usable energy from biomass, the contributions of biomass to our future energy needs are certain to grow.

8.1.1 Carbon Neutrality

One of the most prevalent reasons for the use of biomass as an energy source is its promise of being "carbon-neutral." But what do we really mean by carbon neutrality? Recall the carbon cycle from Chapter 1: the biomass, as it grows, removes carbon from the atmosphere in the form of CO_2 and fixes it into carbohydrates and other carbon-containing material. A tree is a natural carbon sequestration agent. When we then convert that biomass into energy (as in combustion), we generate CO_2 but conservation of matter dictates that we are neither creating nor destroying matter: carbon neutral. But is biomass really carbon neutral? Fertilizers and pesticides (which may be carbon-containing) are often used in growing the material. Machinery, probably powered by fossil fuels, will be used in planting, harvesting, processing, and transporting the biomass. In terms of life-cycle analysis for sustainability, all these factors must be taken into consideration. That said, if the biomass can be produced in a carbon-neutral way and the CO_2 generated during the energy conversion process is sequestered, biomass energy can actually *reduce* the amount of carbon in the atmosphere (Milne and Field 2013). However, an in-depth study of the thermodynamics of energy production from biomass came to the conclusion that biofuel production on an industrial scale is inherently unsustainable (Patzek and Pimentel 2005). Biomass is an important part of the overall sustainable energy solution, but it is not a panacea.

8.1.2 BIOMASS CONSIDERATIONS

8.1.2.1 Energy Density and Land Use

A major consideration for biomass-derived energy is the amount of energy the biomass actually contains. As we saw in Table 1.5, different materials inherently possess different quantities of energy. Crude oil is an excellent source of energy in that it is easily transported and energy rich: at 42 MJ/kg, it is one of the most energy-dense fuels. Methane is even higher, at 55 MJ/kg, and even coal has a respectable energy density (27–32 MJ/kg). Compared to these fossil fuels, the energy density of biomass pales: dry carbohydrate biomass has an energy density in the range of 15–20 MJ/kg (Champagne 2008; da Rosa 2009; Sørensen 2007). This, of course, translates to needing more biomass to produce the same amount of heat or power, which translates into higher transportation costs, processing costs, and so on.

The relatively low energy density of biomass also means that issues associated with land use must be taken into account. Expansion of land use for biofuel production can lead to deforestation (particularly in tropical areas), reducing any potential benefit to using biomass. The best land for agriculture must be used to grow food for a hungry global population. The "ideal" energy crop should be able to be grown on marginal land with little use of fertilizer or pesticides and, potentially, under drought conditions (or at least needing minimal water). Furthermore, energy crops should not be grown at the expense of biodiversity. Given the current level of energy consumption, it would take almost three times all the land currently cultivated for agriculture to satisfy our energy needs through biomass conversions (Barber 2009).

8.1.2.2 Soil and Water

Beyond the enormous area of land needed lies the concern of soil quality. Repeated removal of biomass (as in harvesting of corn stover for energy purposes rather than tilling it back into the soil) may impact the long-term soil quality (Johnson 2013). Water usage, as intimated above, is another major concern: the more biomass-derived energy sources expand, the greater the stress on water supplies. For example, up to six gallons of water are required for every one gallon of bioethanol produced (Aden 2007). Biomass may have a moderate *carbon* footprint, but its *water* footprint is huge. It is true that only a small percentage of the biomass produced by photosynthesis is currently being cultivated, harvested, and used—but how much can be used sustainably? As with any approach to energy generation, the massive demand for energy demand accentuates the need to be careful in considering the use of biomass for energy generation.

8.1.3   WHAT IS BIOMASS?

Just what do we mean by biomass? If we are looking for a material to sustainably replace our fossil fuel supply, it is only natural that we would consider something that we can plant (or birth), grow, and harvest (then repeat). That is the essential definition of *biomass*: organic (as in plant or animal) matter that is available on a renewable basis. Ultimately, biomass is packaged solar energy primarily made up of carbon, hydrogen, nitrogen, and oxygen that has been converted into cellulose, lignin, and other organic molecules by photosynthesis, and into proteins, lipids, nucleic acids, and other biomolecules by other biochemical processes. The more detailed chemical composition of biomass will be examined in Section 8.2.

Where does biomass come from? In terms of renewable energy, we tend to focus on plant matter—trees, grasses, seeds—although animal waste products (e.g., manure, municipal solid waste (MSW), and even waste animal meat) are important biomass resources as well. Crops that are grown for the express purpose of harvesting for energy production are *energy crops* and include plants such as switchgrass, hybrid poplar, and *Camelina sativa* (Benemelis 2012). Residual plant waste (e.g., rice husks, nut shells, or sawdust) is also a ready supply of biomass.

8.1.4   WHAT ARE BIOFUELS?

Biomass can be considered *solar fuel*—after all, it is potential energy created by photosynthesis. There are certainly other kinds of solar fuels, for example, the photosynthetic production of hydrogen covered in Chapter 5. Then there are *biofuels*, those being fuels that are derived from biomass. Biofuels include everything from sugars and fats—simple foodstuffs that power organisms—to biomass-derived ethanol and biodiesel. Biofuels have evolved from "first-generation" biofuels that compete with food production (e.g., ethanol from corn or biodiesel from soybeans) to "second-generation" biofuels that are derived from lignocellulosic biomass (LCB).

Biofuels can also be categorized by their physical properties, much like fossil fuels. Thus, lower-boiling *bioalcohols* include not only biomass-derived methanol and ethanol but also biobutanol. *Bio-oil is* primarily obtained from the pyrolysis of biomass; these are larger molecules that retain the characteristics of higher-boiling organic liquids, that is, oils. *Biodiesel* is a heavier liquid yet. At the other end of the spectrum are gases obtained from biomass, including syngas (from gasification) and *biogas* (from anaerobic digestion or microbial hydrogen generation). Given the similarity of these characteristic groupings to that of fossil fuels, it is no surprise that the *biorefinery* has developed to process biofuels. A biorefinery may have only a single feedstock (e.g., corn) and a single process (e.g., fermentation) to produce a single product (e.g., bioethanol), but as the utilization of biomass for energy generation continues to grow, biorefineries will evolve to take advantage of multiple feedstocks and produce different types of fuels and value-added products from a variety of conversion processes, as depicted in Figure 8.1. This figure also serves as a useful roadmap for the rest of the chapter.

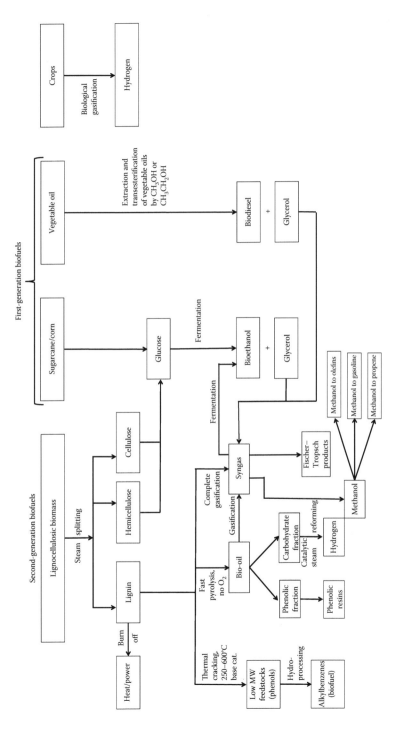

FIGURE 8.1 The various processes of the biorefinery. (From Stocker, M. 2008. Biofuels and biomass-to-liquid fuels in the biorefinery: Catalytic conversion of lignocellulosic biomass using porous materials. *Angew. Chem. Int. Ed.* 47 (47):11. Copyright 2008, Wiley-VCH Verlag GmbH & Co. KGaA, Weinheim.)

8.1.5 Some Basic Biochemistry

Biochemistry is a fascinating extension of the organic chemistry of living organisms but we will examine only the tiniest portion to elucidate the basic processes involved in biomass energy conversions. In order to do so, however, we need to have a basic understanding of some of the nomenclature and concepts, particularly with respect to carbohydrate chemistry.

Carbohydrates, also known as *saccharides*, are "hydrates of carbon," $C_6(H_2O)_6$ $(C_6H_{12}O_6)$. Polymers of sugar units make up carbohydrates, hence the nomenclature polysaccharide, disaccharide, oligosaccharide, and so on. An important feature of a carbohydrate is the *anomeric* carbon, that being the carbon in a carbohydrate that is directly bonded to *two* oxygen substituents (Figure 8.2). The anomeric carbon is assigned C1 for nomenclature purposes. Compounds whose stereochemistry differs only at that stereocenter are known as *anomers*. When the hydroxyl of the anomeric carbon is bonded to some other moiety, it is known as a *glycoside* and the specific bond is known as the *glycosidic linkage* (Figure 8.2). This bond is especially important to recognize since, being an acetal, it is easily hydrolyzed.

The stereochemistry at the anomeric carbon is particularly important and carbohydrate chemistry has its own system of nomenclature for this feature. The relative stereochemistry at the anomeric carbon is indicated by the designators α and β. In the α anomer, the glycosidic substituent is on the side of the ring *opposite* the $-CH_2OH$ group; for the β anomer, the $-CH_2OH$ and glycosidic substituent are on the same side (Figure 8.2).

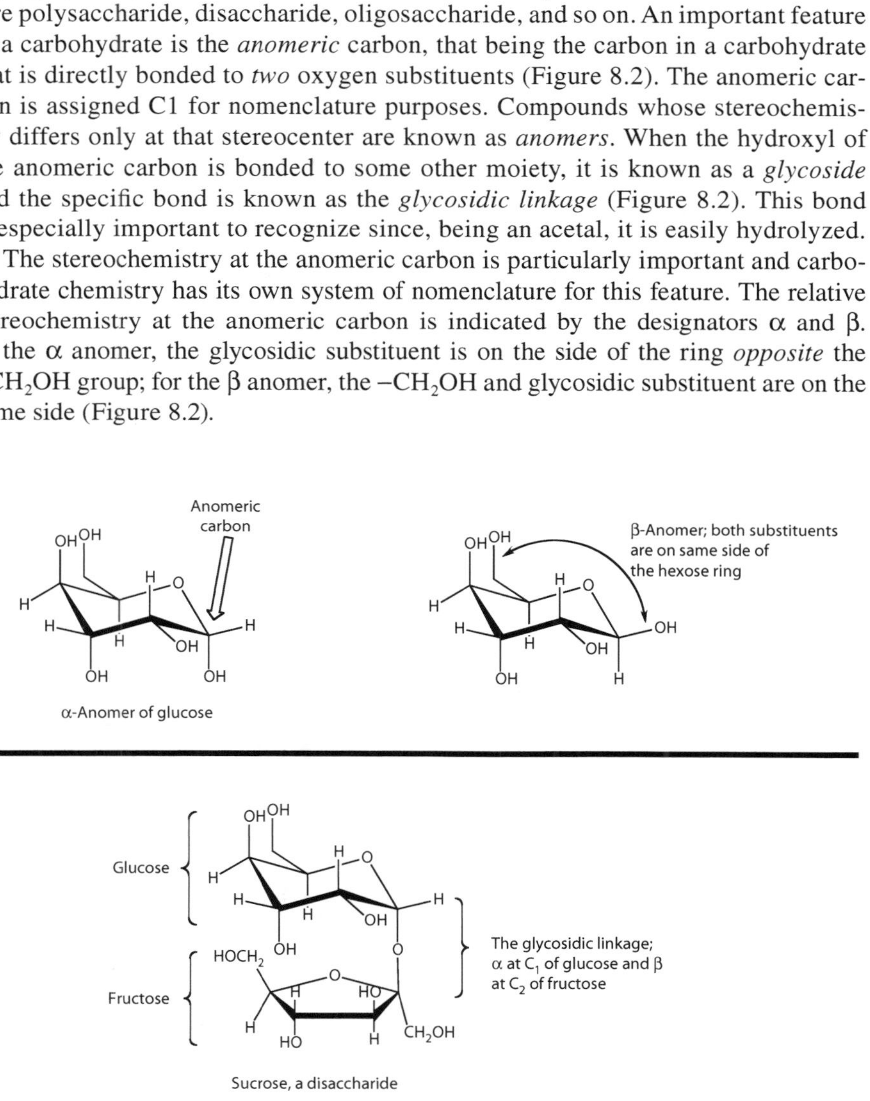

FIGURE 8.2 Basic carbohydrate nomenclature.

8.2 CHEMICAL COMPOSITION OF BIOMASS

While animal waste products, as we have noted, are suitable feedstocks for the production of biofuels of one sort or another, our focus is largely plant biomass. The chemical composition of biomass is very different from fossil fuel feedstocks—plant biomass is loaded with oxygen, with a carbon-to-oxygen ratio of almost one. In contrast, fossil fuels are, for the most part, completely deoxygenated. Because of this high level of oxygen, the energy content of biomass (even discounting the fact that it has low bulk density) is quite poor relative to fossil fuels. Figure 8.3 provides a general overview of the composition of plant biomass. Prior to drying, plant biomass contains up to 95% water, but by dry weight percent, the vast majority of plant biomass is made up of carbon and oxygen in the form of cellulose, hemicellulose, and lignin (hence the name *lignocellulosic biomass*) (Kersten et al. 2007). Small amounts of carboxylic acids, fatty acid esters and alcohols, terpenes, and various other natural products are also found in plants, often lumped together under the heading of "extractives" since they can be extracted out of the plant material by the use of an organic solvent (Alén et al. 1996).

Cellulose (Figure 8.4) accounts for more than 50% of the carbon in the biosphere (Voet and Voet 1990). It is a linear polysaccharide consisting of β-1,4-linked D-glucose units and a degree of polymerization of about 300–15,000, depending on the plant (van Santen 2007). *Cellobiose* is the repeating disaccharide of cellulose. The properties of cellulose are governed by many features:

- Its stable chair conformation leading to the ability to form close-packed linear fibrils with semicrystalline domains
- Its M_w ($\approx 1 \times 10^6$)
- Its intramolecular and intermolecular hydrogen bonding
- Its stereochemistry

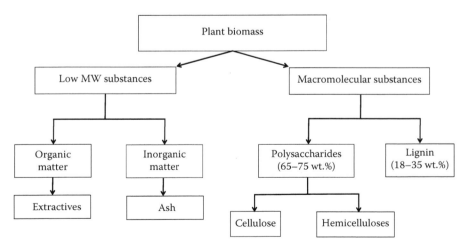

FIGURE 8.3 General chemical composition of plant biomass. (Adapted with permission from Mohan, D., C.U. Pittman, and P.H. Steele. 2006. Pyrolysis of wood/biomass for bio-oil: A critical review. *Energy Fuels* 20:848–889. Copyright 2006, American Chemical Society.)

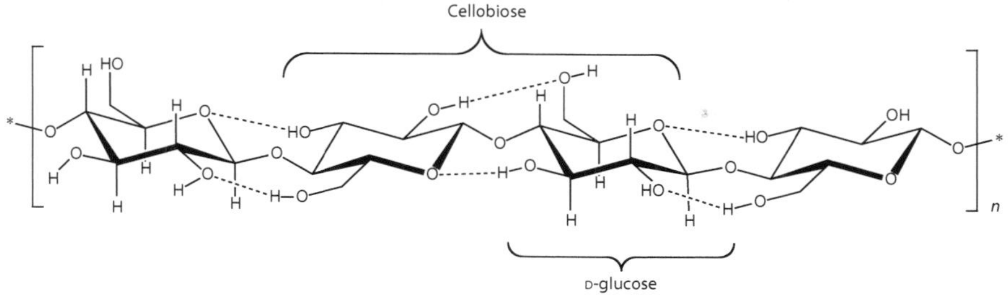

FIGURE 8.4 Cellulose. The dashed lines represent hydrogen bonds.

In contrast to cellulose, *hemicellulose* (occasionally referred to as *polyose*, Figure 8.5) is a diverse polymer made up of a variety of C_5 (pentose) and C_6 (hexose) sugar units, several of which are acetylated (see Figure 8.6). This branched heteropolysaccharide is tightly bound to the cellulose bundles through hydrogen bonding and makes up about 25–33% of most plant materials (Cai and Paszner 1988) (Sun et al. 2004). Hemicellulose is much smaller than cellulose, with a degree of polymerization of 80–200 and a relatively low M_w of $\approx 3 \times 10^4$. The general nature of the hemicellulose structure depends on the type of plant with the result that certain types of lignocellulosic materials are easier to hydrolyze than others (Agbor et al. 2011).

Lignin (Figure 8.7), like hemicellulose, is a diverse conglomerate of a variety of polyphenols that is found predominantly in the cell walls of woody plants. Lignin is embedded within the hemicellulose to provide additional rigidity to the plant. Woody plants are typified by these cellulosic fibers, whereas the fibers in herbaceous plants are more loosely bound, indicating a lower proportion of lignin. Lignin is less polar than cellulose or hemicellulose and is fairly impervious, protecting the polysaccharides from microbial degradation (Vanholme et al. 2010). Lignin is biosynthesized from three basic cinnamyl alcohol units (Figure 8.7) and contains numerous ether linkages,

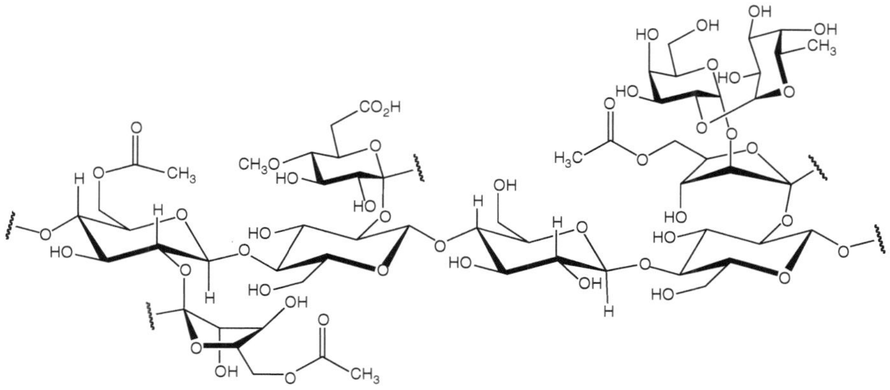

FIGURE 8.5 A representative portion of hemicellulose.

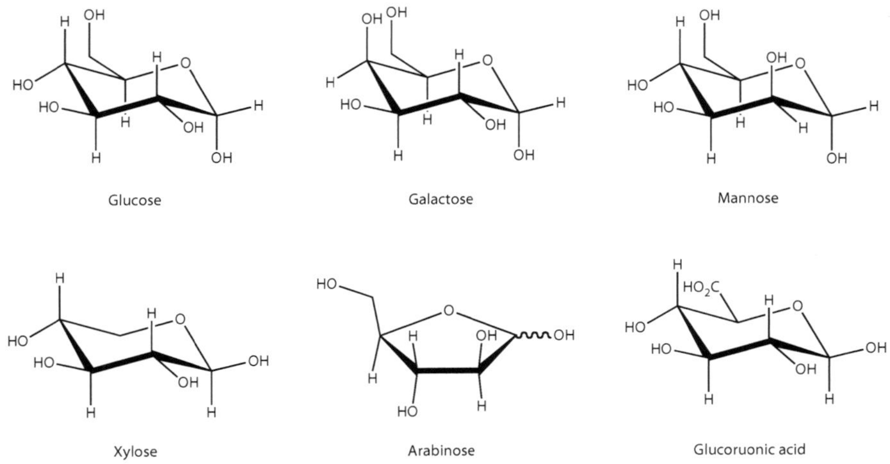

FIGURE 8.6 Structures of the primary sugar components in hemicellulose.

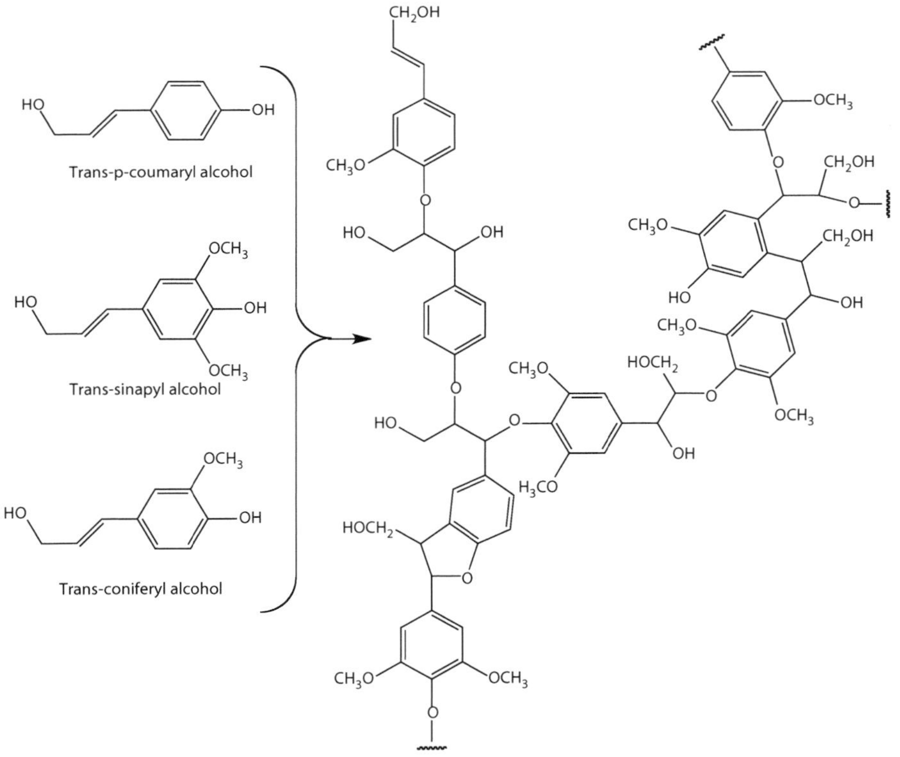

FIGURE 8.7 Lignin and its constituents.

TABLE 8.1

Typical Percent Composition of Inorganic Elements in Plant Biomass

Element	Percent (Dry Matter)
Potassium	0.1
Sodium	0.015
Phosphorus	0.02
Calcium	0.2
Magnesium	0.04

Source: Reprinted with permission from Mohan, D., C.U. Pittman, and P.H. Steele. Pyrolysis of wood/biomass for bio-oil: A critical review. *Energy Fuels* 20. p. 854. Copyright 2006, American Chemical Society.

resulting in a polymer with only about 120 phenolic units and an M_w of roughly 20,000 (Lange 2007). When put together in a plant, lignin, cellulose, and hemicellulose knit together a *lignin–carbohydrate complex* in which the lignin is bonded to hemicellulose, and hemicellulose bonded to cellulose through extensive hydrogen bonding.

Biomass, while largely organic (in the chemical sense), does contain trace amounts of inorganics as well. Silicon, aluminum, titanium, iron, calcium, magnesium, sodium, potassium, phosphorus, sulfur, and chlorine may all be present in the biomass and at different levels depending upon both the environment and the genome; Table 8.1 gives some typical values. Chloride, in particular, can pose problems as thermochemical conversion processes can convert Cl into HCl, although all these contaminants present their own unique challenges depending upon the process. A detailed analysis of the chemical composition of a wide variety of plant materials and their corresponding value as potential feedstocks for biofuel production has recently been published (Godin et al. 2013).

8.3 REACTIVITY AND CONVERSION OPTIONS

8.3.1 CONVERSION OPTIONS

There are essentially three general methods for converting biomass into energy/fuels: thermochemical, biochemical, and mechanical. Table 8.2 provides a very general overview of these processes. The thermochemical processes—pyrolysis, gasification, and combustion—are generally very fast (seconds to minutes) but nondiscriminating, often producing a mixture of products whose composition varies depending upon the particular feedstock and conditions. On the other hand, biochemical methods are relatively slow, taking days or weeks but usually providing 1–2 major product(s). The particular conversion option for which the biomass is best suited depends on the physical and chemical properties of the material. Sewage sludge is more appropriate for digestion than for, say, gasification, due to its moisture content. Each unique method is examined in detail later in this chapter.

TABLE 8.2
Biomass Conversion Options

Type of Conversion	Method	Applications
Thermochemical	Combustion (excess air)	Heat
		Steam → boiler → electricity
	Gasification (partial air)	Heat (CHP)
		Gas turbine → electricity
		Syngas → chemicals, fuels
	Pyrolysis (no air)	Heat, fuel
Biochemical	Digestion	Biogas → electricity
	Fermentation	Fuel (ethanol)
Mechanical	Extraction	Fuel (biodiesel)

Source: Adapted from Turkenburg, W.D. (Ed.), 2000. Renewable energy technologies. In *World Energy Assessment: Energy and the Challenge of Sustainability*, edited by J. Goldemberg, New York: UNDP/UN-DESA/World Energy Council, Chapter 7, Figure 7.1, p. 223.

8.3.2 GENERAL REACTIVITY PATTERNS

Given the complexity of the various materials being used as feedstocks, the reactions involved in biomass energy conversions are far ranging and complex. This diversity is amplified by the huge variability in process conditions, from gas phase to aqueous phase, room temperature to thousands of degrees, and so on. Nevertheless, some general patterns emerge and it is worthwhile to engage in an overview of the reactivity of some typical biomass functionality here.

Understandably, sugars and their derivatives (including the celluloses shown above) play a large role in biomass energy conversion. A simple carbohydrate, such as sucrose, can be considered to be a "polyhydroxyaldehyde," a name that is revealing: the hydroxyaldehyde (or hydroxyketone) open-chain form of the sugar is in equilibrium with the hemiacetal (or hemiketal) closed form (Figures 8.8 and 8.9). The transformations associated with most biomass energy conversion processes are directly related to the presence of the carbonyl and hydroxyl functionality. For example, aldol (Figure 8.10) and conjugate additions (Figure 8.11a and b)—as well as their associated retro reactions—are commonplace and can lead to unwanted side products and degraded products.

Biomass energy conversions are not limited to carbohydrates, however. Plant and animal biomass also contain large amounts of oils (lipids) and proteins. The central reaction relevant to these biomass components is hydrolysis, as shown in Figure 8.12. As we will see later in this chapter, these reactions can be readily accomplished biochemically by enzymatic catalysis.

Because of the harsh operating conditions of thermochemical conversions (gasification and pyrolysis), reactivity is much more erratic. Hemicellulose, cellulose,

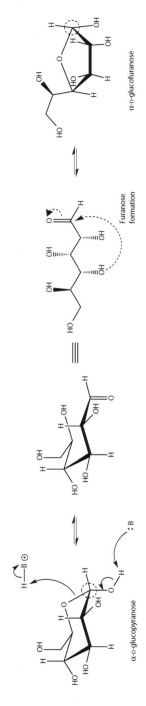

FIGURE 8.8 Equilibrium of glucose illustrating pyranose, furanose, and open-chain forms. The anomeric carbon is circled.

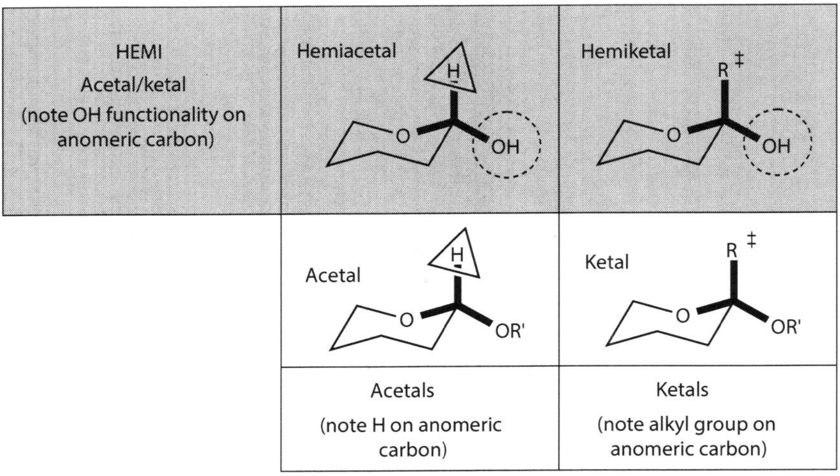

HEMI Acetal/ketal (note OH functionality on anomeric carbon)	Hemiacetal	Hemiketal
	Acetal	Ketal
	Acetals (note H on anomeric carbon)	Ketals (note alkyl group on anomeric carbon)

FIGURE 8.9 The structural differences between acetals, ketals, hemiacetals, and hemiketals.

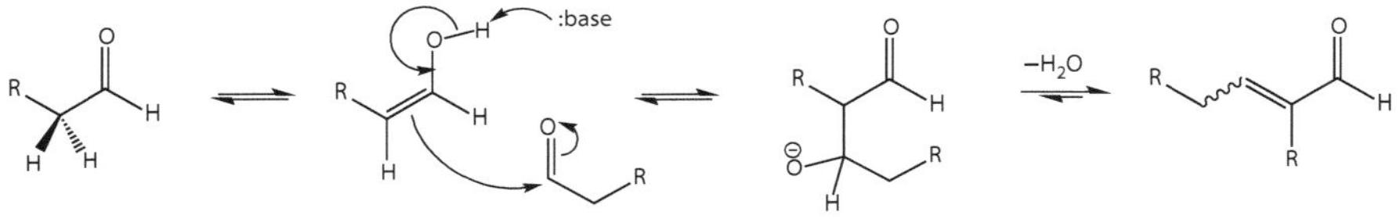

FIGURE 8.10 The aldol condensation.

and lignins decompose to give off mainly small molecules such as water, carbon monoxide, carbon dioxide, and low-molecular-weight oligomers. Cellulose degrades in the range of 240–350°C to give dehydration products and levoglucosan, often by radical processes as depicted in Figure 8.13. Lignins, while not likely to dehydrate, are quite susceptible to radical cleavage reactions at the ether bonds at temperatures of 280–500°C, yielding a complex mixture of phenols (Soltes and Elder 1981).

8.4 BIOMASS BEGINNINGS: HARVESTING AND PROCESSING

We are interested in biomass because of its great potential for sustainability, yet the use of biomass presents some unique challenges. First, it comes in a wide variety of forms, from wood chips to rice husks to animal manure, items with wildly different material handling properties. But we do not grow wood chips, rice husks, or animal manure—we grow *trees*, and *rice*, and … animals. Biomass requires significant *processing* even after it has been *harvested* (not to mention the fact that cultivated biomass needs to be planted, nurtured, and grown prior to harvesting). Furthermore, with regard to harvesting, crops clearly grow in seasonal domains. A constant supply of feedstock is a

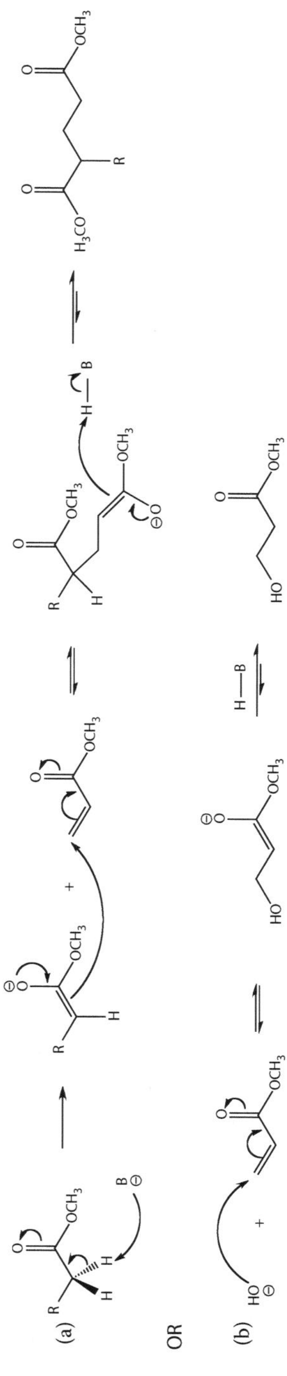

FIGURE 8.11 Conjugate addition to α,β-unsaturated carbonyl compounds. (a) The Michael reaction, (b) conjugate addition.

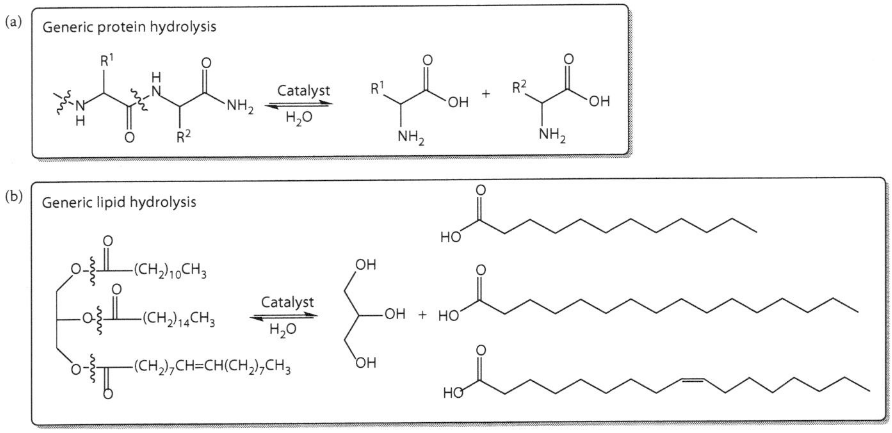

FIGURE 8.12   Hydrolysis of amides (a) and esters (b).

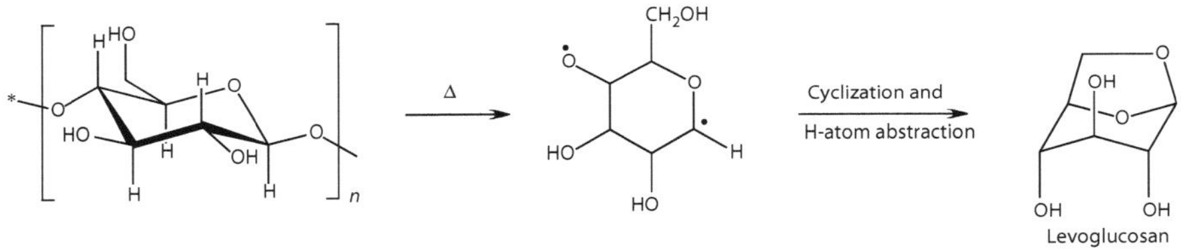

FIGURE 8.13   Thermal decomposition of cellulose with radical pathway to levoglucosan.

desirable, if elusive, ideal for biomass. On the other hand, some waste products (MSW, food processing waste, etc.) can be reliably produced without regard to seasons. All of these factors are pertinent when contemplating biomass as a sustainable energy source.

Almost all biomass feedstocks must undergo some sort of pretreatment before use in an energy conversion process, from harvesting and washing to drying and densification. The goals of pretreatment are to improve the storage, handling, and transport properties as well as the conversion efficiency of the biomass feedstock. First, after harvesting, the desired material must be *separated* from unwanted residues. This could be as simple as washing soil residue from harvested plants or screening to remove gravel, and so on. The next step is to dry the biomass, if necessary.

8.4.1   Drying

The heating value of biomass changes dramatically with moisture content, as we have already seen. Freshly harvested plant biomass may have a moisture content as high as 60%, so for thermochemical energy conversion processes the material must be dried to a moisture content of about 10–15%; otherwise, valuable heat is wasted in the reactor for the evaporation of moisture (recall lower vs. higher heating values as discussed in Chapter 3). The ideal drying agent is solar energy, of course, but drying with waste heat is an efficient alternative. While temperatures above 150°C cause chemical changes in biomass, torrefaction ("roasting" the biomass at a temperature of about 225–300°C in the absence of oxygen) is sometimes desirable—it increases the carbon content of woody biomass as hemicellulose decomposes and volatile compounds evaporate. The material is also made more brittle as the hemicellulose is broken down, making subsequent preconversion treatments easier.

8.4.2   COMMINUTION

Perhaps the most important step in biomass pretreatment is to then *comminute* it: mechanically alter it so that, instead of a wide variety of shapes, sizes, and thicknesses, a conveniently manipulated, uniform solid feedstock is made. This is not done solely for the purpose of convenience in handling. By modifying the particle size and shape and the porosity and surface area, heat and mass transport properties are all changed—all of which greatly impact the behavior (and quality) of the biomass in subsequent energy conversions. Some of the more common processes of comminution are chipping, milling, or grinding until the preferred particle size is obtained.

8.4.3   DENSIFICATION

Once a small particle size is obtained, it can be further modified by compressing or pelletizing in a process that is known as *densification*. The optimum size of biomass pellet, briquette, or particle depends on both the process and the equipment involved and must take into account not only the handling but also the conversion process: a finely distributed powder may be ideal for delivery via a screw auger, but the fine particulate may present major problems later in the conversion process. In any case, transforming the biomass into a pellet or powder is an energy-intensive process that adds to the overall cost of the energy produced.

8.5   THERMOCHEMICAL PROCESSES

8.5.1   INTRODUCTION

Most thermochemical processes use plant biomass as the feedstock, although a mixed feedstock of animal waste (e.g., poultry or feedlot manure) with plant biomass has been used with some success. (Animal waste alone is generally not a suitable feedstock for thermochemical conversion methods because of its high moisture content.) While humans have *combusted* plant biomass for energy generation from our very beginnings, the focus here will be on the two other major thermochemical conversion methods: *gasification* of biomass to generate syngas and *pyrolysis* to make bio-oil. In comparing the specific processes, we will see that there is not a clean line of demarcation between the two; pyrolysis forms some gases and gasification includes pyrolysis reactions. Nevertheless, gasification's goal is syngas, and the goal of modern flash pyrolysis is the production of liquid fuel. Furthermore, two similarities in these processes stand out: first, the complexity of the biomass feedstock going *in* leads to complex product mixtures coming *out*. Second, the many variables associated with each process (reactor design, temperature, rate of heating, biomass particle size, gasification agent, catalyst, moisture level, residence time, etc.) have a profound impact on the composition of the product. While many of the basic reactions during the thermochemical conversion are known, control of the reaction output is still very much an empirical science.

8.5.2 PYROLYSIS

8.5.2.1 Introduction

Pyrolysis is ancient technology: heating biomass in the absence of oxygen was used in ancient Egypt to produce pitch for embalming purposes and for waterproofing boats (Ringer et al. 2006). Pine was once "distilled" to obtain turpentine, and methanol is also known as wood alcohol for a reason: wood heated in the absence of oxygen produces methanol. *Slow pyrolysis* (over a period of days) is still the main method for the production of charcoal, but the primary goal of modern *flash pyrolysis*—with a residence time of less than 2 s—is to convert solid, low-density biomass into a convenient liquid fuel, usually referred to as *bio-oil* (not to be confused with oil *expressed* out of other biomass, as in oilseeds). The *bio-oil* (or *pyrolysis oil*) so obtained can be upgraded for subsequent use in diesel engines, turbines, and blending with other fuels. (N.B. Another process for conversion of biomass into liquids is *liquefaction*, which is often confused with pyrolysis. Liquefaction is carried out at a lower temperature and under a pressurized hydrogen atmosphere—the key difference being the reducing medium. Our focus is on pyrolysis, favored for commercialization due to the lower capital costs (Graça et al. 2012).)

8.5.2.2 Process

The pyrolysis process is quite straightforward: in a hot ($\approx$500–600°C) oxygen-free environment, a flash of thermal energy infuses the biomass feedstock, pyrolyzing it (literally "cleaving with heat"), then the pyrolysis product is cooled as quickly as possible (quenched) to prevent further reaction of the components that would result in the formation of tar. The precise nature of the product (i.e., the proportion of solid, liquid, and gas) depends a great deal on the process variables.

The short residence time is key: this is *fast* (or *flash*) *pyrolysis*. Just how fast is fast? The rate of heating has been estimated to be >1000°C/s amounting to a rate of heat transfer between 600 and 1000 W/cm^2 (Reed et al. 1980). Ideally, the residence time should be only a few hundred milliseconds (Mohan et al. 2006). The bio-oil produced condenses out as a greenish, dark brown to reddish oil; most char that is formed is collected at the bottom of the reactor. The high rate of heat transfer is needed to chemically shatter the biomass, cleaving the chemical bonds to make smaller molecules from MW of 2 (H_2) to 300–400 (more detail on the reaction chemistry is presented in Section 8.5.4.2). About 70% of the product is bio-oil (note that this includes water), 10–15% is gaseous, and the remainder is solid char (Ringer et al. 2006).

Because the heat transfer must be extremely fast, the feedstock for pyrolysis must be particularly fine (particle size of about 2 mm) and the convection good (Bridgewater et al. 2001). Upon entering the pyrolysis zone any moisture content in

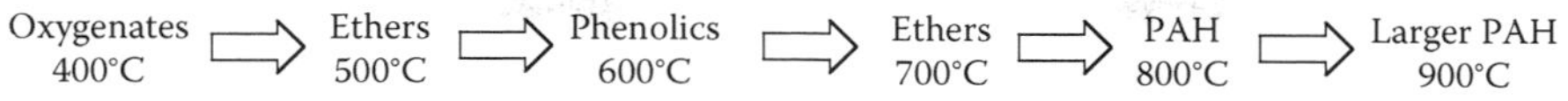

| Oxygenates 400°C | $\Longrightarrow$ | Ethers 500°C | $\Longrightarrow$ | Phenolics 600°C | $\Longrightarrow$ | Ethers 700°C | $\Longrightarrow$ | PAH 800°C | $\Longrightarrow$ | Larger PAH 900°C |

FIGURE 8.14   The changes in pyrolysis composition with time. PAH refers to polyaromatic hydrocarbons. (Adapted from Elliot, D.C. 1986. Analysis and comparison of biomass pyrolysis/gasification condensates. PNL-5943, Final Report. Pacific Northwest Laboratory. Richland, WA.)

the biomass explosively vaporizes, shredding the biomass and releasing the volatile organics to initiate the pyrolysis reactions that form the primary products. These volatiles can condense out or react with one another to produce the secondary products, including tars. The constitution of the pyrolysis product changes with both time and temperature as shown in Figure 8.14 (Elliot 1986). Even though the residence time is remarkably short, there is continuing interaction between the primary products formed from the initial pyrolysis and the secondary products produced between and among the primary and secondary products. If the pyrolysis temperature is too high, the product distribution will tend toward char via oligomerization, dehydration, decarbonylation, and decarboxylation reactions. Temperatures above 700°C can lead to gasification and a low-quality syngas.

Another consideration with respect to the need for fast heat transfer is the reactor design. Pyrolysis reactors incorporate some mechanism by which the fast heat transfer can be achieved at a uniform temperature, something that is generally accomplished by good mixing. A fluidized bed is one example of a reactor that works effectively for pyrolysis. In a fluidized bed, the materials are vigorously mixed by suspending them (under pressure) in a gaseous flow via bubbling or some other form of circulation. As a result, the particulate within the reactor behaves like a fluid, hence the name. This almost always requires the addition of an inert additive such as sand. One problem that fluidized bed reactors exhibit is that their high velocity carries char fines and ash with the product as it condenses.

8.5.2.3   Product

Pyrolysis of biomass gives three products: the *permanent gases*, the bio-oil, and char. The process conditions are optimized to obtain the highest yield possible of the desired bio-oil. It is a delicate balancing act: higher temperatures and/or longer residence times can lead to tar formation or the opposite—cracking to form more gases. The physical properties of the pyrolysis oil depend to some degree on the feedstock (in terms of *type* of feedstock but even its growth environment), but the heating value (about 17 MJ/kg) is essentially the same regardless of the source (provided that the water content of product is held constant (Ringer et al. 2006)). It is worth noting two comparisons: (1) the heating value of the starting biomass is roughly 18 MJ/kg, hence pyrolysis simply converts the biomass from a low bulk density solid fuel into a more convenient liquid fuel, not a fuel with a significantly higher heating value, and (2) the heating value of crude bio-oil is only about 40–45% (by weight) that of liquid hydrocarbon fuels (Mohan et al. 2006).

The bio-oil product is a free-flowing liquid that is a microemulsion containing as much as 50% water (although typically more like 15–30%). It is not miscible with nonpolar organic solvents, making it unsuitable for blending with hydrocarbon fuels. It is corrosive due to the presence of carboxylic acids in the product, giving a pH of 2–2.5. Furthermore, because the product components are formed without being allowed to reach equilibrium, the resultant oil is unstable: further reactions take place upon aging, leading to a higher-viscosity product. These reactions (aldol and other condensation reactions leading to phase separation) are catalyzed by acid and the presence of char fines in the bio-oil. As a result, the shelf life is weeks to a few months at most. Finally, bio-oil suffers from low volatility, cold-flow problems, and poor combustion characteristics (Wornat et al. 1994). As a result, upgrading of bio-oil is required (*vide infra*).

The chemical makeup of bio-oil is, in a word, complex: over 400 compounds have been identified, typically by GC-MS, but to try and characterize every compound in bio-oil would be pointless (Graça et al. 2012; Soltes and Elder 1981). Volatiles (CO, CO_2, methanol, acetaldehyde, acetic and formic acids, and other small molecules) are formed from the pyrolysis of all three major components of plant biomass (cellulose, hemicellulose, and lignin). The breakdown of cellulose and hemicellulose leads to mostly levoglucosan (roughly 35–45% by weight) and a wide variety of furans, hydroxyaldehydes, hydroxyketones, and sugars. Lignin breaks down into a huge variety of complex phenols (see Table 8.3) and aromatics (Alén et al. 1996). Some representative mechanisms for the formation of a few of these compounds will be examined in the next section.

8.5.2.4 Pyrolysis Reactions

As can be gleaned from the description of the chemical makeup of the pyrolysis product, the pyrolysis reactor is a den of wild abandon with respect to cleavage of carbon–carbon and, particularly, carbon–oxygen bonds. The energy needed to break these bonds is relatively low (compared to, e.g., the reforming of methane to form hydrogen (Ringer et al. 2006)). A plethora of possible reaction pathways exist, from retro-aldols, intramolecular acetal and ketal formation, dehydrations, depolymerizations, and repolymerizations via condensation or radical reactions. Many reaction mechanisms have been proposed for the formation of the molecules that make up bio-oil, although a complete mechanistic unraveling is neither realistic nor necessary—the formation of these products follows the same pathways seen in other organic fragmentation processes, both polar and radical. For example, a plausible mechanistic pathway for the formation of furans from cellulose (via levoglucosan) is proposed in Figure 8.15 (recall the formation of levoglucosan from cellulose shown in Figure 8.13) (Thangalazhy-Gopakumar et al. 2011). Retro-aldol reactions can lead to any number of smaller molecules, as shown in Figure 8.16a and b; a related fragmentation mechanism is shown in Figure 8.17 (Lomax et al. 1991). Any of these smaller molecules could re-condense into higher-molecular-weight compounds, so it is not difficult to envisage how a complex mixture of molecules can result under the harsh conditions of pyrolysis.

8.5.2.5 Upgrading Bio-Oil

Given crude bio-oil's undesirable properties (corrosiveness, instability, low heating value, and poor combustion properties), upgrading is required to make a

TABLE 8.3
Some Phenolic Lignin Breakdown Products

Catechols

Guiacols

Syringol/als

Source: Adapted from Thangalazhy-Gopakumar, S. et al. 2011. *Energy Fuels* 25 (3):1191–1199.

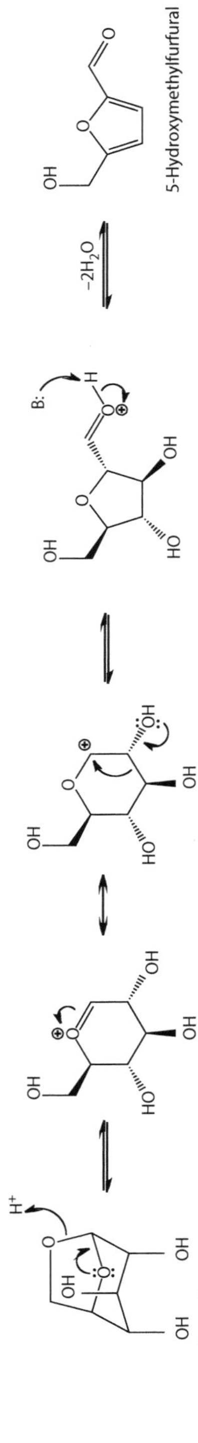

FIGURE 8.15 Formation of 5-hydroxymethylfurfural from cellulose via levoglucosan.

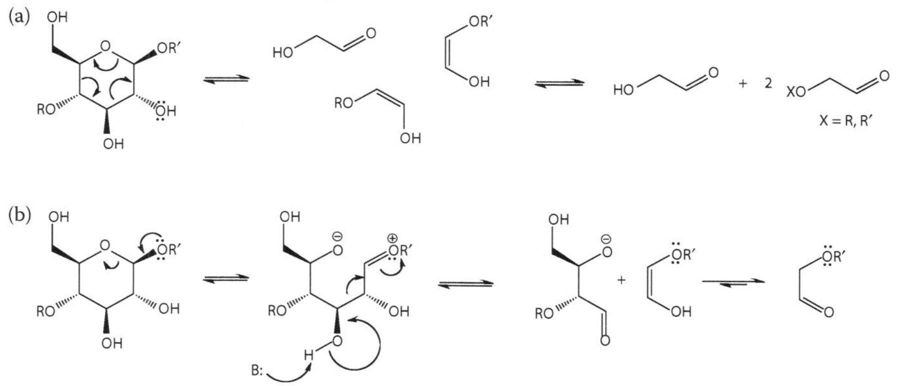

FIGURE 8.16   Fragmentation mechanisms in pyrolysis. (a) A retro [2 + 2 + 2] cycloaddition and (b) a retro-aldol reaction.

commercially viable product. Cleanup by extractive separation is one approach; for example, the National Renewable Energy Laboratory (NREL) solvent method (Figure 8.18) was developed for deriving a useful adhesive. But upgrading to make a high-quality fuel has required extensive research into developing suitable catalytic processes to convert the bio-oil into a less viscous, more stable material that more closely mimics crude oil. The presence of aldehydes, ketones, and other reactive oxygenated species is primarily responsible for the instability of the bio-oil, so deoxygenation is required as a part of the upgrading. Viscosity reduction requires lowering in the proportion of higher-molecular-weight compounds. The general reactions associated with upgrading are given in Table 8.4.

Upgrading can take place *in situ* (i.e., during the pyrolysis process) or after the product formation. Two of the more commonly used methods to achieve these goals are hydrodeoxygenation (HDO) and cracking (specifically cracking with a zeolite catalyst) (Mortensen et al. 2011). The HDO method borrows heavily from progress made in hydrodesulfurization of diesel, gasoline, and other petroleum products. The catalysts most frequently used are Co- or Ni-promoted molybdenum sulfide

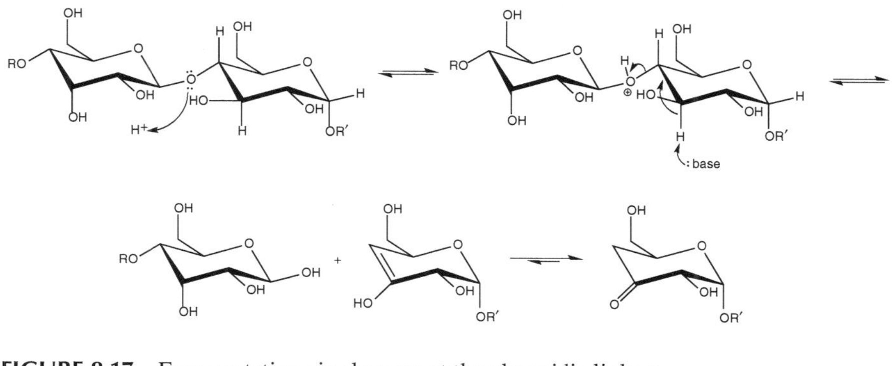

FIGURE 8.17   Fragmentation via cleavage at the glycosidic linkage.

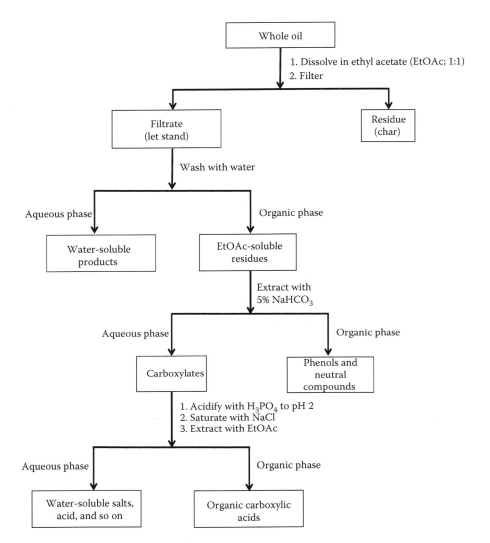

FIGURE 8.18 Cleanup of bio-oil via solvent extraction. (Adapted with permission from Chum, H. et al. 1989. Biomass pyrolysis oil feedstocks for phenolic adhesives. In *Adhesives from Renewable Resources*, edited by R.W. Hemingway, A.H. Conner and S.J. Branham. Washington, D.C.: Copyright 1989, American Chemical Society: 135–151.)

(MoS_2) on Al_2O_3. The process is run under an atmosphere of hydrogen (ca. 1000–12,000 kPa) at 250–450°C to reductively deoxygenate the bio-oil and generate water as the by-product.

Elucidating the mechanism of the HDO process has been a matter of intense research effort, particularly with regard to theoretical studies. While the specifics are not yet well understood, several key steps can be reasonably anticipated (see Figure 8.19):

TABLE 8.4
Reactions in the Catalytic Upgrading of Bio-Oil

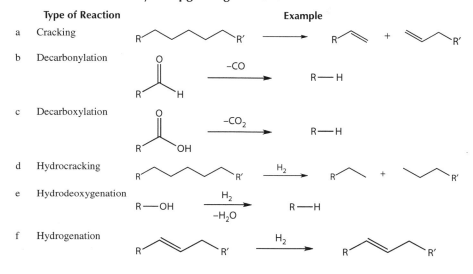

	Type of Reaction	Example
a	Cracking	
b	Decarbonylation	
c	Decarboxylation	
d	Hydrocracking	
e	Hydrodeoxygenation	
f	Hydrogenation	

Source: Reprinted from *Appl. Catal. A Gen.* 407 (1–2), Mortensen, P.M. et al. A review of catalytic upgrading of bio-oil to engine fuels. 1–19, Copyright 2011 with permission from Elsevier.

- The activation of the molybdenum–sulfur catalyst (1) presumably takes place by oxidative addition of hydrogen gas followed by loss of hydrogen sulfide (H_2S) to form a vacant coordination site (2).
- If the oxygenated species is an alcohol, the compound coordinates in a σ fashion to give intermediate (3).
- The actual deoxygenation step is not well understood; it has been postulated to proceed by a carbocation intermediate (Romero et al. 2010) or by bimolecular nucleophilic substitution (Dupont et al. 2011). In the latter case, an adjacent thiol attacks the coordinated alcohol (3), transferring the organic group to the sulfur (4). Subsequent desulfurization (presumably by reductive elimination) yields the deoxygenated species.
- Regeneration of the catalyst likely takes place by hydrogenation and loss of water (5).

Unraveling a basic understanding of the HDO mechanism should lead to the design of improved catalysts for the process—a high priority, since the conversion of bio-oil to higher-value fuels requires thorough deoxygenation.

Zeolite cracking of bio-oils is different from the HDO method in that hydrogenation is unnecessary so that the process can be run at atmospheric pressure. Recall from Section 2.3 that the cage-like structure of the zeolites allows molecules to enter the pores and be "cracked" by acid catalysis at moderately high

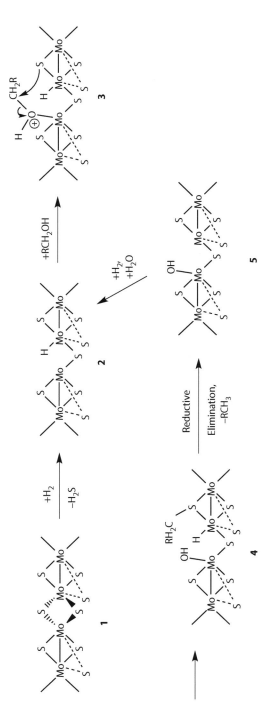

FIGURE 8.19 Proposed steps in the deoxygenation of bio-oil by a molybdenum–sulfur catalyst.

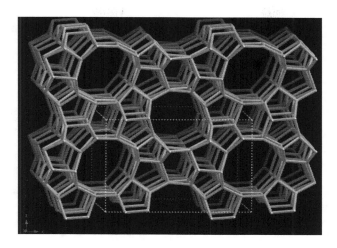

FIGURE 8.20 The ZSM-5 zeolite framework. (Reprinted with permission from C. Baerlocher and L.B. McCusker, Database of Zeolite Structures: http://izasc.ethz.ch/fmi/xsl/ IZA-SC/Credits.htm.)

temperatures (usually between 300°C and 600°C) (Mortensen et al. 2011). Bio-oil cracking with zeolites is most often carried out with an activated ZSM-5 catalyst (Figure 8.20) leading to carbocation-based fragmentations and loss of oxygen via acid-catalyzed dehydration reactions along with decarbonylations and decarboxylations. The average pore size of ZSM-5 and its high number of both Lewis and Brønsted acid sites make it a fairly active cracking catalyst, and *in situ* cracking with a variety of modified ZSM-5 catalysts (nickel-, cobalt-, iron-, or gallium-substituted ZSM-5) has produced as much as 16 wt.% hydrocarbons (French and Czernik 2010). A significant problem with zeolite cracking, however, is coking, a side reaction that is also enhanced by the number of acidic sites. The formation of these carbon deposits (primarily polyaromatics) blocks the pores of the zeolite, deactivating the catalyst.

A kinetic description of the bio-oil cracking process illustrates its complexity: the initial separation of the bio-oil into a volatile and a nonvolatile mixture is followed by cracking of the nonvolatiles into further volatiles *or* polymerization/condensation to form char. The volatiles are distributed between oil, aqueous, and gaseous fractions and can also polymerize or condense to form more carbon residue (Adjaye and Bakhshi 1995). Unfortunately, the yields from zeolite cracking of bio-oil are quite low (≈20%), largely a result of extensive coking (Balat et al. 2009). A one-step upgrading process was recently described in which supercritical ethanol was used as both reagent and solvent. Hydrogen gas and a palladium/zirconium catalyst were mixed with the bio-oil/ethanol mixture and heated in an autoclave to 280°C under a pressure of about 10 MPa. As the GC-MS chromatographs in Figure 8.21 illustrate, this treatment resulted in a significant cleanup of the crude bio-oil (note that the entire analysis time was 58 min, hence the entire scan is divided into three intervals for clarity of presentation). The pH and heating value of the upgraded product increased while the viscosity decreased. The problematic reactive species

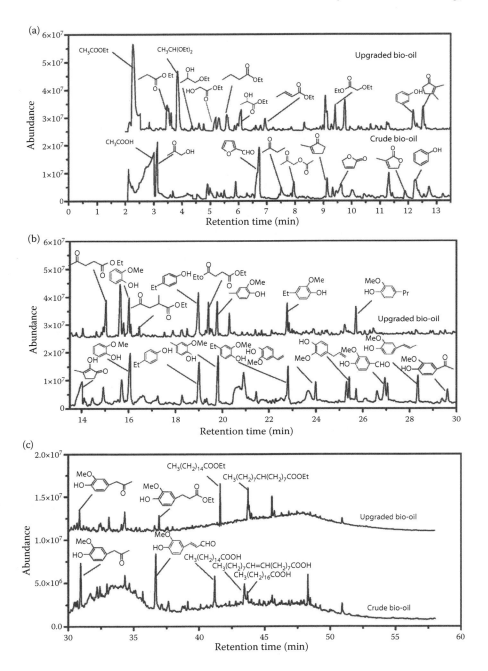

FIGURE 8.21 Comparison of crude and upgraded bio-oils by GC-MS. (a) Retention time $t_R = 0-13$ min; (b) $t_R = 14-30$ min; and (c) $t_R = 30-60$ min. (Reprinted with permission from Tang, Z. et al. 2009. One step bio-oil upgrading through hydrotreatment, esterification, and cracking. *Ind. Eng. Chem. Res.* 48:6923–6929. Copyright 2009, American Chemical Society.)

in crude bio-oil (aldehydes and ketones) were reduced (presumably to alcohols) and carboxylic acids converted into esters, resulting in an overall diminishing of the compounds that lead to premature aging of the product (Tang 2009). Overall, the hydrodeoxygenation route currently appears to be more promising in terms of preparing a higher-value bio-oil. But in the end analysis, bio-oil upgrading is far from commercially viable and, as a result, so is the prospect of bio-oil as a large-scale renewable fuel source (Mortensen et al. 2011).

8.5.3 GASIFICATION

8.5.3.1 Introduction

As we saw in Section 5.2.2, gasification of coal is a well-established technology, having been carried out in the 1800s to produce "town gas" for lighting and heating of boilers and long used in the production of hydrogen as a component of coal-derived syngas. Other more recent applications of syngas are electricity generation and the synthesis of liquid chemicals via the Fischer–Tropsch (Section 5.2.2) and other processes. Biomass gasification, however, is not as well established and is only now approaching large-scale commercialization with the primary use being combined heat and power applications. Several <100 MW$_{th}$ (where the subscript *th* simply means we are referring to megawatts of *thermal* energy) biomass gasification plants exist in Europe and India with a total syngas production capacity of well over 2 million Nm^3/day (normal cubic meters; measured at standard temperature and pressure) (National Energy Technology Laboratory 2010) (Kirkels and Verbong 2011). Countries around the globe are very actively engaged in biomass gasification research.

How does gasification differ from pyrolysis? The obvious difference is in the product: gasification strives for a gaseous product whereas pyrolysis produces the liquid bio-oil. The difference in process comes in the presence or absence of oxygen. Gasification requires controlled combustion in the presence of a substoichiometric amount of oxygen; pyrolysis is conducted in the absence of oxygen. Gasification is considered by many to be the most promising of the biomass conversion processes. However, there are three major problems linked to biomass gasification:

1. Biomass ash has a low melting point—as a result, there is a significant formation of tar and ash.
2. The biomass feedstock requires extensive energy-intensive pretreatment, including drying.
3. The product gas may need further purification if required for a subsequent application.

In addition, from a strictly thermodynamic point of view, the high oxygen content of biomass means that it is not an ideal feedstock for gasification, although pretreatments such as torrefaction (to increase the C:O ratio) improve the energy balance. That being said, among the thermochemical conversion processes for biomass, gasification is the most efficient and biomass gasification has filled an important niche in small-scale renewable energy technologies.

8.5.3.2   Process Parameters and Reactor Design

As we saw in the section on gasification of coal, the reaction chemistry taking place in a gasifier is complex and varied. Some of the most important reactions seen in the steam gasification of biomass are presented in Table 8.5. There is significant overlap between the two feedstocks with respect to these reactions, although the wide variety of oxygenated compounds in biomass makes the entire palette of chemistry much more complex. Gasification of biomass requires taking a messy mixture of highly oxygenated polymers and converting it into as much CO and H_2 as possible with minimal formation of by-products. A broadly oversimplified description would be that the biomass is combusted (oxidized to CO_2), pyrolyzed (to form char and to degrade the cellulose/hemicellulose/lignin), and reduced (to form CO and H_2). Several process parameters make a large difference in the product composition.

8.5.3.2.1   Feedstock and Pretreatment

The feedstock for biomass gasification is typically woody biomass with a 5–30% moisture content, although rice husks and "black liquor," a lignin-rich by-product from paper manufacture, have also been gasified. MSW and animal manures can also be used, often in conjunction with drier feedstocks and on a smaller scale (Ro et al. 2007). In any case, the desired product gas is a mixture of H_2 and CO (syngas), with other components consisting of CO_2, CH_4, alkenes, steam, and (if air is used as the gasification agent) nitrogen gas.

Pretreatment of lignocellulosic biomass (LCB) (as described in Section 8.4) has a large impact on the composition of the product gas and contaminants. Particle

---

TABLE 8.5
Reactions in the Gasification of Biomass

	Equation	Name of Reaction	$\Delta H^\circ_{f(298)}$ (kJ/mol)
a	$CH_4 + CO_2 \rightarrow 2CO + 2H_2$	Dry reforming of methane	123.8
b	$CH_4 + H_2O \rightarrow CO + 3H_2$	Steam reforming of methane	205.3
c	$CO + H_2O \rightarrow H_2 + CO_2$	Water gas shift (WGS) reaction	−42.2
d	$C + H_2O \rightarrow H_2 + CO$	Heterogeneous WGS reaction	130.4
e	$C + CO_2 \rightarrow 2CO$	Boudouard equilibrium	172.6
f	$C + 2H_2 \rightarrow CH_4$	Hydrogenation	−74.9
g	$C + O_2 \rightarrow CO_2$	Combustion of char	−393.8
h	$C + \frac{1}{2}O_2 \rightarrow CO$	Combustion of char	−111
i	$H_2 + \frac{1}{2}O_2 \rightarrow H_2O$	Formation of water	−242
j	$CO + 3H_2 \rightarrow CH_4 + H_2O$	Steam reforming	−206

Source: Reprinted with permission from Salaices, E., B. Serrano, and H. de Lasa. Biomass catalytic steam gasification thermodynamics analysis and reaction experiments in a CREC riser simulator. *Industrial & Engineering Chemistry Research* 49 (15):6837. Copyright 2010, American Chemical Society.

size in particular is a critical concern because gasification requires fast heat transfer and good convection to give a higher-quality product. For example, a fluidized bed gasifier requires finely divided fuel in order to attain the necessary convection. In general, the smaller the biomass particle size, the better the heat transfer, but with the trade-off that the feedstock is more costly due to the higher pretreatment costs. If a higher-quality gas is not required for subsequent application, a larger particle feedstock (such as corn cobs or wood chips) can be used effectively, although the use of these feedstocks requires a longer residence time in the gasifier.

8.5.3.2.2 Temperature

At the most basic level, gasification of biomass is the partial oxidation of organic material at elevated temperature to yield a combustible gas. In essence, it consists of four processes in one unit:

1. Drying (evaporation of moisture at ≤120°C)
2. Pyrolysis (volatilization; up to ca. 350°C)
3. Cracking/reforming of the volatilized vapors
4. Gasification

As usual, a happy medium must be reached in optimizing the temperature range for gasification of biomass and the optimum temperature range varies with the type of gasifier. Temperatures that are too low increase the production of contaminants (especially tar), while temperatures that are higher (>950°C), while producing higher-quality syngas, favor the formation of molten ash, a.k.a. *slag*, that can severely foul the reactor. The typical temperature range for most biomass gasification is 750–900°C.

8.5.3.2.3 Gasification Agent

As is the case for coal gasification (Section 5.2.2), a number of different compounds can be used as the actual oxidant in biomass gasification, including pure oxygen, air, water (steam, liquid water, or supercritical water), or mixtures of any of the above. The nature of the gasification agent impacts the efficiency of the process by altering the ratio of the product gases. For example, if air (inexpensive) is used in lieu of pure oxygen (more costly), costs are reduced at the front end but the energy value of the producer gas is reduced more than twofold due to dilution by N_2 (the LHV of this so-called "producer gas" is only about 5–6 MJ/Nm³) (Overend 2004). The use of air leads to an H_2:CO ratio of ≤1, while the use of steam increases the proportion of H_2, as shown in Equations 8.1 and 8.2.

$$C_6H_8O_4 + 2H_2O \rightarrow 6CO + 6H_2 \tag{8.1}$$

$$C_6H_8O_4 + 8H_2O \rightarrow 6CO_2 + 12H_2 \tag{8.2}$$

As usual, the product gas composition is further altered toward a higher proportion of hydrogen by the water gas shift reaction (entry c, Table 8.5).

8.5.3.2.4 Reactor Design

Although the design and engineering of gasification reactors is well beyond the scope of this book, the reactor has a large influence on the chemistry and product outcome (Reed 2002). There are two basic types of gasifier designs—*fixed beds* and *fluidized beds* (fluidized beds were described in Section 8.5.2.2). Two specific types of fixed bed gasifiers are shown in Figure 8.22a and b with air as the gasification agent.

1. *Updraft (countercurrent)*. This and the downdraft (2) designs are fixed bed gasifiers in which the physical and chemical changes occur inside the reactor in a series of zones. As can be seen in Figure 8.22a, in the updraft gasifier the air/oxidant flows in from the bottom of the gasifier as the biomass fuel enters in from the top, in a counter-current flow. Updraft gasifiers are notorious for tar production.
2. *Downdraft (cocurrent; Figure 8.22b)*. In this reactor configuration, the gasification takes place across a constricted throat resulting in better mixing, some cracking, and a cleaner gas with much less tar than in an updraft gasifier.

These fixed bed gasifiers are essentially stationary blast furnaces. Their simplicity of design is their main advantage, but the produced gas usually needs secondary cleanup (removal of tar and ash) if used for anything other than direct heating (NETL 2013). For both fixed bed gasifiers, the typical gas composition is (de Lasa et al. 2011):

Gas	Percent by Volume
Carbon monoxide	20–30
Hydrogen	5–15
Methane	1–3
Carbon dioxide	5–15

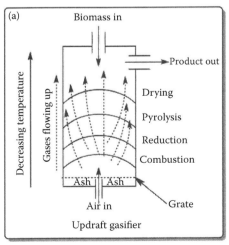
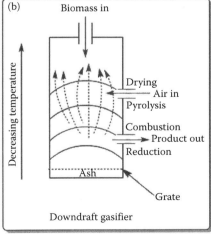

FIGURE 8.22 Common fixed bed gasifier designs. (a) Updraft and (b) downdraft.

The fluidized bed reactor described in the pyrolysis section can be used for biomass gasification and an even more complicated gasifier design, the *entrained-flow* reactor, shows promise despite its higher-temperature operation. These reactors require a higher initial capital investment but ultimately result in a lower fuel production cost, provided that the waste heat from the syngas is captured and used efficiently (Swanson et al. 2010). Entrained-flow gasifiers operate at high temperature and pressure and add the oxidant and the biomass fuel in a turbulent flow through the top of the gasifier simultaneously, entraining the fine particles of fuel as they move in a very turbulent flow through the gasifier. Since these gasifiers operate at 1100–1300°C, the ash melts but the turbulence prevents the formation of large chunks of slag, so the molten ash is entrained with the gas flow (NETL 2013). Furthermore, the high operating temperature results in a much lower tar formation. An entrained-flow oxygen-blown pressurized biomass gasification pilot plant using wood powder as fuel was able to produce high-quality syngas with a product ratio of H_2 (19–23%), N_2 (18–25%), CO (35–40%), and CO_2 (16–19%) (Weiland et al. 2013).

8.5.3.3 Gasification Reactions

Different types of chemistry take place within certain temperature ranges within the gasifier; we will focus on the process in an updraft fixed bed gasifier for simplicity's sake. At the lowest and hottest zone, oxidation takes place to form CO_2 from char (entry g, Table 8.5). This is the rate-limiting step in gasification (Bronson et al. 2012). The CO_2 so formed, along with steam, travels up through the gasifier to provide the heat needed to drive the highly endothermic water gas shift and Boudouard reactions (entries d and e). The CO_2 and steam are reduced to CO and H_2 in the reduction zone by the excess carbon in the char. The product gases travel upward and dry the biomass. As the temperature cools to below about 600°C, pyrolysis of the dried biomass particulate occurs and low-molecular-weight compounds are formed, for example vaporized or liquid oxygenated compounds such as levoglucosan, hydroxyaldehydes, and various phenols, along with small molecules including methane and hydrogen. Some pyrolysis vapors travel up and condense on the cooled fuel, going out with the product gas and resulting in the formation of tar. Biochar forms by pyrolysis as well, largely by dehydration and cross-linking reactions. The char is gasified while the other primary products react at higher temperatures (700–850°C) to make alkenes, condensable oils, and tars. At even higher temperatures (>850°C), polynuclear aromatics and methylated aromatics can be formed and can condense into tar. At these higher temperatures, equilibrium is established so that the product composition is determined primarily by thermodynamics.

Much study has been devoted to understanding the formation of tar during gasification since its removal adds cost to the manufacturing process. As we saw in Section 8.5.3.2, the gasifier configuration plays a large role in the amount of tar that is formed and can range from as little as 0.5 g/m³ to as much as 100 g/m³ product (Devi et al. 2003)! Tar is mostly a messy mixture of aldehydes and ketones that have undergone condensation reactions plus various complex phenols, furans, and other aromatic compounds, including toluene, naphthalene, anthracene, pyrenes, and so on (Maniatis and Beenackers 2000). As we will see in the next section, its removal presents a sticky challenge for biomass energy conversion.

8.5.3.4   Contaminants and Catalysis

Gasification of biomass invariably results in the formation of by-products such as tar, biochar, ash, and inorganic volatiles such as ammonia and HCl. The proportion of these contaminants is dependent upon the feedstock, gasifier configuration, and operating conditions. Ash consists of a mélange of oxides (CaO, K_2O, P_2O_5, MgO, SiO_2, Na_2O) plus carbon and can be used as a soil amendment. It can be removed from the product stream by mechanical means *if* it does not melt inside the reactor and form slag, a major annoyance.

Tar is an even bigger problem: it fouls equipment, deactivates catalysts, clogs pipes and can ruin compressors, pumps, and turbines. If the end use of the gasified biomass is strictly for heating, the gas formed likely needs no further purification. However, for high-quality (read: tar-free) syngas, there are two approaches: optimize the gasification so that little tar is formed, or purify the product stream. For either approach, it is catalysis to the rescue. We will focus on post-gasification tar removal (so-called *secondary* methods).

Removal of tar by secondary methods is basically revisiting the realm of catalytic cracking (Section 2.3.3.2) where a catalyst takes the higher-molecular-weight compounds found in tar and converts them into smaller, more volatile components that may be captured for heating fuel. Not only does the catalyst need to be effective at removing tar, it should also be able to reform methane to hydrogen and CO (recall entries a and b of Table 8.5) and, of course, be robust and inexpensive. Catalysts used for tar removal are subject to the same sorts of failure pathways as other catalysts: poisoning by impurities, sintering, coking and so on, and the ideal tar removal catalyst does not yet exist. A wide variety of catalysts have been examined for use in tar removal; even the char naturally formed in the gasifier is catalytic in the destruction of tar but deactivates readily. We will focus on just two types of tar removal catalysts: basic catalysts (in particular, the naturally occurring minerals dolomite and olivine) and nickel-based catalysts. Many other materials have been studied (clay minerals, ferrous metal oxides, zeolites, many other transition metals, etc.) with varying degrees of success so that the search is ongoing.

The catalyst standard for tar removal and upgrading the product stream to form a higher-quality (more H_2-rich) product is a nickel-based catalyst on an Al_2O_3 support. A number of nickel-based catalysts that are commercially available have been used effectively, often with a promoter such as molybdenum, cerium oxide, magnesium oxide, or aluminum (Anis and Azinal 2011). While these nickel catalysts are the most effective at tar removal, they suffer from the same drawbacks we have seen previously: high cost, deactivation by coking (a result of the steam reforming process) and poisoning by sulfur, chlorine, and other contaminants.

A low-cost alternative to nickel is the naturally occurring mineral dolomite $[CaMg(CO_3)_2]$. Dolomite is a particularly brittle solid whose effectiveness as a tar-removal catalyst can be enhanced by calcining, a process whereby the material is heated to drive off any volatiles without sintering or fusing the particles together. Nevertheless, its physical fragility is such that it cannot be used in a fluidized bed. Olivine $[(Mg, Fe^{2+})_2 SiO_4]$ has also been used for secondary cleanup of producer gas and is more robust and less prone to coking than dolomite (Ammendola et al. 2010). Unfortunately, however, it is also less effective than dolomite, even after calcining (Corella et al. 2004). While both these minerals are effective, the nickel-based catalysts are considerably more so. An attractive compromise is the combination of dolomite with nickel. Such a catalyst was prepared by combining an aqueous solution of nickel nitrate with a high-Fe_2O_3-content dolomite, and then drying and calcining the catalyst. The performance of this catalyst (4.1 wt.% nickel) compared favorably to two commercial catalysts (ICI-46-1 at 24 wt.% nickel and Z409 at 21 wt.% nickel), removing up to 97% of the tar and producing nearly 70% H_2 in the product gas. Given the significant difference in weight percent Ni, the Ni/dolomite combination appears to be a promising, less expensive option (Wang et al. 2004).

8.5.4 CONCLUSIONS

The thermochemical processes for converting biomass into energy fuels is complicated by the messiness and variety (in terms of both physical form and chemical composition) of biomass, but both pyrolysis and gasification have a long history of proven technology. Thermochemical conversion of biomass remains an extremely active research area, particularly with respect to catalytic upgrading of the product, be it bio-oil or biomass-derived syngas.

8.6 BIOCHEMICAL PROCESSES

In this section, three major methods for the biochemical transformation of biomass to a useful fuel will be examined: fermentation (alcohol formation), anaerobic digestion (biogas formation), and extraction/transesterification (biodiesel formation). Recall that other biochemical methods of producing energy (in the form of hydrogen gas) have already been examined in Section 5.2.3. Many of the concerns that apply to thermochemical processes apply in the case of biochemical transformations as well. For example, some kind of pretreatment—mechanical, chemical, or both—is often necessary (particularly for lignocellulosic feedstocks). But in contrast to thermochemical conversion processes, biochemical conversion processes are slow and are carried out by enzymatic (or whole organism) catalysis in aqueous solutions. We will once again see that while research continues intensely in all of these areas, this, too, is largely an empirical science, although advances in genetic engineering, directed evolution, and proteomics are advancing our understanding immensely.

8.6.1 FERMENTATION

Probably not too long after humans started burning biomass for cooking and warmth they discovered fermentation. Fermented beverages were known in the ancient Near East, with biomolecular archaeologists determining that wine was produced as long ago as 8500–4000 BCE (Hutkins 2006). Today's fermentation is based on the same basic biochemical processes: converting glucose into ethanol via enzymatic degradation. Because of the need for renewable energy sources and the convenience of ethanol as a blendable liquid transportation fuel, the volume of ethanol produced annually has skyrocketed in the last decade, particularly in the United States where almost 14,000 million gallons were produced in 2011, a level that dropped slightly in 2012 due to drought (Figure 8.23).

Three main biomass sources have served as *first-generation* feedstocks for ethanol: sugarcane, corn, and wheat (and other cereal grains). These starch-based plant materials are relatively easy to convert into ethanol—the breakdown of starch (see Figure 8.24) can be readily carried out by α-amylase (*vide infra*). Cellulose, on the other hand, is fairly inert to decomposition unless by a termite or a fungus. But while it is easier (and therefore less expensive) to ferment starch- and sucrose-based plant materials into ethanol, their use as a fuel feedstock can compete with the food supply. Hence the driving force in fermentation research is to optimize the production of *second-generation* bioethanol by fermenting lignocellulosic biomass. The 200 million dry tons of lignocellulosic biomass produced in the United States each year *could* be converted into over 16 billion gallons of ethanol (Perlack et al. 2005). By late 2012, a combination of companies in the United States produced about 20,000 gallons of fuels using cellulosic biomass with estimates that this amount will grow to over five million gallons in 2013 (U.S. Energy Information Administration 2013a). But something as simple as stereochemistry presents an obstinate obstacle to lignocellulosic ethanol formation. We will look at the fermentation of starch-based biomass as an introduction to the biochemistry

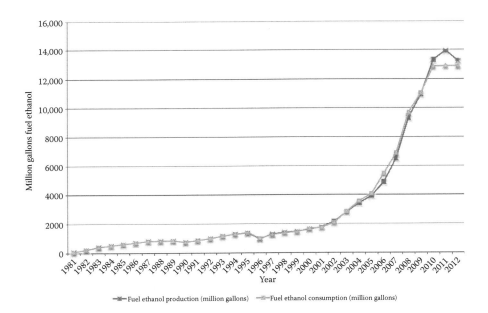

FIGURE 8.23 Production and consumption of fuel ethanol (the United States), 1981–2012. *Annual Energy Review.* Available from http://www.eia.gov/totalenergy/data/annual/#renewable

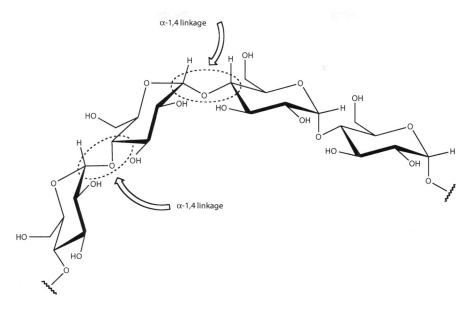

FIGURE 8.24 α-Amylose, a major structural component of starch.

of fermentation and then turn to the problems associated with the fermentation of lignocellulosic biomass.

8.6.1.1 Fermentation of Starch

As noted above, sucrose-laden biomass such as sugarcane and sugar beets, or starchy feedstocks such as barley, corn, and wheat, are good fodder for fermentation. The overall process of isolating ethanol from fermentation is outlined in Figure 8.25. Just as for other processes, the biomass may need to be comminuted prior to chemical processing. In the case of fermentation, however, the additional step of chemical hydrolysis is required in order to convert the carbohydrates (starch, sucrose) into glucose for fermentation via anaerobic glycolysis. This hydrolysis step is frequently referred to as *saccharification.*

Sucrose (Figure 8.26a and b) is a simple disaccharide of glucose and fructose. Note that sucrose is frequently shown in a Haworth projection (Figure 8.26b) for convenience in highlighting the stereochemistry at the glycoside linkage—an unfortunate shortcut for those who are familiar with line structure notation in organic

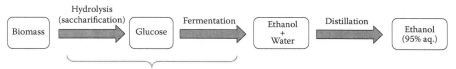

FIGURE 8.25 Overview of the fermentation process.

(a)   (b)

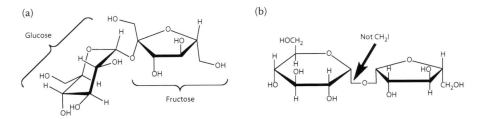

FIGURE 8.26   Sucrose in a normal representation (a) versus a Haworth projection (b).

chemistry! Starch found in grains is a mixture of polysaccharides (α-amylose and amylopectin) possessing α-1,4- and α-1,6-linkages between the individual glucose units (Figure 8.27a). α-Amylose (Figure 8.27b), in contrast to its isomer cellulose, can be cleaved with α-amylase that hydrolyzes the internal α-1,4-linkages to give several glucose units linked together (instead of the original polymer consisting of several thousand linked glucose units). Additional hydrolysis eventually converts these oligosaccharides into glucose that then undergoes fermentation via an organism of the species *Saccharomyces cerevisae,* otherwise known as baker's yeast.

The metabolic pathway of glycolysis is primarily about conversion of glucose into energy for living organisms, but fermentation hijacks the process for the purpose of ethanol production. The mechanistic pathway whereby the C6 sugar glucose is broken down to ethanol and carbon dioxide is well understood and outlined in the following. (Note that there are numerous phosphorylations and dephosphorylations that take place but these mechanisms are not shown for the sake of clarity.)

Glucose-6-phosphate is first isomerized to *fructose-1,6-bisphosphate* as shown in Figure 8.28. The next key step in glycolysis is a retro-aldol reaction that cleaves the six-carbon chain of glucose/fructose into the two three-carbon fragments *dihydroxyacetone phosphate* and *glyceraldehyde-3-phosphate*. This reaction, shown in Figure 8.29, is catalyzed by an aldolase enzyme and proceeds via the intermediacy of a Schiff base (otherwise known as an imine) from a lysine residue.

So it is a retro-aldol reaction that takes a six-carbon chain down to two three-carbon pieces, but how do we get to ethanol? First, glyceraldehyde-3-phosphate is converted into pyruvate through a series of enzyme-catalyzed reactions, including oxidation to the carboxylate, more phosphorylations, dehydration and hydrolysis. The product of all these machinations, *pyruvate*, is decarboxylated to make acetaldehyde and then reduced to make ethanol as summarized in Figure 8.30. How does this all happen? The catalytic cycle whereby the *coenzyme thiamine pyrophosphate* (TPP) converts pyruvate into acetaldehyde is shown in Figure 8.31. The *thiazolium ring* of TPP plays a central role because it is an ylide: stabilization by the neighboring positively charged quaternary nitrogen allows for the formation of a vinyl carbanion (**1**, Figure 8.31). This anion can then bond with the carbonyl of pyruvate followed by decarboxylation, proton transfer, and regeneration of the catalyst with concomitant formation of acetaldehyde, the ethanol precursor. Reduction to ethanol takes place by way of *nicotinamide adenine dinucleotide* (NADH; Figure 8.32).

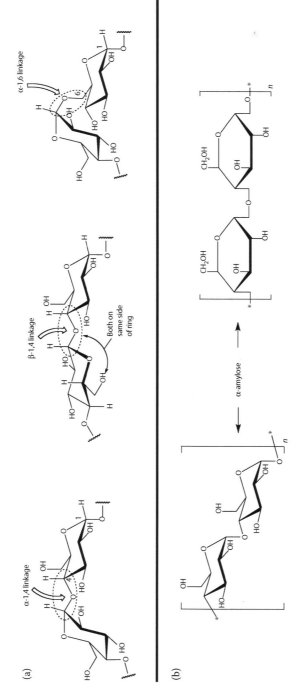

FIGURE 8.27 (a) Starch's α-1,4- versus 1,6-glycosidic linkages; (b) the α-amylose substructure.

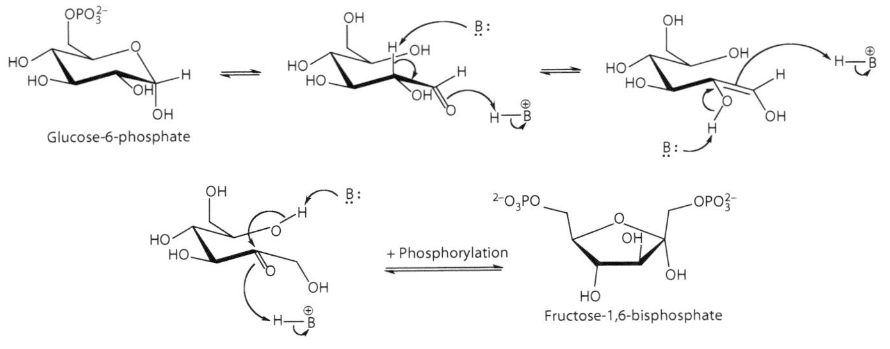

FIGURE 8.28   Formation of fructose-6-bisphosphate.

An important aspect of fermentation is that as the amount of ethanol produced increases, a level of alcohol will be reached that is lethal to the organisms doing the fermentation. Thus, the ethanol must be distilled out of the fermentation broth. Unfortunately, it forms an azeotrope with water so that the distillate is approximately 95% ethanol. To obtain dry (absolute) ethanol requires additional drying through molecular sieves. The residue from the fermentation process is called *dry distillers grains* (DDG) and is used as an animal feed.

The actual process of ethanol production for biofuel use is very well established, particularly in the United States (where the primary feedstock is corn) and Brazil (sugarcane). However, the production of bioethanol from corn is controversial. First, the energy value of ethanol is, of course, considerably less per mole than a comparable hydrocarbon. In addition, the sustainability of bioethanol production is questionable at best (Patzek and Pimentel 2005). The impacts of bioethanol production on the global food system, greenhouse-gas emissions (from growing and harvesting to combustion as a fuel), soil, water and air quality, and biodiversity must all be considered. Bluntly put, the future of corn ethanol is bleak as the focus turns to the development of cellulosic ethanol.

8.6.1.2   Fermentation of Lignocellulosic Biomass

Given that lignocellulosic biomass must replace food-based plant feedstocks for ethanol production, the technology development is well underway. The biggest problem facing the large-scale production of LCB-based bioethanol is the pretreatment required to break down the LCB into fermentable sugars.

8.6.1.2.1   Pretreatment

As we have seen, the lignocellulosic matrix is a complex physicochemical composite that is, for obvious reasons, structurally sound. To break it down so that microorganisms can have access to the sugars necessary for fermentation into ethanol is nontrivial and includes physical, chemical, and biological methods. Lignin again is a particular problem, since it protects the cellulose from breakdown by absorbing and inactivating the enzymes (up to 60–70% of the total enzyme content can be

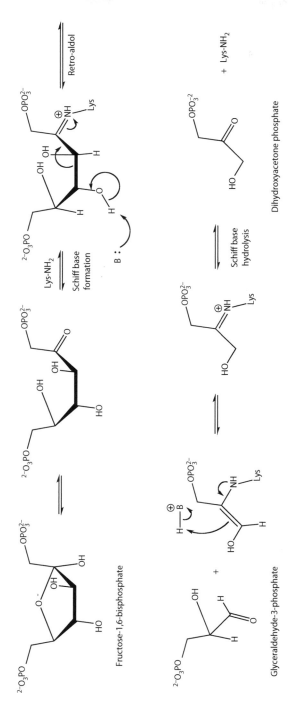

FIGURE 8.29 The aldolase-catalyzed formation of dihydroxyacetone phosphate and glyceraldehyde-3-phosphate.

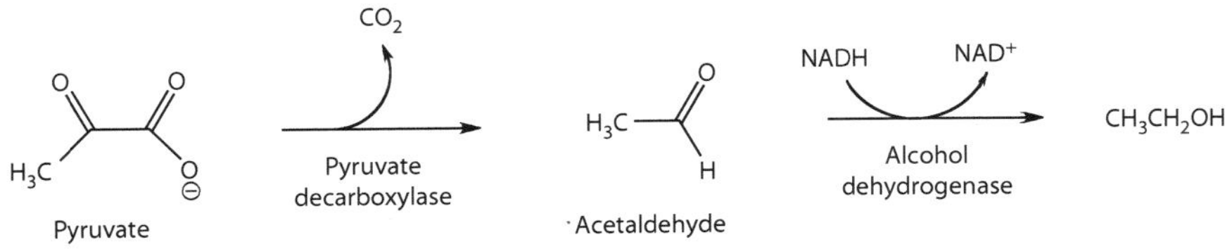

FIGURE 8.30   Conversion of pyruvate to ethanol.

adsorbed onto lignin (Jørgensen et al. 2007)). Furthermore, pretreatment is a balancing act: deconstruction of the lignocellulose matrix requires harsh conditions, but if the conditions are too severe, inhibitory byproducts can be formed and/or valuable compounds in the feedstock lost to the waste stream.

Feedstock for lignocellulosic ethanol production can be quite varied, from agricultural residue such as corn stover (the stalks left behind after the ears of corn are harvested) or straw to energy crops such as switchgrass (*Panicum virgatum* L.) and *Miscanthus × giganteus*. These perennial grasses are now the more encouraging option for bioenergy production. Switchgrass, considered the "model perennial grass for bioenergy production," is the leading candidate for bioethanol production from LCB in the midwestern United States (U.S. Department of Energy 2011). It is a warm-weather perennial grass that consists of about 30 dry weight percent cellulose with less ash and moisture than wood (Yang et al. 2009). *M. giganteus* is a particularly high-yielding energy crop with low maintenance, good tolerance to drought, and good growth in varied settings. However, its growth may be too good, as it is invasive toward native species (U.S. Department of Energy 2011). Like any other, this feedstock requires pretreatment.

The feedstock is first reduced in size by some sort of chopping or grinding in order to facilitate the next step of hydrolysis. LCB can be further broken down by biological fungi, as any forest floor reveals. However, the use of these fungi to reduce the LCB is far too time-consuming for an industrial process, so rather severe chemical treatments are necessary. Strong alkalis (NaOH, KOH) can make modest progress toward the breakdown of lignin, and the carbohydrates in the cellulose and hemicellulose are hydrolyzed to mono- and oligosaccharides. Dilute H_2SO_4 is successful at solubilizing most of the hemicellulose, exposing the cellulose for further decomposition (Agbor et al. 2011). However, fermentation inhibitors may be formed during the pretreatment process as well, requiring careful attention to process conditions.

The robust nature of LCB has led to the development of other, more effective pretreatments. For example, *steam explosion* (treatment of the comminuted biomass with saturated steam at high pressure) solubilizes the hemicellulose by hydrolysis of the acetyl groups. The acetic acid that is formed is thought to catalyze further hydrolysis of the hemicellulose (in fact, sometimes dilute H_2SO_4 is part of the steam pretreatment). However, this steam pretreatment method does little to disrupt the lignin or cellulose and results in the loss of some portion of the soluble sugars, reducing the overall fermentation yield.

Another pretreatment option is AFEX, which stands for the *ammonia fiber/ freeze explosion* method, an alkaline pretreatment method that makes use of liquid

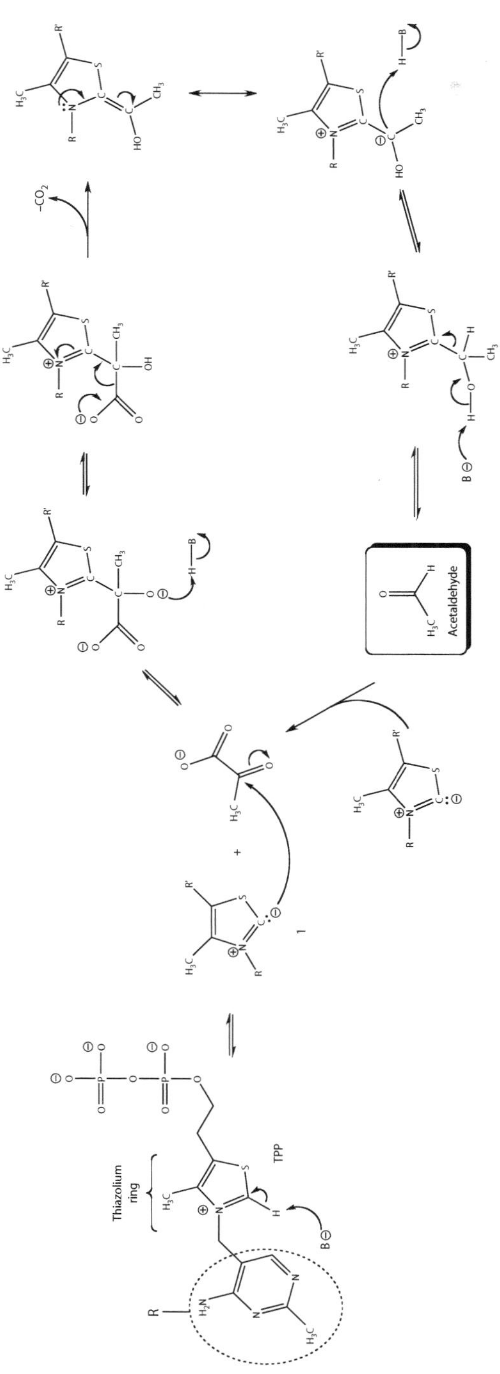

FIGURE 8.31 Mechanism of the TPP-catalyzed decarboxylation of pyruvate.

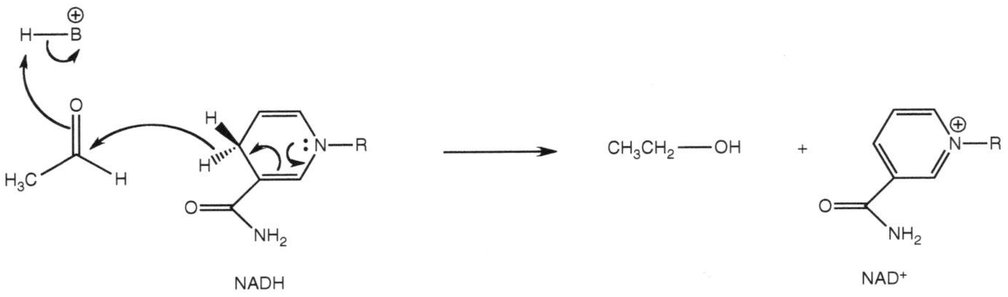

FIGURE 8.32   Reduction of acetalydehyde by NADH.

ammonia to swell the cellulose and cleave and remove the lignin. In so doing, the treated biomass is then more amenable to hydrolysis. AFEX is carried out at 60–90°C and at pressures >3 MPa for about 30 min, then the reactor vessel is vented to explosively release the pressure. Because of these relatively mild conditions, few by-products (like furans from sugar decomposition that can inhibit fermentation) form (Palmqvist and Hahn-Hägerdal 2000).

8.6.1.2.2 Fermentation

Given the carbohydrate complexity of LCB a number of different enzymes are needed for efficient fermentation. Intensive effort has been devoted to the discovery of biological systems that can optimize this process. The enzymes necessary for biochemical conversion of cellulose to ethanol fall into three classes (Jørgensen et al. 2007):

- Exo-1,4-β-glucanases: These enzymes cleave cellobiose units from the end of the cellulose chain.
- Endo-1,4-β-D-glucanases: These are responsible for random hydrolysis of β-1,4-glucosidic bonds in the cellulose chain.
- 1,4-β-D-glucosidases: These enzymes convert cellobiose into glucose (Figure 8.33) and cleave individual glucose units from oligosaccharides.

Other enzymes are necessary for the fermentation of the hexose and pentose sugars obtained from hydrolysis of hemicellulose. And, as noted above, lignin is a real barrier to enzymatic decomposition of LCB. Even discounting these challenges, the enzymatic hydrolysis of treated LCB is not straightforward. Substrates for one enzyme

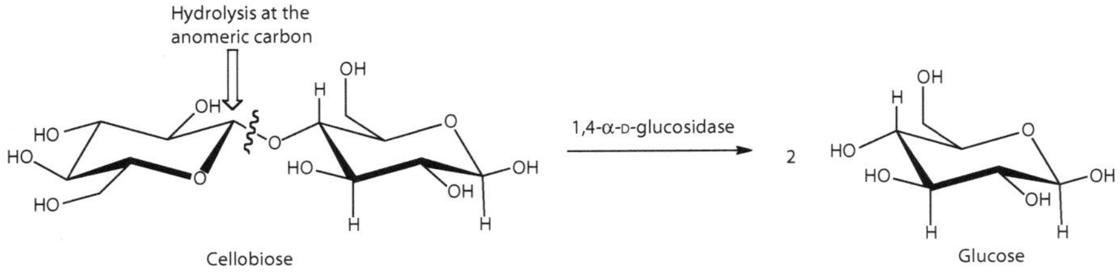

FIGURE 8.33   Enzymatic cleavage of cellobiose.

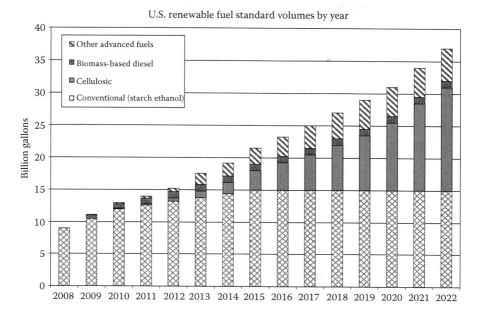

FIGURE 8.34 Renewable fuel standards for bioethanol and biodiesel (the United States). (From EIA Monthly Energy Review, Table 10.3 and 10.4 and the EPA Renewable Fuels Standard EMTS Informational Data. Available at: http://www.afdc.energy.gov/data/10421.)

may inhibit another, accumulation of one hydrolytic product may inhibit a different enzyme, and other degradation products may be inhibitors. As a result, the effectiveness of cellulase enzymes is limited. Some of the research being done to address these issues includes development of new enzymes that have a greater tolerance for products and by-products, a greater operating temperature range as well as a wider pH tolerance. Nevertheless, development of new cellulase enzymes is a major barrier to scale-up of cellulosic ethanol and is of particular concern given the push to move from starch-based ethanol. As the renewable fuel standards graph in Figure 8.34 shows, cellulosic ethanol production must increase dramatically in order to meet these volume requirements. Enormous research effort will be required in order to meet this demand.

8.6.2 ANAEROBIC DIGESTION

"Waste-to-energy" sounds like an ideal renewable, sustainable energy conversion and *anaerobic digestion* (AD) can do just that. It can be a real "one person's trash is another one's treasure" story, where anaerobic digestion turns organic waste into *biogas*, a methane-rich fuel that can be used for CHP application, for electricity generation, or for reforming into hydrogen for use in fuel cells. We are awash in waste, from restaurant and cafeteria food wastes to brewery and winery waste, slaughterhouse and food-processing plant waste to MSW, sewage sludge, feedlot waste, industrial wastewater, and agricultural residues. We do not need to grow an energy crop or divert the food supply to use this source of energy that is available every day all around us.

Anaerobic digestion of waste streams consisting of <12% dry solids is a mature technology that can produce biogas consisting of 35–75% methane by volume. Biogas so generated has a reasonably high energy content of 15–30 MJ/Nm^3 compared to that of pure methane at 38 MJ/Nm^3 (Abatzoglou and Boivin 2009; Nizami and Murphy 2010). Biogas generated from waste is increasingly being recognized as a valuable component of sustainable energy generation—over 30 million small-scale (household) digesters and 3000 large-scale digesters have already been built in China (Li et al. 2009).

8.6.2.1 Biochemistry of Digestion

After pretreatment (depending upon the feedstock, *vide infra*), the biomass is subjected to biochemical transformation carried out by microorganisms (*methanoarchaea*) in the absence of oxygen, that is, *anaerobic digestion*. The bacteria that do the work can be divided into three groups: (1) those that carry out the fermentation, (2) the *acetogenic* bacteria, and (3) the *methanogenic* bacteria. This mixed community of microbes carry out four essential steps in the anaerobic digestion of organic matter (Liu et al. 2010):

1. *Hydrolysis.* The feedstocks used for anaerobic digestion may vary widely in their chemical composition, from LCB to animal proteins and lipids. In order to liquify or solubilize the material for digestion, proteins must be broken down into amino acids, carbohydrates to sugars, and lipids to glycerol and fatty acids (Figure 8.35). This is all carried out by the diversity of enzymes present in microorganisms. This hydrolytic solubilization is the rate-determining step of the anaerobic digestion process.
2. *Acidogenesis.* In this step, the water-soluble hydrolysis products (sugars, amino acids, etc.) are fermented into a variety of smaller molecules, including ethanol, acetate, and low-molecular-weight (C3 and C4) fatty acids.
3. *Acetogenesis.* At this stage, acetogenic bacteria take butanoate, propanonate, lactate [$CH_3CH(OH)CO_2^-$], succinate ($^-O_2CCH_2CH_2CO_2^-$), and ethanol and

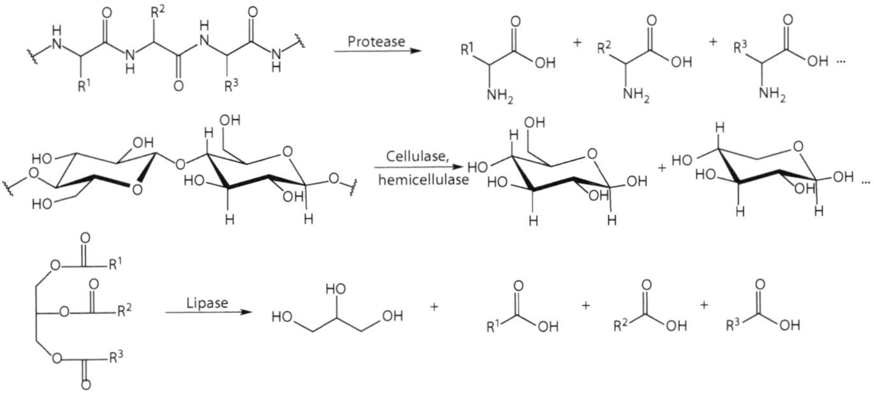

FIGURE 8.35 The various hydrolysis pathways of anaerobic digestion.

convert these substrates into acetate, H_2, and CO_2, the substrates for conversion into methane.

4. *Methanogenesis.* The final stage of anaerobic digestion takes place as methanogenic archaea cleave acetate to form methane and CO_2 and use hydrogen gas and bicarbonate to generate additional methane (Willey et al. 2013).

The biochemistry of methanogenesis is another interesting example of nature's use of metals. The greatly oversimplified summary is that acetate is cleaved to provide CO_2 followed by reduction of CO_2 into CH_4 with hydrogen gas, all of which is catalyzed by metalloenzymes (Figure 8.36). Methyl coenzyme reductase is a nickel-containing enzyme common to all methanogenic pathways, for example, and all methanoarchaeal F_{420}-dependent hydrogenases contain iron–sulfur clusters and nickel, with some also containing selenium (recall the discussion of hydrogenases in Section 5.2.3.2) (Ferry 2002). An abbreviated catalytic scheme in which these hydrogenases catalyze the transformation of CO_2 to CH_4 is given in Figure 8.37 with the key to the biomolecular structures in Figure 8.38. In step (1), methanofuran (a) is formylated with CO_2 to generate formylmethanofuran (b), which then transfers the formyl group to tetrahydromethanopterin [H_4MPT, (c)] in step (2), regenerating methanofuran. The formylated H_4MPT (d) can then undergo an intramolecular ring closure (step 3) to form the dihydroimidazolium intermediate (e) in step (4). Reduction of this iminium ion takes place stereospecifically, presumably by hydride transfer from an F_{420}-dependent enzyme. The next step, (5), is another F_{420}-dependent catalysis to reductively cleave the N–C bond in f and set up the methane precursor (g). This compound delivers a methyl group to coenzyme M in step (6). A cobalt–porphyrin cofactor is an intermediary in the regeneration of H_4MPT and transfer of the methyl group to coenzyme B (not shown in Figure 8.37). Oxidative coupling of coenzymes M and B with reduction of the methyl group to methane completes the methanogenesis cycle (step 7) (Ferry 2002).

The product—biogas—is mostly a mixture of methane and CO_2 with other minor components such as hydrogen sulfide, ammonia, oxygen, nitrogen, and so on with other minor components such as hydrogen sulfide, ammonia, oxygen, nitrogen, and so on. CO_2 makes up the vast majority of the biogas balance, thus reducing its energy value. Siloxanes (oligomeric Si–O–Si compounds) may be present in minute amounts in landfill gas from MSW disposal, since silicon-containing compounds are present in a large number of personal care products, pharmaceuticals, and so on (Mota et al. 2011).

8.6.2.2 Process and Parameters

As with other biomass energy conversion processes, the process for anaerobic digestion is under continual optimization. Variables such as reactor design, pH,

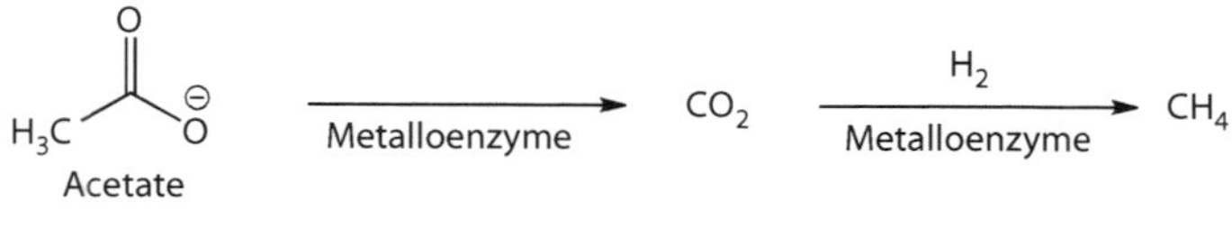

FIGURE 8.36   An overview of methanogenesis.

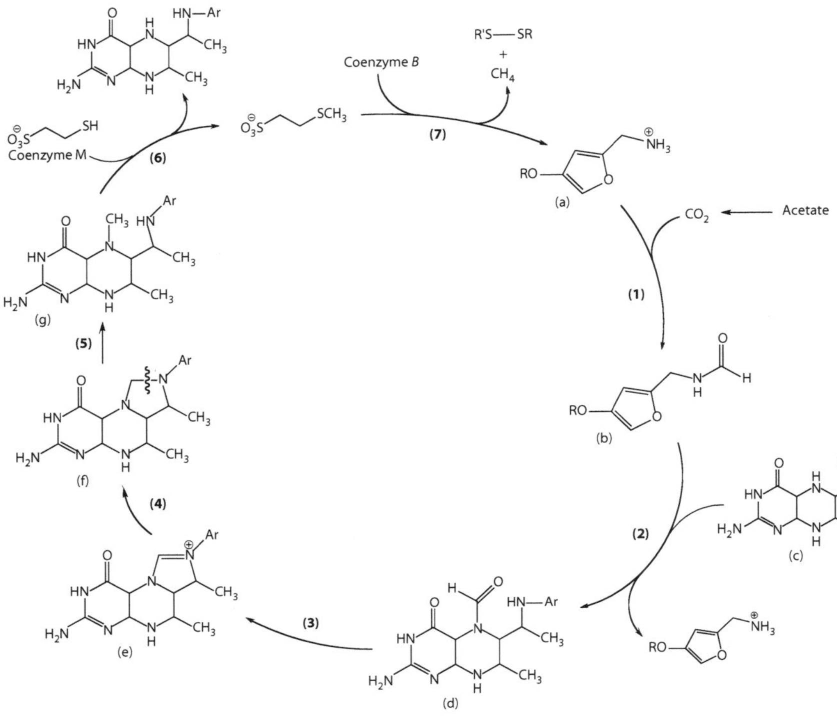

FIGURE 8.37 The catalytic cycle of anaerobic digestion (see text).

temperature, the nature of the pretreatment, and the nature of the feedstock all strongly impact the yield and quality of the biogas produced. An anaerobic digester may be a simple single-stage system consisting of a tank with an inlet for sludge and a vent for biogas collection wherein solids settle to the bottom of the tank while methane and CO_2 bubble out (essentially a glorified septic tank). Two-stage digesters that separate out the steps and/or the microorganisms have been studied in attempts to improve the rate and yield of biogas formation, *vide infra* (Schievano et al. 2012).

8.6.2.2.1 Pretreatment

Pretreatment may or may not be completely removed from the fermentation/digestion process. The ultimate goal of pretreatment of the feedstock is to reduce particle size and enhance solubilization and may include mechanical, biochemical, or physicochemical methods such as freeze–thaw cycling or ozonoloysis. As expected, pretreatment can drastically impact the rate and yield of biogas formation.

8.6.2.2.2 pH

The long-term stability of the digester relies upon careful control of pH—a complicated matter when fatty acids, acetates, ammonia, and so on are being continually generated. Buffering with alkali such as lime (CaO), sodium hydroxide, or ammonium carbonate is required, yet different pH ranges are needed for the optimum

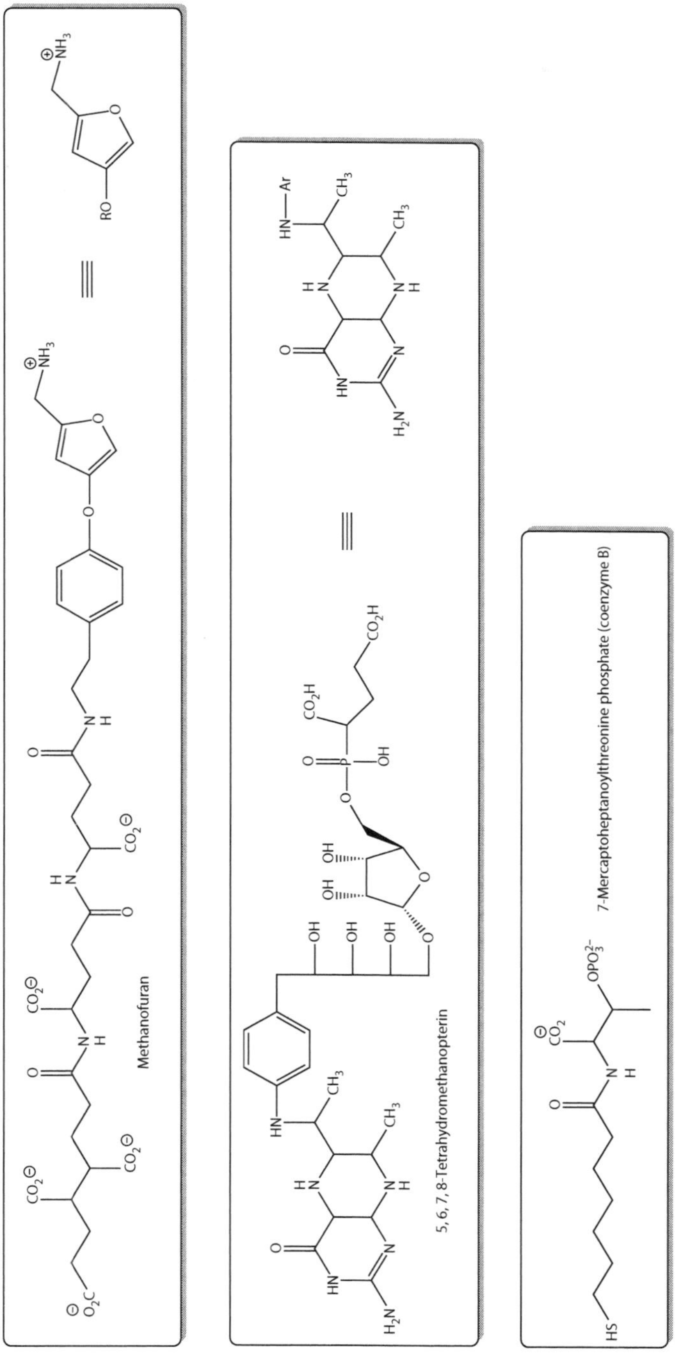

FIGURE 8.38 Key to the abbreviations in Figure 8.37.

activity of each of the various bacterial/archaeal colonies (Liu et al. 2010). The methanogens, for example, perform optimally in a narrow pH range around 7 and will die if the pH drops below 5, whereas the optimal pH for the acidogenesis step is ≈5–6 (Elefsiniotis and Oldham 1994). A recent study found that by initially allowing hydrolysis to occur in the digester for 8 days at pH 10, then adjusting the pH down to ≈7.0 the yield of biogas was more than four times greater than untreated sludge. This striking improvement in yield was attributed to enhanced hydrolysis in concert with an increase in microbial activity (Zhang et al. 2010).

8.6.2.2.3 Temperature
The temperature of the digester is another important variable—a higher temperature can improve the microbial activity and lead to a greater destruction of pathogens, but there is a limit. Most anaerobic processes are *mesophilic*, being carried out around 35–38°C (Banerji et al. 2010). Fermentation at higher temperatures (*thermophilic* fermentation, usually around 40–50°C) can increase the yield of the biogas and decrease the time in the reactor. The use of a two-stage digester to carry out biogas formation allows the two stages to be run at different temperatures. In one example studying the digestion of sludge from wastewater treatment, the first stage (hydrolysis/fermentation) was carried out at 70°C followed by batch anaerobic digestion of this prefermented sludge at 37°C. This method showed a 30–50% increase in biogas formation although accumulation of 3-methylbutanoic acid, a branched isomer that is more difficult to degrade, was observed (Bolzonella et al. 2007).

8.6.2.3 Landfill Gas
Anaerobic digestion is a well-established technology that is reliable and can make use of a feedstock that can otherwise be a problem. However, there are additional considerations when contemplating the use of anaerobic digestion for bioenergy production:

- Bacteria and archaea have nutritional requirements that may or may not be met by the feedstock. For example, the use of olive oil factory wastewater as a feedstock required the addition of ammonium chloride as a nitrogen source, an issue that can be resolved by codigestion with another (nitrogen-containing) feedstock, for example, cattle manure (Gannoun et al. 2007).
- The "waste" from the waste feedstock (the leftover sludge from biogas generation), if not treated properly, may be concentrated in toxic compounds, including heavy metals, phosphates, and a variety of unsavory inorganic and organic compounds. Disposal of this residue becomes a new issue.

The decomposition of MSW (otherwise known as landfill trash) is yet another example of anaerobic digestion and *landfill-gas-to-energy* projects are enticing. Landfill gas is 40–60% methane by volume, 20–40% CO_2, and, like the sludge-generated version, contaminated with very low percentages of other volatiles (Demirbaş 2001). Capture and use of landfill gas can provide significant environmental benefits since methane is a greenhouse gas with a much greater global warming capacity than CO_2. The potential is huge: MSW landfills are the second-largest source of human-related methane emissions in the United States (Pelley 2009); other sources put the production of biogas from landfills at 80% of the total global output (Kothari et al. 2010). In any case, the capture of methane from properly engineered landfills can be as high as 90–99%, although a value of about 75% is more typical (Barlaz 2009).

Despite the obvious attraction of capturing landfill gas, the concept is not without controversy. Methane collection efficiencies at landfills vary widely, so some amount of methane still escapes into the atmosphere. Alternatives to landfills, such as diverting organic waste for composting, form less methane; by burying compostable organics in landfills levels of methane in the atmosphere could actually be inadvertently increased. But overall, anaerobic digestion to form biogas from waste—no matter the waste source—is a very promising waste-to-energy conversion process that has even captured the interest of large technology companies such as Google and Microsoft. Because data transfer can take place from anywhere anytime, the idea of building large data centers in less expensive rural locations near large dairy farms that can generate biogas to power the technology is an appealing prospect (Vance 2010). Anaerobic digestion adds one more option to the renewable energy palette, and further research on reactor design, codigestion options, and pretreatment optimization should lead to more effective generation and capture of biogas in the future.

8.6.3 Biodiesel

8.6.3.1 Introduction

The potential of a biologically based diesel transportation fuel (*biodiesel*) has been recognized for decades, as Rudolph Diesel (of engine fame) reportedly used vegetable oil in his engine even in 1900 (Peterson 1986). The American Society for Testing and Materials defines biodiesel as a transportation fuel made up of "mono-alkyl esters of long chain fatty acids derived from vegetable oils or animal fats" (ASTM 2013). In other words, biodiesel is produced from the transesterification of triacylglycerides (TAGs) obtained from various plant and animal feedstocks, a first-generation process that is far and away the most common production method. Most biodiesel is prepared by transesterification with methanol, hence the biodiesel that results is known as *FAME biodiesel*, for fatty acid methyl ester (Figure 8.39). We will focus on this method and take care not to confuse this "FAE biodiesel" (obtained from fatty

FIGURE 8.39 Transesterification with methanol to produce FAME biodiesel.

acids and fatty acid esters) with other types of biodiesel or bio-oil (obtained from pyrolysis of biomass; Section 8.5.2).

FAE biodiesel, as its name implies, is used to replace petroleum-based diesel fuels either as a stand-alone product or as a blend. Blends consisting of up to 20% biodiesel ("B20") can be used in many diesel engines and, in fact, blends are mandated in many European countries (Bezergianni and Dimitriadis 2013). It is a "carbon neutral" (with the same caveats previously discussed), biodegradable, and relatively clean-combusting fuel, with very little sulfur and considerably fewer particulate emissions than petroleum diesel, as well as lower hydrocarbon, CO, and CO_2 pollutants. Additionally, FAE biodiesel exhibits some favorable performance characteristics relative to regular diesel: it has a higher heating value (37–40 MJ/kg for biodiesel vs. 35 for petroleum diesel) and a higher cetane number than petroleum diesel (45–70 vs. ≈50; the diesel standard is a minimum of 51)[*] (Bezergianni and Dimitriadis 2013).

The 2012 production of FAE biodiesel in the United States was 969 million gallons, an increase of 2 million gallons over 2011 and a huge jump from 2010's total of 343 million gallons (U.S. Energy Information Administration 2013b). The use of FAE biodiesel is not without limitations, however. The cold-temperature performance of FAE biodiesel precludes its use in higher-proportion blends in cold climates. FAE biodiesel also contains a significant amount of moisture that leads to corrosion issues. And, as we will see, the presence of unsaturated fatty acids leads to stability problems. Nevertheless, the use of this renewable biofuel has caught on and the volume of FAE biodiesel production is expected to increase globally.

8.6.3.2 Feedstocks

One of the most important advantages of the production and use of FAE biodiesel is the wide variety of renewable feedstocks that are available for conversion into the fuel. Virtually any source of fatty acid esters is a potential feedstock, from oilseed crops to microalgae (Table 8.6). In the United States, the 2012 production of FAE biodiesel was primarily from soybean oil, with corn oil and greases important secondary feedstocks. The production of FAE biodiesel thus again raises the "food versus fuel" conundrum so that many nonfood-related oil crops have been identified as promising feedstocks, including oilseeds from *Jatropha curcas* L. (a drought-resistant shrub in the Euphorbiaceae family), *Madhuca indica* (a tropical tree), and *Calophyllum inophyllum* L. (a leafy evergreen tree whose fruit kernels have an oil content of 75%), among others (Atabani et al. 2013; Venkanna and Venkataramana 2009).

Since the feedstock itself is a large component of FAE biodiesel cost (roughly 80% or more), biodiesel made from virgin oils is much more expensive than, say, that made from waste oils and fats collected from restaurants. However, turning that disgusting used French fry grease into fuel for your truck is not as simple as it seems: waste greases have a higher percentage of free fatty acids (FFA). Virtually every potential FAE biodiesel feedstock possesses some amount of FFAs, but waste oils

---

[*] The cetane number is a parallel to the octane measurement for gasoline; the higher the cetane number, the better the engine performance. (Baig and Ng 2010; Speight 2007).

TABLE 8.6
Some Major Biodiesel Feedstocks

Source	Typical Oil Content (%)
Algae	15–75 (dry weight)
Castor	44–45
Coconut	65–69
Corn	≈4
Cottonseed	18–25
Jatropha nut	40–60
Jojoba	45–55
Linseed	39–42
Palm kernel	21–25
Palm fruit	46–57
Peanut	≈50
Rapeseed/canola[a]	≈40
Soybean	16–18
Sunflower	40–50
Waste oils	–

[a] Canola stands for Canadian oil low acid and refers to a hybrid of rapeseed that has a low percentage of erucic acid making it more suitable for consumption.

and greases have an FFA content of 15% or above. Jatropha oil consists of roughly 28% FFA and some microalgae oil contains as much as 50% FFAs (Atadashi et al. 2012). The presence of these fatty acids wreaks havoc on the production process as will be explained in Section 8.6.3.4.

One of the more promising feedstocks for biodiesel production is microalgae. More than 50,000 species of microalgae are known and many have a high oil content (Table 8.6). Microalgae have two distinct advantages as a potential feedstock: they do not compete with food production, and they do not require large tracts of land for growth. In fact, microalgae potentially provide 132 times the yield per acre as soybean (see Table 8.7). For each acre of water surface used to grow microalgae, up to 15,000 gallons of feedstock oil could be produced, compared with 18, 48, 102, 127, or 635 gallons for corn, soybean, safflower, sunflower, rapeseed and oil palm, respectively (Chisti 2007). Furthermore, algae is relatively fast growing and could potentially grow on nitrate- and phosphate-rich wastewater since the only required nutrients are nitrogen, phosphorus, and CO_2. Upon extraction of the lipids from microalgae, the biomass residue remains high in protein and carbohydrates (among other nutrients) and has the potential to be used as animal feed or as feedstock for other biomass-to-energy conversion processes (Spolaore et al. 2006). The latter use, in fact, has been suggested as part of a life-cycle assessment of the use of algae for biodiesel production. This study found that, while microalgae production requires less land and needs less pesticide than conventional crops, there are bottlenecks to

TABLE 8.7
Biodiesel Feedstock Acreage Demands

Crop	Oil Yield (L/hectare/ year)	Crop Area (hectare × 10^6)	% U.S. Cropland
Soybean	446	1481	326
Oil palm	5950	45	24
Microalgae (@ 30%)	58,700	4.5	2.5

Source: Reprinted from *Biotechnol. Adv.* 25 (3), Chisti, Y. 2007. Biodiesel from microalgae, p. 296, with permission from Elsevier.

its sustainability, namely, the levels of fertilizer used and the high rate of energy consumption for production, harvesting, and oil extraction (Lardon et al. 2009).

How does algae, corn, or animal matter get converted into feedstock for biodiesel? We will not go into great detail, but the processing figures importantly into the cost and sustainability of the fuel. *Rendering* refers to the process whereby lard, grease, or tallow (among other products) are obtained from an animal carcass. For plant sources, processing will include cleaning and comminution followed by some sort of solvent extraction or mechanical expression of the oil from the nuts, seeds, or the like (Ribeiro et al. 2011).

8.6.3.3 Biochemistry of Fatty Acids

Once the crude oils and fats are obtained from the feedstock the key transformation is the transesterification of the triacylglycerides therein. As much of 98% of the oil in these feedstocks is triacylgycerides, that is, a triester made up of glycerol and three fatty acids (Figure 8.40). As noted above, some amount of FFAs also accompanies the TAGs, a reality that complicates biodiesel processing (see below). Fatty acids almost always contain an even number of carbons due to their biosynthesis from the essential building block acetyl CoA, shown in the simplified overview in Figure 8.41. A Claisen condensation with concomitant decarboxylation of the intermediate yields

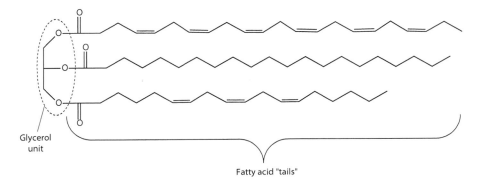

FIGURE 8.40   A typical triacylglyceride.

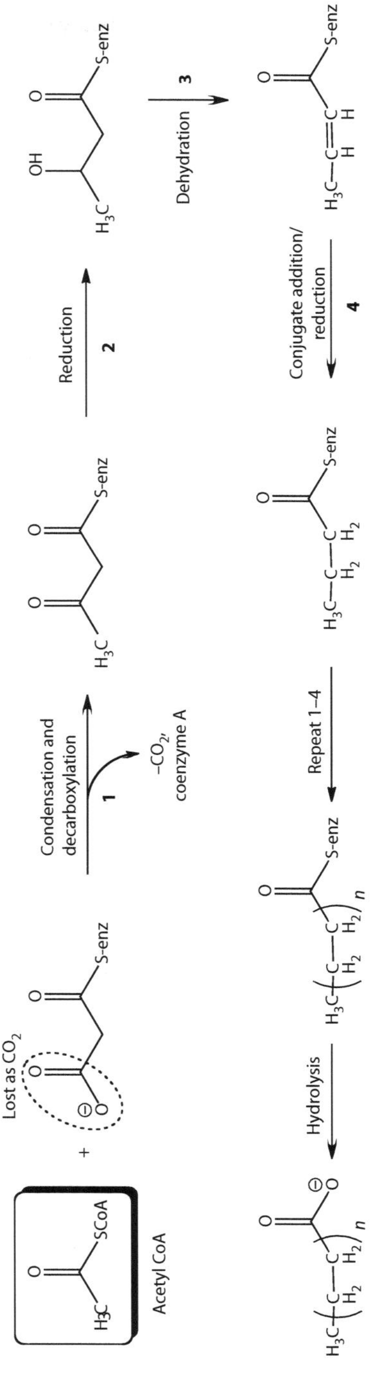

FIGURE 8.41 The biosynthetic pathway for fatty acids.

the four-carbon β-ketoacyl skeleton (1). Reduction of the ketone functionality (2) followed by dehydration (3) and conjugate reduction (4) yields the four-carbon acyl building block that can again undergo condensation and decarboxylation to lengthen the chain. Of course, all of these steps are catalyzed by enzymes and the fatty acid carboxylate is ultimately freed from its enzyme tether. Subsequent enzyme-catalyzed transformations produce the variety of unsaturated and polyunsaturated fatty acids listed in Table 8.8. Also note that almost all naturally occurring unsaturated fatty acids have the *cis* double bond geometry.

The fatty acids found in TAGs vary depending on the source and climate. In algae, for example, the fatty acids *cis*-5,8,11,14,17-eicosapentenoic acid (EPA) and *cis*-4,7,10,13,16,19-docosahexenoic acid (DHA) predominate. In tropical areas, shorter-chain fatty acids are more common (C12–C14), while in cooler climates, longer chains (≥C18) dominate. A representative sampling of fatty acid components in various oils and fats is shown in Table 8.9.

The fatty acids that make up the biodiesel feedstock strongly influence the properties of the biodiesel, as one might expect. In general, carbon chains of 18 carbons or fewer are needed to achieve the desired viscosity properties. Given that biodiesel may be used in a wide range of climates, several performance specifications must be met in its manufacture. For example, the *cloud point* (the temperature at which wax crystals begin to appear) and the *pour point* (the temperature at which the material can no longer be poured) may limit the use of a specific biodiesel in cold climates. These values are intimately related to the conformational properties of the TAGs

TABLE 8.8
Common Fatty Acids

# Carbons: # Double Bonds	Common Name	Scientific Name
12:0	Lauric acid	Dodecanoic acid
14:0	Myristic acid	Tetradecanoic acid
16:0	Palmitic acid	Hexadecanoic acid
16:1	Palmitoleic	9-Hexadecenoic acid
18:0	Stearic acid	Octadecanoic acid
18:1	Oleic acid	9-Octadeceneoic acid
18:1	Ricinoleic acid	12-Hydroxy-9-octadecenoic acid
18:1	Vaccenic acid	11-Octadecenoic acid
18:2	Linoleic acid	9,12-Octadecadienoic acid
18:3	α-Linolenic acid	9,12,15-Octadecatrienoic acid
18:3	γ-Linolenic acid	6,9,12-Octadecatrienoic acid
20:0	Arachidic acid	Eicosanoic acid
20:1	Gadoleic acid	9-Eicosenoic acid
20:4	Arachidonic acid	5,8,11,14-Eicosatetraenoic acid
20:5	EPA	5,8,11,14,17-Eicosapentaenoic acid
22:0	Behenic acid	Docosanoic acid
22:1	Erucic acid	13-Docosenoic acid
22:6	DHA	4,7,10,13,16,19-Docosahexenoic acid

TABLE 8.9

Fatty Acid Components of Some Representative Oils and Fats

Source	Myristic 14:0	Palmitic 16:0	Palmitoleic 16:1	Stearic 18:0	Oleic 18:1	Linoleic 18:2	Linolenic 18:3	Gadoleic 20:1	Erucic 22:1
Canola	–	4	–	2	56	26	10	–	–
Soybean	–	11	–	4	22	53	8	–	–
Palm	–	44	–	4	39	11	–	–	–
Rapeseed (*Brassica napus*)	–	3	–	1	17	14	9	11	45
Chicken	1	24	6	6	40	17	1	–	–

Source: Adapted from DeMan, J.M. 1999. *Principles of Food Chemistry*, 3rd Ed., Food Science Text Series. New York: Springer.

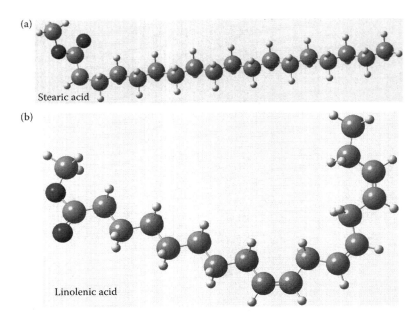

(a)

Stearic acid

(b)

Linolenic acid

FIGURE 8.42 (a) Fully saturated stearic acid. (b) Bent linoleic acid with three *cis* double bonds.

and FFAs in the feedstock. Straight-chain, saturated fatty acids have higher melting points due to increased intermolecular interactions, whereas those TAGs and FFA containing unsaturation are lower-melting—the "kink" from the *cis* double bond(s) in the fatty acid chain prevents close contact between molecules, with a resultant decrease in the intermolecular forces responsible for melting point (Figure 8.42a and b). Unsaturation also plays an important role with respect to the stability of the biodiesel product and biodiesel from common feedstocks contains a large proportion (60–85%) of unsaturated esters (Schumacher et al. 2009). The *iodine value* (or iodine number) of a fat or oil (including FAE biodiesel) is a measure of the amount of unsaturation present, an important consideration in that double bonds add reactivity. Unsaturation can lead to polymerization and unsaturated fatty acids/esters are especially prone to allylic radical oxidation—problems when it comes to biodiesel degradation and engine performance. The typical iodine value for petroleum diesel is negligible, while for biodiesel the typical *limit* is 120 (Hart Energy Consulting 2008).

8.6.3.4 Production and Catalysis

8.6.3.4.1 Introduction

In an ideal world, FAE biodiesel would be simple to produce: simply throw a basic catalyst into the feedstock with an alcohol, heat and stir, and violá—FAE biodiesel! Unfortunately, it is considerably more complicated. First, transesterification is an equilibrium reaction, hence a large excess of a fairly volatile alcohol such as methanol or ethanol must be used to push the equilibrium to the product side as per Le Chatelier's principle (Figure 8.39). This results in the need to remove a large excess of the unreacted alcohol, adding to the cost of manufacture. While ethanol is attractive due to its potential origin from renewable resources, less-expensive methanol is the more popular choice. It is critical that the process is forced toward completion in order to ensure the highest possible yield of the transesterified product, meaning a three-step, one-pot sequence: conversion of the triglyceride into the diglyceride, then the monoglyceride and all, ultimately, into FAME biodiesel, as shown in Figure 8.43. While somewhat complicated at the molecular level, high yields (>90%) of the FAME product can be obtained, but large-scale manufacture requires an inexpensive catalyst that can carry out the desired transformation under mild conditions with minimal by-product formation and a simple workup. The search for this ideal catalyst is ongoing.

Catalysis of the transesterification reaction with a homogeneous catalyst such as CH_3ONa is the predominant method for carrying out FAME biodiesel production, but homogeneous catalysts present problems with respect to their eventual reclamation. The removal of the stoichiometric amount of glycerol by-product, required neutralization of the catalyst, and washing (and subsequent drying) of the product to remove all traces of the residual salts make the production process complex. When FFAs are present, the picture is even more complicated because the basic catalyst will, of course, preferentially react with the FFA to produce soap (Figure 8.44). Soap, being an excellent emulsifier, makes product separation difficult at best. Furthermore, unless the reaction conditions are scrupulously anhydrous, transesterification under basic conditions can generate additional FFA. Thus, while catalysis with alkali is speedier and leads to higher yields, an acid catalyst is required when a high concentration of FFA are present, and a two-step process—acid-catalyzed *esterification* of the FFA,

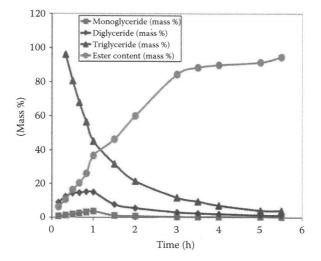

FIGURE 8.43 Transesterification of soybean oil with 3 wt.% acid catalyst at 200°C; 1:24 oil/methanol. (Reprinted with permission from Baig, A. and F. T. T. Ng. A single-step solid acid-catalyzed process for the production of biodiesel from high free fatty acid feedstocks. *Energy Fuels* 24 (9):4712–4720. Copyright 2010, American Chemical Society.)

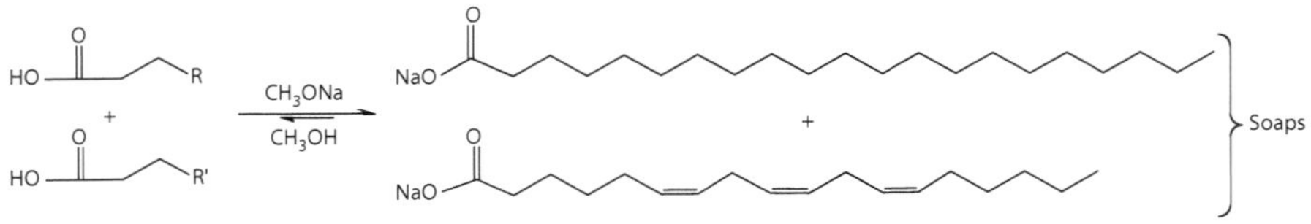

FIGURE 8.44 Acid–base reaction with free fatty acids to make soap.

followed by base-catalyzed *transesterification*—is frequently used. The alternative—removal of FFA prior to transesterification—is wasteful and adds expense.

8.6.3.4.2 Heterogeneous Basic Catalysis

FAE biodiesel formation with heterogeneous catalysts is preferable due to the stream-lined catalyst reclamation yet the problems associated with the nature of the catalyst—acidic or basic—persist. Acid catalysts are better for the esterification of the FFA (but lead to corrosion concerns) and basic catalysts are the better choice for transesterification. Using about 4 wt.% of a basic catalyst with a low-FFA-content feedstock, FAME biodiesel can be produced in good yields using a large excess of methanol at ≈70°C for, on average, about 4 h (Borges and Díaz 2012). A recent review provides a detailed summary of these heterogeneous catalysts, including bifunctional catalysts that, in effect, "do it all" with some success (Borges and Díaz 2012). An overview follows.

Compounds that have been found to be promising heterogeneous basic catalysts are some minerals, metal oxides, and derivatives thereof. For example, dolomite can be calcined at 850°C and used at 1.5 wt.% to catalyze the formation of FAME biodiesel in over 90% yield (Ngamcharussrivichai et al. 2010). Recall that dolomite is $CaMg(CO_3)_2$ and has been tested as a catalyst in the tar cleanup of syngas produced in biomass gasification (Section 8.5.3.4). Magnesium, strontium, and barium oxides are also suitable catalysts: the use of strontium oxide to transesterify TAGs from a variety of sources in a one-stage process (i.e., without a separate extraction step) at 60°C using microwave irradiation as the heat source gave essentially quantitative conversion of the feedstocks to FAME (Koberg and Gedanken 2012). The currently prevailing solid basic catalyst for FAME biodiesel formation, however, is calcium oxide (CaO), being obtained from the calcining of calcium carbonate. A particularly attractive aspect of CaO catalysis is that CaO-precursor waste products such as eggshells or sea shells can be used to catalyze the reaction (Viriya-empikul et al. 2010).

A wide variety of modified basic metal oxides have also been shown to be good heterogeneous catalysts for the transesterification reaction, including potassium-, zinc-, and iron-modified calcium oxides, or CaO supported on alumina or zinc oxide. Similarly, alumina loaded with alkali metal salts also showed good activity as a transesterification catalyst. Hydrotalcite ($[Mg_{(1-x)}Al_x(OH)_2]^{x+}(CO_3)_{x/n}^{2-}$, where $x = 0.25$–0.55) or hydrotalcite-like compounds are also promising transesterification catalysts in that they can be readily tailored (Figure 8.45). By changing the metal ion ratios, the base strength can be altered, plus these compounds possess good surface area—favorable attributes for good catalytic activity.

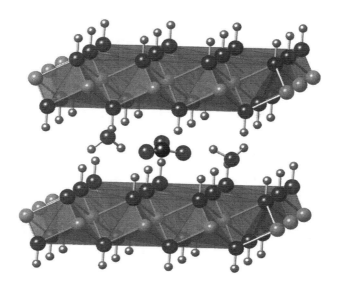

FIGURE 8.45   The hydrotalcite structure with carbonate counterion. (From Pérez-Ramírez, J., S. Abelló, and N.M. van der Pers. 2007. Influence of the divalent cation on the thermal activation and reconstruction of hydrotalcite-like compounds. *J. Phys. Chem.* C 111 (9):3642–3650. Copyright 2007, American Chemical Society.)

8.6.3.4.3   Heterogeneous Acidic Catalysis

When the feedstock contains significant amounts of FFAs, a basic catalyst will not suffice. Homogeneous acidic catalysts such as sulfuric acid will certainly do the job but again present serious contamination and separation problems, and trace acid in the FAE biodiesel may result in the product failing to pass corrosion standards. Thus, numerous heterogeneous acid catalysts have been tested for the esterification reaction which typically requires a reaction temperature >150°C (Atadashi et al. 2012; Borges and Díaz 2012; Jothiramalingam and Wang 2009). Thus, metal oxides of titanium and tin are both effective as acidic esterification catalysts, but zirconium oxide (and modified variants) has been the most extensively studied. Unfortunately, zirconium is also an expensive and increasingly utilized element, making its sustainable use a concern. Zeolites and acid-activated kaolinite, halloysite, or montmorillonite clay minerals have shown some success, as have acidic ion-exchange resins such as Amberlyst-15 (a sulfonated styrene-divinylbenzene copolymer) and even Nafion®, the commonly used fuel cell membrane. However, for some of these sulfonated materials, the acidic functionality can be leached out resulting in diminished catalyst activity. A recent attempt to identify a stable, highly active solid acid catalyst made use of a mesoporous carbon material. The highly porous catalyst was prepared by depositing furfuryl alcohol on a mesoporous silica template. Several steps of thermal treatment gave the carbonized platform that was then oxidized with hydrogen peroxide to give the mesoporous carbon with extensive hydroxyl and carboxyl functionality. Sulfonation with concentrated H_2SO_4 at 150°C yielded the final solid acid catalyst, which was shown to have a surface area of 475 m²/g. This catalyst performed well when compared to other solid acid catalysts including acid-activated kaolin and sulfonated mesoporous silica. Furthermore, its activity was undiminished through five cycles (Chang et al. 2013).

8.6.3.4.4   Bifunctional Catalysis

The best possible outcome for catalytic production of FAE biodiesel would be to identify a catalyst that performs well in both the esterification and the transesterification stages, allowing for a single-step process that can be used with even high-FFA-content feedstocks. These catalysts must exhibit both acidic and basic tendencies in just the right balance in order to perform well. Research efforts to develop such a catalyst are making inroads and have largely been based on empirical studies. In but one example, the mineral Quintinite 3T ($Mg_4Al_2(OH)_{12}CO_3 \cdot 4H_2O$), a *layered double hydroxide* compound was able to convert a feedstock with up to 30% FFA to biodiesel under relatively mild conditions (10 wt.% catalyst, 75°C, 2 h). More impressively, the yields of biodiesel were always at least 96% and the catalyst activity remained high (95%) even after five cycles. Given this catalyst's ability to convert both FFA and TAGs into biodiesel in a single process, its promise for lower-cost biodiesel production is excellent (Kondamudi et al. 2011).

8.6.3.4.5   Enzymatic Catalysis

Given that the production of FAE biodiesel is based on a biological feedstock it stands to reason that biological catalysis is an option. Indeed, enzymatic catalysis of FAE biodiesel production is a well-studied phenomenon for the usual good reasons: enzymatic catalysis leads to a much milder, greener chemical transformation given a lower temperature of reaction and absence of harsh acids and bases. The issues associated with enzymatic catalysis are also unsurprising, that is, high cost and competitive inhibition by impurities in the feedstock.

8.6.3.4.6   Lipases

Triacylglycerides are lipids and the enzymes known to catalyze lipid esterification, hydrolysis, and transesterification (and, hence, biodiesel production) are, logically enough, *lipases*. These enzymes are widely distributed in nature, being produced by microorganisms as well as plants and animals. Microorganismal lipases from the genera *Candida*, *Rhizopus*, and *Pseudomonas* are the primary sources for industrial applications. Lipases are especially active for lipophilic substrates, so fatty acid esters of over 10 carbon atoms are good substrates. They do not require *cofactors* (a cofactor is a small molecule that must associate with the enzyme to activate it) and they are stable in the presence of organic solvents. In addition, some lipases are relatively nonspecific: they will cleave any triacyl-, diacyl-, or monoacylglyceride. All of these features make them attractive candidates for use in industry (Ribeiro et al. 2011). A reasonable general mechanism for the lipase-catalyzed transesterification is shown in Figure 8.46, where Ser is serine and His is histidine.

8.6.3.4.7   Immobilization

Because of the high cost associated with these enzymes their practical application requires immobilization in or on some solid material that allows their easy recovery. Immobilization of the lipase on a solid inert material not only results in cost

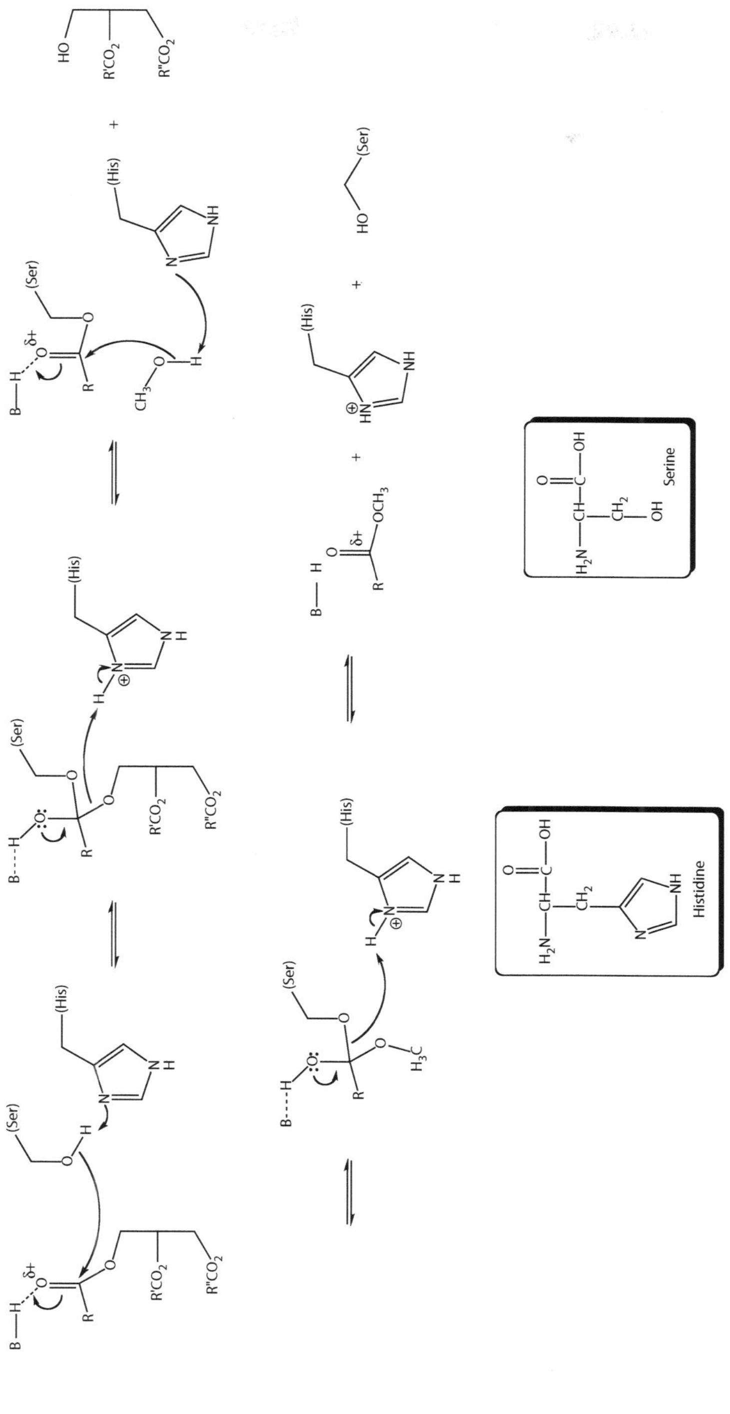

FIGURE 8.46 General mechanism for lipase-catalyzed transesterification of triacylglyceride.

savings due to recycling, it also stabilizes the lipase. As a result, fixed bed reactors (such as described in Section 8.5.3.2) are preferred for enzyme-catalyzed continuous biodiesel manufacture (as opposed to a reactor that would subject the lipase to too much shear stress) (Ribeiro et al. 2011). While lipases have been successfully immobilized on ceramic or polymer resin beads, diatomaceous earth or entrapped in sol–gels, the current most common method of immobilization is adsorption on textile: a pretreated textile is placed in a solution of the enzyme then dried, analyzed for activity, and used in biodiesel production (Lu et al. 2007). *Pseudozyma antartica, Thermomyces lanuginosus, Candida* sp. 99-125, and commercially available *Candida antartica* lipase (Novozym® 435) have all proven to be very effective immobilized enzymes for biodiesel production and yields of >90% for a wide variety of feedstocks can be reliably obtained.

8.6.3.4.8   Process Conditions
Lipase catalysis takes place, as expected, at lower temperatures and in general must be carried out at temperatures below 60°C, at which point the lipase is deactivated. However, catalysis of FAEE biodiesel produced from soybean oil with Novozym 435 gave the optimum yield of 92% when carried out at 65°C (Brusamarelo et al. 2010). Just as for the catalysts described in the previous sections, a large excess of the alcohol must be used in order to present a favorable equilibrium. The alcohol is also a necessary reaction component in that it acts as a solvent and reduces the viscosity of the reaction mixture. However, some lipases are inhibited by alcohols, particularly shorter-chain alcohols such as methanol or ethanol. Two approaches to address these concerns are (1) to add the alcohol in portions and monitor the reaction, adding more alcohol as it is consumed, and (2) to use other organic solvents that do not inhibit the enzyme. Petroleum distillates (e.g., hexanes or petroleum ether) are completely compatible with lipase-catalyzed biodiesel production.

8.6.3.5   Conclusions
The production of FAE biodiesel has well-established and reliable technology behind it and biofuel plays a minor, but significant, role in many areas of the globe. Industrial-scale applications have much room for improvement, however, and in the end analysis, FAE biodiesel is a first-generation biofuel that, in many instances, competes with food production. New methods of biodiesel production have been developed that are based on crop residues and LCB such as switchgrass. These methods—Fischer–Tropsch diesel, hydrotreated vegetable oils, and hydrotreating of waste cooking oils—offer additional promise in that they do not play into the "food versus fuel" debate and they produce a higher-performance diesel (Bezergianni and Dimitriadis 2013). Nevertheless, these new technologies are likewise in the early stages of development and bring their own biomass energy issues with them.

8.7   SUMMARY

The concept of utilizing biomass for energy has many very attractive positive aspects. Biomass is well distributed across the globe, it is renewable, and its conversion into energy is particularly appropriate in rural areas where the feedstock does not need to be trucked or transported by pipeline over hundreds or thousands of miles. Biochemical processes for conversion are currently more cost effective than thermochemical processes, but continuing research could further reduce costs.

Biomass energy crops invariably impact agricultural land. Including reforestation efforts to help minimize the increase in CO_2, this competition for food-producing land has been forecast to raise food prices by as much as 80% (Reilly et al. 2012). Biomass energy conversion processes are best suited for small, local installations—no biomass conversion process is truly industrial scale. Ultimately, there is much skepticism that biomass will ever become a major contributor to sustainable energy production: it has been estimated that biomass produced in a *sustainable* manner can contribute at most 10–15% of our future energy needs (Doornbosch and Steenblik 2007). As Jerald L. Schnoor, editor-in-chief of the journal *Environmental Science and Technology* and Allen S. Henry Chair in Engineering at the University of Iowa, states (emphasis his): "BIOFUELS ARE NOT SUSTAINABLE. At least not in the way we practice row-crop agriculture today…. Biofuels are one element of a rational and strategic response to the problem" (Schnoor 2006).

OTHER RESOURCES

BOOKS

Centi, G. and R.A. van Santen. 2007. *Catalysis for Renewables: From Feedstock to Energy Production*. Weinheim, FRG: Wiley-VCH.
Clark, J.H. and F.E.I. Deswarte. 2008. *Introduction to Chemicals from Biomass*. Chichester, West Sussex, UK: Wiley & Sons.

ONLINE RESOURCES

International Energy Agency/Bioenergy: http://www.iea.org/topics/bioenergy/
U.S. Department of Energy/Energy Efficiency and Renewable Energy/Biomass: http://www.eere.energy.gov/topics/biomass.html

REFERENCES

Abatzoglou, N. and S. Boivin. 2009. A review of biogas purification processes. *Biofuels, Bioprod. Biorefin.* 3 (1):42–71.
Aden, A. 2007. Water usage for current and future ethanol production. *Southwest Hydrology* 5:22–23.
Adjaye, J.D. and N.N. Bakhshi. 1995. Catalytic conversion of a biomass-derived oil to fuels and chemicals I: Model compound studies and reaction pathways. *Biomass Bioenergy* 8 (3):131–149.
Agbor, V.B., N. Cicek, R. Sparling et al. 2011. Biomass pretreatment: Fundamentals toward application. *Biotechnol. Adv.* 29:675–685.
Alén, R., E. Kuoppala, and P. Oesch. 1996. Formation of the main degradation compound groups from wood and its components during pyrolysis. *J. Anal. Appl. Pyrolysis* 36 (2):137–148.
Ammendola, P., B. Piriou, L. Lisi et al. 2010. Dual bed reactor for the study of catalytic biomass tars conversion. *Exp. Therm. Fluid Sci.* 34 (3):269–274.

Anis, S. and Z.A. Azinal. 2011. Tar reduction in biomass producer gas via mechanical, catalytic and thermal methods: A review. *Renew. Sustain. Energy Rev.* 15:2355–2377.

ASTM. 2013. *ASTM D 6751 Specification for Biodiesel Fuel Blend Stock (B100) for Middle Distillate Fuels.* West Conshohocken, PA: ASTM International.

Atabani, A.E., T.M.I. Mahlia, I. Anjum Badruddin et al. 2013. Investigation of physical and chemical properties of potential edible and non-edible feedstocks for biodiesel production, a comparative analysis. *Renew. Sustain. Energy Rev.* 21 (0):749–755.

Atadashi, I.M., M.K. Aroua, A.R. Abdul Aziz et al. 2012. Production of biodiesel using high free fatty acid feedstocks. *Renew. Sustain. Energy Rev.* 16 (5):3275–3285.

Baerlocher, C. and L.B. McCusker, Database of Zeolite Structures: http://izasc.ethz.ch/fmi/xsl/IZA-SC/Credits.htm.

Baig, A. and F.T.T. Ng. 2010. A single-step solid acid-catalyzed process for the production of biodiesel from high free fatty acid feedstocks. *Energy Fuels* 24 (9):4712–4720.

Balat, M., M. Balat, E. Kirtay et al. 2009. Main routes for the thermo-conversion of biomass into fuels and chemicals. Part 1: Pyrolysis systems. *Energy Conv. Mgmt.* 50 (12):3147–3157.

Banerji, S.K., R.Y. Surampalli, C.M. Kao et al. 2010. High strength wastewater to bioenergy. In *Bioenergy and Biofuel from Biowastes and Biomass*, edited by S.K. Khanal, 23–42. Reston, VA, USA: American Society of Civil Engineers.

Barber, J. 2009. Photosynthetic energy conversion: Natural and artificial. *Chem. Soc. Rev.* 38 (1):185–196.

Barlaz, M.A. 2009. Landfill gas recovery. *Environ. Sci. Technol.* 43 (9):2995.

Benemelis, I. 2012. USDA Announces Additional 9,000 Acres for Non-Food Energy Crop Production (Release No. 0195.10). June 13, 2012.

Bezergianni, S. and A. Dimitriadis. 2013. Comparison between different types of renewable diesel. *Renew. Sustain. Energy Rev.* 21 (0):110–116.

Bolzonella, D., P. Pavan, M. Zanette et al. 2007. Two-phase anaerobic digestion of waste activated sludge: Effect of an extreme thermophilic prefermentation. *Ind. Eng. Chem. Res.* 46 (21):6650–6655.

Borges, M.E. and L. Díaz. 2012. Recent developments on heterogeneous catalysts for biodiesel production by oil esterification and transesterification reactions: A review. *Renew. Sustain. Energy Rev.* 16 (5):2839–2849.

Bridgewater, A.V., S. Czernik, and J. Piskorz. 2001. An overview of fast pyrolysis. In *Progress in Thermochemical Biomass Conversion*, edited by A.V. Bridgwater, 977–997. Oxford: Blackwell Science.

Bronson, B., F. Preto, and P. Mehrani. 2012. Effect of pretreatment on the physical properties of biomass and its relation to fluidized bed gasification. *Environ. Prog. Sustain. Energy* 31 (3):335–339.

Brusamarelo, C.Z., E. Rosset, A. de Césaro et al. 2010. Kinetics of lipase-catalyzed synthesis of soybean fatty acid ethyl esters in pressurized propane. *J. Biotechnol.* 147 (2):108–115.

Cai, Z.S., and L. Paszner. 1988. Salt catalyzed wood bonding with hemicellulose. *Holzforschung* 42:11–20.

Champagne, P. 2008. Biomass. In *Future Energy*, edited by T.M. Letcher, 151–170. Oxford, UK: Elsevier.

Chang, B., J. Fu, Y. Tian et al. 2013. Multifunctionalized ordered mesoporous carbon as an efficient and stable solid acid catalyst for biodiesel preparation. *J. Phys. Chem. C* 117 (12):6252–6258.

Chisti, Y. 2007. Biodiesel from microalgae. *Biotechnol. Adv.* 25 (3):294–306.

Chum, H. et al. 1989. Biomass pyrolysis oil feedstocks for phenolic adhesives. In *Adhesives from Renewable Resources*, edited by R.W. Hemingway, A.H. Conner and S.J. Branham, 135–151. Washington, D.C.: American Chemical Society.

Corella, J., J.M. Toledo, and R. Padilla. 2004. Olivine or dolomite as in-bed additive in biomass gasification with air in a fluidized bed: Which is better? *Energy Fuels* 18:713–720.

da Rosa, A.V. 2009. *Fundamentals of Renewable Energy Processes*. Burlington, MA: Academic Press/Elsevier.

de Lasa, H., E. Salaices, J. Mazumder et al. 2011. Catalytic steam gasification of biomass: Catalysts, thermodynamics and kinetics. *Chem. Rev.* 111 (9):5404–5433.

Demirbaş, A. 2001. Biomass resource facilities and biomass conversion processing for fuels and chemicals. *Energy Conv. Mgmt.* 42 (11):1357–1378.

DeMan, J.M. 1999. *Principles of Food Chemistry*, 3rd Ed., Food Science Text Series. New York: Springer.

Devi, L., K.J. Ptasinski, and F.J.J.G. Janssen. 2003. A review of the primary measures for tar elimination in biomass gasification processes. *Biomass Bioenergy* 24 (2):125–140.

Doornbosch, R. and R. Steenblik. Biofuels: Is the Cure Worse Than the Disease? Report No. SG/SD/RT(2007)3. Organisation for Economic Co-operation and Development: Paris. 2007.

Dupont, C., R. Lemeur, A. Daudin et al. 2011. Hydrodeoxygenation pathways catalyzed by MoS_2 and NiMoS active phases: A DFT study. *J. Catal.* 279 (2):276–286.

Elefsiniotis, P. and W.K. Oldham. 1994. Influence of pH on the acid-phase anaerobic digestion of primary sludge. *J. Chem. Technol. Biotechnol.* 60 (1):89–96.

Elliot, D.C. 1986. Analysis and Comparison of Biomass Pyrolysis/Gasification Condensates. PNL-5943, Final Report. Pacific Northwest Laboratory. Richland, WA.

Ferry, J.G. 2002. Methanogenesis biochemistry. In *Encyclopedia of Life Sciences*. London/New York: Nature Publishing Group.

French, R. and S. Czernik. 2010. Catalytic pyrolysis of biomass for biofuels production. *Fuel Process Technol.* 91 (1):25–32.

Gannoun, H., N.B. Othman, H. Bouallagui et al. 2007. Mesophilic and thermophilic anaerobic co-digestion of Olive Mill wastewaters and Abattoir wastewaters in an upflow anaerobic filter. *Ind. Eng. Chem. Res.* 46:6737–6743.

Godin, B., S. Lamaudière, R. Agneessens et al. 2013. Chemical composition and biofuel potentials of a wide diversity of plant biomasses. *Energy Fuels* 27 (5):2588–2598.

Graça, I., J.M. Lopes, H.S. Cerqueira et al. 2012. Bio-oils upgrading for second generation biofuels. *Ind. Eng. Chem. Res.* 52 (1):275–287.

Hart Energy Consulting. 2008. Establishment of the guidelines for the development of biodiesel standards in the APEC Region; EWG 02/2007a.

Hutkins, R.W. 2006. *Microbiology and Technology of Fermented Foods*. Ames, IA: Blackwell Publishing Professional.

Johnson, J. 2013. *Regional Corn Stover Removal Impact Study—Morris (II)*. Morris, MN: United State Department of Agriculture Agricultural Research Service. Original edition, 3645-11610-001-05.

Jørgensen, H., J.B. Kristensen, and C. Felby. 2007. Enzymatic conversion of lignocellulose into fermentable sugars: Challenges and opportunities. *Biofuels Bioprod. Biorefin.* 1 (2):119–134.

Jothiramalingam, R. and M.K. Wang. 2009. Review of recent developments in solid acid, base, and enzyme catalysts (heterogeneous) for biodiesel production via transesterification. *Ind. Eng. Chem. Res.* 48 (13):6162–6172.

Kersten, S.R.A., W.P.M. van Swaaij, L. Lefferts et al. 2007. Options for catalysis in the thermochemical conversion of biomass into fuels. In *Catalysis for Renewables: From Feedstock to Energy Production*, edited by G. Centi and R.A. van Santen, 119–145. Weinheim, FRG: Wiley-VCH.

Kirkels, A.F. and G.P.J. Verbong. 2011. Biomass gasification: Still promising? A 30-year global overview. *Renew. Sustain. Energy Rev.* 15 (1):471–481.

Koberg, M. and A. Gedanken. 2012. Optimization of bio-diesel production from oils, cooking oils, microalgae, and castor and Jatropha seeds: Probing various heating sources and catalysts. *Energy Environ. Sci.* 5:7460–7469.

Kondamudi, N., S.K. Mohapatra, and M. Misra. 2011. Quintinite as a bifunctional heterogeneous catalyst for biodiesel synthesis. *Appl. Catal. A Gen.* 393 (1–2):36–43.

Kothari, R., V.V. Tyagi, and A. Pathak. 2010. Waste-to-energy: A way from renewable energy sources to sustainable development. *Renew. Sustain. Energy Rev.* 14 (9): 3164–3170.

Lange, J. 2007. Lignocellulose conversion: An introduction to chemistry, process and economics. In *Catalysis for Renewables: From Feedstock to Energy Production*, edited by G. Centi and R.A. van Santen, 21–51. Weinheim, FRG: Wiley-VCH.

Lardon, L., A. Hélias, B. Sialve et al. 2009. Life-cycle assessment of biodiesel production from microalgae. *Environ. Sci. Technol.* 43 (17):6475–6481.

Li, R., S. Chen, X. Li et al. 2009. Anaerobic codigestion of kitchen waste with cattle manure for biogas production. *Energy Fuels* 23 (4):2225–2228.

Liu, X.-Y., H.-B. Ding, and J.-Y. Wang. 2010. Food waste to bioenergy. In *Bioenergy and Biofuel from Biowastes and Biomass*, edited by S.K. Khanal and R.Y. Surampalli, 43–70. Reston, VA, USA: American Society of Civil Engineers.

Lomax, J.A., J.M. Commandeur, P.W. Arisz et al. 1991. Characterisation of oligomers and sugar ring-cleavage products in the pyrolysate of cellulose. *J. Anal. Appl. Pyrolysis* 19:65–79.

Lu, J., K. Nie, F. Xie et al. 2007. Enzymatic synthesis of fatty acid methyl esters from lard with immobilized *Candida* sp. 99–125. *Process Biochem.* 42 (9):1367–1370.

Maniatis, K. and A.A.C.M. Beenackers. 2000. Tar protocols. IEA bioenergy gasification task. *Biomass Bioenergy* 18 (1):1–4.

Milne, J.L. and C.B. Field. Assessment Report from the GCEP Workshop on Energy Supply with Negative Carbon Emissions. 2013.

Mohan, D., C.U. Pittman, and P.H. Steele. 2006. Pyrolysis of wood/biomass for bio-oil: A critical review. *Energy Fuels* 20:848–889.

Mortensen, P.M., J.D. Grunwaldt, P.A. Jensen et al. 2011. A review of catalytic upgrading of bio-oil to engine fuels. *Appl. Catal. A Gen.* 407 (1–2):1–19.

Mota, N., C. Alvarez-Galvan, R.M. Navarro et al. 2011. Biogas as a source of renewable syngas production: Advances and challenges. *Biofuels* 2 (3):325–343.

National Energy Technology Laboratory 2010. 2010 worldwide gasification database, edited by U.S. Department of Energy. http://www.netl.doe.gov/technologies/coalpower/gasification/worlddatabase/summary.html

National Energy Technology Laboratory. 2013. *Gasifipedia: Gasification in Detail. Types of Gasifiers.* U.S. Department of Energy [cited February 22 2013]. Available from http://www.netl.doe.gov/technologies/coalpower/gasification/gasifipedia/4-gasifiers/.

Ngamcharussrivichai, C., P. Nunthasanti, S. Tanachai et al. 2010. Biodiesel production through transesterification over natural calciums. *Fuel Processing Technology* 91:1409–1415.

Nizami, A.S. and J.D. Murphy. 2010. What type of digester configurations should be employed to produce biomethane from grass silage? *Renew. Sustain. Energy Rev.* 14 (6):1558–1568.

Overend, R.P. 2004. Thermochemical conversion of biomass in renewable energy sources charged with energy from the sun and originated from earth-moon interaction. *Encyclopedia of Life Support Systems (EOLSS)*, edited by E.E. Shpilrain. Oxford, UK: Developed under the auspices of the UNESCO, EOLSS Publishers.

Palmqvist, E. and B. Hahn-Hägerdal. 2000. Fermentation of lignocellulosic hydrolysates. II: Inhibitors and mechanisms of inhibition. *Bioresour. Technol.* 74 (1):25–33.

Patzek, T.W. and D. Pimentel. 2005. Thermodynamics of energy production from biomass. *CRC Crit. Rev. Plant. Sci.* 24 (5–6):327–364.

Pelley, J. 2009. Is converting landfill gas to energy the best option? *Env. Sci. Technol.* 43 (3):555–555.

Pereira, E.G., J.N. da Silva, J.L. de Oliveira et al. 2012. Sustainable energy: A review of gasification technologies. *Renew. Sustain. Energy Rev.* 16 (7):4753–4762.

Pérez-Ramírez, J., S. Abelló, and N.M. van der Pers. 2007. Influence of the divalent cation on the thermal activation and reconstruction of hydrotalcite-like compounds. *J. Phys. Chem.* C 111 (9):3642–3650. American Chemical Society.

Perlack, R.D., L.L. Wright, A.F. Turhollow et al. 2005. *Biomass as Feedstock for a Bioenergy and Bioproducts Industry: The Technical Feasibility of a Billion-Ton Annual Supply.* Oak Ridge TN: Oak Ridge National Laboratory.

Peterson, C.L. 1986. Vegetable oil as a diesel fuel: Status and research priorities. *Trans. ASAE* 29 (5):1413–1422.

Reed, T.B. 2002. *Encyclopedia of Biomass Thermal Conversion. The Principles and Technology of Pyrolysis, Gasification & Combustion.* 3rd ed. Golden, CO Biomass Energy Foundation Press.

Reed, T.B., J.P. Diebold, and R. Desrosiers. 1980. Perspectives in Heat Transfer Requirements and Mechanisms for Fast Pyrolysis (SERI/CP-622–1096). Paper read at Specialists Workshop on Fast Pyrolysis of Biomass, 1980, at Golden, CO: Solar Energy Research Institute.

Reilly, J., J. Melillo, Y. Cai et al. 2012. Using land to mitigate climate change: Hitting the target, recognizing the trade-offs. *Env. Sci. Technol.* 46 (11):5672–5679.

Ribeiro, B.D., A.M. de Castro, M.A.Z. Coelho et al. 2011. Production and use of lipases in bioenergy: A review from the feedstocks to biodiesel production. *Enzyme Res.* 2100:1–16.

Ringer, M., V. Putsche, and J. Scahill. 2006. *Large-Scale Pyrolysis Oil Production: A Technology Assessment and Economic Analysis,* edited by U.S. Department of Energy; National Renewable Energy Laboratory, Technical Report NREL/TP-510-37779. Oak Ridge, TN: U.S. Department of Energy.

Ro, K.S., K. Cantrell, D. Elliott et al. 2007. Catalytic wet gasification of municipal and animal wastes. *Ind. Eng. Chem. Res.* 46 (26):8839–8845.

Romero, Y., F. Richard, and S. Brunet. 2010. Hydrodeoxygenation of 2-ethylphenol as a model compound of bio-crude over sulfided Mo-based catalysts: Promoting effect and reaction mechanism. *Appl. Catal. B* 98 (3–4):213–223.

Salaices, E., B. Serrano, and H. de Lasa. Biomass catalytic steam gasification thermodynamics analysis and reaction experiments in a CREC riser simulator. *Indust. Eng. Chem. Res.* 49 (15):6837. American Chemical Society.

Schievano, A., A. Tenca, B. Scaglia et al. 2012. Two-stage vs. single-stage thermophilic anaerobic digestion: Comparison of energy production and biodegradation efficiencies. *Env. Sci. Technol.* 46 (15):8502–8510.

Schnoor, J.L. 2006. Biofuels & the environment. *Environ. Sci. Technol.* 40 (13):4042.

Schumacher, L.G., J. Van Gerpen, and B. Adams. 2009. Biodiesel fuels. In *Renewable Energy Focus Handbook,* 483–493. Amsterdam: Elsevier.

Soltes, E.J., and T.J. Elder. 1981. Pyrolysis. In *Organic Chemicals from Biomass,* edited by I.S. Goldstein, 63–95. Boca Raton, FL: CRC Press.

Sørensen, B. 2007. *Renewable Energy Conversion, Transmission & Storage.* Burlington MA: Academic Press.

Speight, J.G. 2007. *The Chemistry and Technology of Petroleum.* 4th ed. Boca Raton, FL: CRC Press.

Spolaore, P., C. Joannis-Cassan, E. Duran et al. 2006. Commercial applications of microalgae. *J. Biosci. Bioeng.* 101:87–96.

Stocker, M. 2008. Biofuels and biomass-to-liquid fuels in the biorefinery: Catalytic conversion of lignocellulosic biomass using porous materials. *Angew. Chem. Int. Ed.* 47 (47):11, Wiley-VCH Verlag GmbH & Co. KGaA, Weinheim.

Sun, R., X.F. Sun, and J. Tomkinson. 2004. Hemicelluloses and their derivatives. In *Hemicelluloses: Science & Technology*, edited by P. Gatenholm and M. Tenkanen. Washington D.C.: American Chemical Society.

Swanson, R.M., A. Platon, J.A. Satrio et al. 2010. Techno-economic analysis of biomass-to-liquids production based on gasification. *Fuel* 89 (suppl.1):S11–S19.

Tang, Z., Q. Lu, Y. Zhang et al. 2009. One step bio-oil upgrading through hydrotreatment, esterification, and cracking. *Ind. Eng. Chem. Res.* 48:6923–6929.

Thangalazhy-Gopakumar, S., S. Adhikari, R.B. Gupta et al. 2011. Influence of pyrolysis operating conditions on bio-oil components: A microscale study in a pyroprobe. *Energy Fuels* 25 (3):1191–1199.

Turkenburg, W.D. (Ed.), 2000. Renewable energy technologies. In *World Energy Assessment: Energy and the Challenge of Sustainability*, edited by J. Goldemberg, New York: UNDP/UN-DESA/World Energy Council, Chapter 7, Figure 7.1, p. 223.

U.S. Department of Energy. 2011. *U.S. Billion-Ton Update: Biomass Supply for a Bioenergy and Bioproducts Industry*. Oak Ridge TN: U.S. Department of Energy, 227.

U.S. Energy Information Administration. 2013a. *Cellulosic Biofuels Begin to Flow But in Lower Volumes than Foreseen by Statutory Targets* [cited February 28 2013]. Available from http://www.eia.gov/todayinenergy/detail.cfm?id=10131

U.S. Energy Information Administration. 2013b. *Petroleum and Other Liquids: Monthly Biodiesel Production Report*. Energy Information Administration 2013b [cited February 29 2013].

U.S. Energy Information Administration. 2012. *Annual Energy Review, 2011*. U.S. Energy Information Administration. Available from http://www.eia.gov/totalenergy/data/annual

van Santen, R.A. 2007. Renewable catalytic technologies—A perspective. In *Catalysis for Renewables: From Feedstock to Energy Production*, edited by G. Centi and R.A. van Santen, 1–19. Weinheim, FRG: Wiley-VCH.

Vance, A. 2010. One moos and one hums, but they could help power Google. *The New York Times*, 18 May 2010, B1.

Vanholme, R., B. Demedts, K. Morreel et al. 2010. Lignin biosynthesis and structure. *Plant Physiol.* 153 (3):894–905.

Venkanna, B.K. and R.C. Venkataramana. 2009. Biodiesel production and optimization from *Calophyllum inophyllum* Linn oil (honne oil)—A three stage method. *Bioresour. Technol.* 100 (21):5122–5125.

Viriya-empikul, N., P. Krasae, B. Puttasawat et al. 2010. Waste shells of Mollusk and egg as biodiesel production catalysts. *Bioresour. Technol.* 101 (10):3765–3767.

Voet, D. and J.G. Voet. 1990. *Biochemistry*. New York: John Wiley & Sons.

Wang, T., J. Chang, P. Lv et al. 2004. Novel catalyst for cracking of biomass tar. *Energy Fuels* 19 (1):22–27.

Weiland, F., H. Hedman, M. Marklund et al. 2013. Pressurized oxygen blown entrained-flow gasification of wood powder. *Energy Fuels* 27 (2):932–941.

Willey, J.M., L.M. Sherwood, and C.J. Woolverton. 2013. *Prescott's Microbiology*, 9th Ed. New York City, NY, USA: McGraw-Hill.

Wornat, M.J., B.G. Porter, and N.Y.C. Yang. 1994. Single droplet combustion of biomass pyrolysis oils. *Energy Fuels* 8 (5):1131–1142.

Yang, Y., R.R. Sharma-Shivappa, J.C. Burns et al. 2009. Saccharification and fermentation of dilute-acid-pretreated freeze-dried switchgrass. *Energy Fuels* 23 (11):5626–5635.

Zhang, D., Y. Chen, Y. Zhao et al. 2010. New sludge pretreatment method to improve methane production in waste activated sludge digestion. *Env. Sci. Technol.* 44 (12):4802–4808.

9 Nuclear Energy

9.1 INTRODUCTION

As noted in the introduction to this book, there are several relevant topics that would not be covered even though there is legitimate claim to their inclusion. Given that limitation, why include nuclear power? After all, it is a topic that is highly controversial with respect to sustainability and certainly one with a clouded history. Yet the following two quotes from expert scientists well summarize the need to consider nuclear energy:

> To the extent that we live in a hydrocarbon-limited world, generate too much CO_2, and major hydropower opportunities have been exhausted worldwide, new nuclear power stations must be considered.
>
> Tad W. Patzek, Cockrell Family Regents Chair, Petroleum and Geosystems Engineering, University of Texas at Austin
>
> *and* David Pimentel, Professor Emeritus, College of Agriculture and Life Sciences, Cornell University. (Patzek and Pimentel 2005)

> ... (T)here is probably nothing in this world that challenges the notion of sustainability as much as the safeguarding of nuclear waste repositories continuously for 40,000 years or more .... Still, I conclude that the alternative of rampant climate change is even more risky. Thus, low carbon energy sources are desperately needed. Nuclear power in the 21st century promises to be more modular and much safer than existing nuclear plants. And I must concede that the existing nuclear power industry has established a good safety record (Chernobyl and Fukushima notwithstanding), especially when compared to the coal-mining industry.
>
> **Jerald L. Schnoor, Editor-in-Chief**
> *Environmental Science & Technology (2013)*

Generation of electricity by nuclear power plants started in the 1950s, the heyday of nuclear power. The jump in oil prices in the 1970s gave nuclear energy an additional boost, but infamous nuclear power plant accidents—Three Mile Island (1979), Chernobyl (1986), and Fukushima (2011)—have loomed large in the general public's consciousness with the result that construction of new nuclear power plants has decreased noticeably since peaking in the 1980s (Davis et al. 2010). There is no new construction in the United States and limited construction globally. Nevertheless, the tide is turning for including nuclear power as part of the energy mix for several reasons. Nuclear energy is comparable to wind and solar energy in terms of its low contribution to carbon emissions (nuclear produces between 40 and 110 g of CO_2/kWh, compared with 900 g CO_2/kWh for coal, 20–60 g CO_2/kWh for photovoltaics and 10 g CO_2/kWh for wind (Armaroli and Balzani 2011)). It is both reliable and powerful: the energy released per gram of U-235 that undergoes fission is equivalent to 2.5 million times the energy released in burning 1 g of coal (Ansolabehere et al. 2003). And nuclear energy already provides a significant portion of the world's primary energy—5.7% of the world's supply in 2010, with the United States producing roughly 30% of the world total and France leading the way in terms of proportion of domestic energy production (75%) (IEA/International Energy Agency 2013). Thus nuclear energy, despite the serious concerns surrounding its use, is almost certain to continue to be a source of final energy for our growing energy consumption.

9.2 NUCLEAR CHEMISTRY BASICS

9.2.1 GENERAL CHEMISTRY REVIEW

The energy conversions we have examined thus far have relied primarily upon capturing the energy of chemical reactions that center on electron transfer or excitation. We have been able to ignore, if you will, the rest of the atom—until now. Nuclear power comes about from release of the binding energy of elemental particles that make up the nucleus of atoms. Because of the fundamental relationship between matter and energy ($E = mc^2$), the loss of a tiny amount of mass in a nuclear reaction translates into a huge amount of energy. Understanding nuclear power requires a quick review of some nuclear nomenclature and basic general chemistry.

A *nucleon* refers to either a proton or a neutron in the nucleus of an atom made up, of course, of positively charged protons (p) represented as $_1^1 p$ (or $_1^1 H$), negatively charged electrons represented as $_{-1}^0 e$ (or $_{-1}^0 \beta$, the *beta particle*), and neutrons depicted as $_0^1 n$. (Recall that superscript denotes the unit mass of the particle while the subscript denotes the charge.) All elements beyond lead-208 (atomic number > 83) are unstable and spontaneously undergo *radioactive decay*. For example, uranium-238 decays with loss of an alpha particle ($_2^4 He$) to give thorium-234; thorium-234, in turn, decays with emission of a beta particle to give protactinium-234 (plus a beta particle), and so on (Equation 9.1):

$$_{92}^{238} U \xrightarrow{\text{alpha decay}} {}_{90}^{234} Th + {}_2^4 He \xrightarrow{\text{beta decay}} {}_{91}^{234} Pa + {}_{-1}^0 \beta \qquad (9.1)$$

In contrast to radioactive decay, the nucleus of an element is made larger by *neutron capture* wherein a neutron is absorbed by a nucleus, as shown in Equation 9.2 for indium-115:

$$_{49}^{115} In + {}_0^1 n \xrightarrow{\text{neutron capture}} {}_{49}^{116} In \qquad (9.2)$$

As above, subsequent beta decay of the indium-116 isotope leads to the *transmutation* to tin-116 (Equation 9.3):

$$_{49}^{116} In \xrightarrow{\text{beta decay}} {}_{50}^{116} Sn + {}_{-1}^0 \beta \qquad (9.3)$$

Nuclear reactions between two nuclei may proceed by *fusion* or *fission*. In either case, huge amounts of energy are produced. For example, the fusion of hydrogen nuclei in the Sun's core produces helium and energy as shown in Equations 9.4 through 9.6. The energy (15,000,000°C) and density (150 g/cm³) in the Sun's core is necessary to force two hydrogen atoms to fuse and make the deuterium isotope with the release of a beta particle (Equation 9.4). Fusion of a deuterium nucleus with an additional proton yields helium-3 nuclei (Equation 9.5) which fuse, to generate helium-4 plus two additional protons (Equation 9.6); the overall process gives off unfathomable amounts of energy and is the reaction underlying the hydrogen bomb.

$$2\,_1^1\mathrm{H} \rightarrow {}_1^2\mathrm{H} + {}_{-1}^{0}\beta \tag{9.4}$$

$$_1^1\mathrm{H} + {}_1^2\mathrm{H} \rightarrow {}_2^3\mathrm{He} \tag{9.5}$$

$$2\,_2^3\mathrm{He} \rightarrow {}_2^4\mathrm{He} + 2\,_1^1\mathrm{H} \tag{9.6}$$

While fusion energy has many attractive features for the generation of final energy (Section 9.4.4), the nuclear chemistry behind conventional nuclear power is based on fission. A material that undergoes an appreciable amount of fission products when bombarded with a thermal neutron (*vide infra*) is termed *fissile*. (A material that does not undergo fission itself, but can be converted into fissile material, is known as a *fertile* material; for example, fertile U-238 can be converted into fissile Pu-239, *vide infra*.) Whether an element undergoes fusion or fission is determined entirely by the binding energy of the nuclei involved. Figure 9.1 is a plot of the average nuclear binding energy versus mass number and shows that iron-56 is the tipping point between fusion and fission: elements of lower atomic weight will undergo *fusion* and release energy as a more stable nucleus with greater binding energy is formed. Elements with a higher atomic weight will undergo *fission* to produce fragment nuclei with higher binding energies. It is helpful to think of nuclear binding energy as analogous to Gibbs free energy—a nuclear transition will occur if energy is released, which will be the case if the binding energy is greater on the product side of the equation. These fundamental nuclear transitions underlie all nuclear energy, and the *lanthanides* and *actinides*, Table 9.1, are the key radioactive nuclei in the discussion of nuclear power.

9.2.2 BIRTH OF NUCLEAR ENERGY

The alpha decay of U-238 occurs naturally, but in the 1930s Otto Hahn, Lise Meitner, and Fritz Strassman began to study the *neutron-induced* radioactive decay of uranium. Bombardment of a target nucleus with neutrons can result in elastic collisions (the neutron just bounces off with a loss of energy), inelastic scattering (leaving the target nucleus in an excited state and the neutron with less energy), capture of the

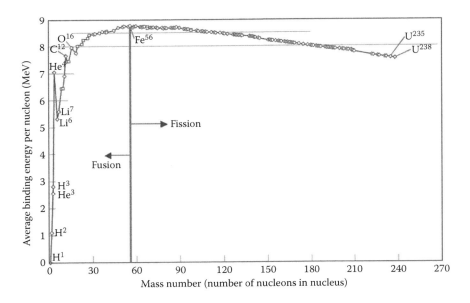

FIGURE 9.1   Binding energy per nucleon versus mass number. (From Wikimedia Commons. http://en.wikipedia.org/wiki/File:Binding_energy_curve_-_common_isotopes.svg. Accessed June 10, 2013.)

neutron by the target nucleus (this gives an excited state *compound nucleus* that then decays), or fragmentation (fission). Hahn and his coworkers found that bombardment of the uranium nucleus resulted not in radioactive decay but instead led to production of two large (relatively speaking) fragments, that is, fission of the atomic nucleus. Enrico Fermi determined that this neutron bombardment resulted in the release of additional neutrons. With the buildup to World War II occurring, these discoveries rapidly led to the basic understanding of nuclear chain reactions, as illustrated in Figure 9.2. When a high-energy neutron collides with a U-235 particle, about 200 MeV of energy is released as fragmentation occurs to give the more stable atomic nuclei xenon ($^{143}_{54}$Xe) and strontium ($^{90}_{38}$Sr) *plus* two additional neutrons. These additional neutrons can, in turn, collide with another U-235 nucleus and continue the chain reaction. The amount of material needed to sustain a nuclear chain reaction for U-235 (the *critical mass*) is 1 kg; if less than this amount is present, the reaction will fizzle out because of the natural inefficiency of neutrons escaping the core, for example. Unleashing the energy of a nuclear chain reaction in an uncontrolled manner is the basis of the atomic bomb, but in a nuclear power plant the rate of the sustained reaction is controlled to provide a steady level of heat for powering a turbine generator.

Naturally occurring uranium is a mixture of U-235 and U-238 and both isotopes are present in a conventional nuclear reactor. Those neutrons captured by U-238 result in transmutation, forming plutonium-239 by a sequence of beta particle emissions (*beta decay*) as shown in Equation 9.7.

TABLE 9.1
Lanthanides and Actinides

Lanthanides			Actinides				
Mass Number	Symbol	Name	Mass Number	Symbol	Name	Principal Isotope	Half-Life (years)
57	Ln	Lanthanum	89	Ac	Actinium	Ac-227	21.7
58	Ce	Derium	90	Th	**Thorium**	*Th-232*	1.39×10^{10}
59	Pr	Praseodymium	91	Pa	**Protactinium**	Pa-231	3.28×10^{5}
60	Nd	Neodymium	92	U	**Uranium**	U-235	7.13×10^{8}
						U-238	4.50×10^{9}
61	Pm	Promethium	93	Np	Neptunium	Np-237	2.20×10^{6}
						Pu-238	86.4
62	Sm	Samarium	94	Pu	Plutonium	Pu-239	24,360
						Pu-242	3.79×10^{5}
						Pu-244	8.28×10^{7}
63	Eu	Europium	95	Am	Americium	Am-241	433
						Am-243	7650
64	Gd	Gadolinium	96	Cm	Curium	Cm-242	4.56×10^{-1}
						Cm-244	18.12
65	Tb	Terbium	97	Bk	Berkelium	Bk-249	9.13×10^{-1}
66	Dy	Dysprosium	89	Cf	Californium	Cf-252	2.57
67	Ho	Holmium	99	Es	Einsteinium		
68	Er	Erbium	100	Fm	Fermium		
69	Tm	Thulium	101	Md	Mendelevium		
70	Yb	Ytterbium	102	No	Nobelium		
71	Lu	Lutetium	103	Lr	Lawrencium		

Note: Elements in bold are present in nature in practical amounts for extraction. Isotopes in italics are fertile.

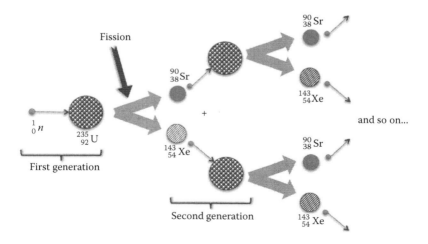

FIGURE 9.2   A nuclear chain reaction.

$$^{238}_{92}\text{U} + ^1_0n \xrightarrow{\textit{neutron capture}} {}^{239}_{92}\text{U} \xrightarrow{\textit{beta decay}} {}^{239}_{93}\text{Np} + {}^{\;0}_{-1}\beta \xrightarrow{\textit{beta decay}} {}^{239}_{94}\text{Pu} + {}^{\;0}_{-1}\beta$$

$$(9.7)$$

Plutonium-239, unlike uranium-239 or neptunium-239, does not decay appreciably while in the reactor: it has a half-life of about 24,000 years. It, too, is a fissile material and plays a key role in *breeder reactors* and, unfortunately, nuclear weapons.

The first nuclear power plants (*Generation I*) were built in the 1950s and by the year 1960, 17 nuclear plants were in operation. Slightly more advanced nuclear reactors built in the 1970s and 1980s are known as *Generation II* reactors and are the topic of the next section (Char and Csik 1987). Advanced variants of the Generation II reactors, having improved safety features and more modular architecture, are *Generation III/III+* reactors. We will examine the potential of *Generation IV* reactors in Section 9.4.3.

9.2.3 NUCLEAR REACTORS

Some of the key components of a nuclear reactor in terms of chemistry are the *fuel*, the *moderator*, the *control* (or poison), and the *coolant*.

- The fuel is any fissionable nucleus; for conventional nuclear power this means an isotope of uranium or plutonium.
- The moderator—water, heavy water (D_2O), or graphite—is the material within the reactor core that reduces the energy of the bombarding neutrons through elastic and inelastic collisions. A moderator is needed for the successful operation of *thermal* reactors, as we will see in the next section. It should *not* be a good neutron absorber lest the chain reaction be slowed too much.
- The control is a material that *is* a good neutron absorber—this material absorbs neutrons as needed to keep the reaction rate under control.
- The coolant is exactly what one would expect: some fluid that can act as a heat sink to carry excess heat away from the reactor core. In conventional nuclear reactors the coolant is water, but for advanced reactors operating at very high temperatures a molten salt or molten metal may be necessary.

Table 9.2 summarizes most of the nuclear reactor designs including those in use and those in the design and development phase (Generation IV). The pressurized light water reactor will be described in some detail as a basis for understanding other types of nuclear reactors.

9.2.3.1 Conventional Nuclear Power

Electricity from nuclear energy is generated in the same manner as in a coal-fired power plant except that the source of heat for firing the steam-powered turbine is from nuclear reactions. Most nuclear power plants in operation today are

TABLE 9.2
Some Types of Nuclear Reactors

Type of Reactor	Acronym	Typical Fuel	Coolant	Description
Light water reactor (LWR), pressurized	PWR	Enriched UO_2	Water	The most prevalent design in current use. Water is also the moderator. T_{in}/T_{out} = 290/320°C. Contains 200–300 fuel rod assemblies.
Boiling (light) water reactor	BWR			Similar to PWR but runs at a lower pressure (7 MPa vs. 16 MPa). T_{in}/T_{out} = 280/288°C. Contains around 700–800 fuel rod assemblies
Heavy water reactor	CANDU	Natural or enriched U, Th-233	Heavy water (D_2O)	Heavy water serves as the moderator and coolant. The heavy water analogue to the PWR = PHWR
Fast breeder	FBR	UO_2 and PuO_2	Liquid Na°	No moderator is required and operating temperatures inside the reactor core are higher than LWRs
European pressurized reactor	EPR	UO_2 or MOX	Water	Essentially a "new and improved" PWR. Construction of the first EPR has begun in Olikiluoto, Finland, but has been stalled by many delays. Two more are under construction in China
Westinghouse Advanced PWR	AP1000	UO_2	Water	Similar to EPR—a Generation III PWR
Sodium fast reactor	SFR	MOX	Liquid Na°	Uses specialized stainless-steel cladding. T_{in}/T_{out} = 370/550°C

continued

TABLE 9.2 (continued)
Some Types of Nuclear Reactors

Type of Reactor	Acronym	Typical Fuel	Coolant	Description
Lead fast reactor	LFR	U/Pu nitride mixture	Lead or Pb/Bismuth	$T_{in}/T_{out} = 600/800°C$
Molten salt reactor	MSR	$^{233}UF_4$ or $^{235}UF_4$	Molten salt (e.g., FLiNaK)	$T_{in}/T_{out} = 700/1000°C$
Very-high-temperature reactor	VHTR	UO_2	High-pressure helium gas	$T_{in}/T_{out} = 600/1000$ °C; 50% efficiency. Electricity is generated via a gas turbine driven by hot helium. Designed for the co-production of hydrogen. Graphite moderated
Pebble bed modular reactor	PBMR	U° or UOX distributed in porous graphite	Helium	A version of the VHTR. The core uses tennis ball-sized spherical graphite "pebbles" containing ceramic-coated fuel particles
Gas fast reactor	GFR	MOX	He + supercritical CO_2	Operating temperature of 850°C; 48% efficiency
Supercritical water-cooled reactor	SCWR	UO_2	Supercritical water	$T_{in}/T_{out} = 290/600°C$. Can operate at very high pressures because supercritical water has good heat capacity and therefore can be used in smaller volumes

Sources: Adapted from International Energy Agency. 2007. Nuclear Power. ETE04. http://www.iea.org/publications/freepublications/publication/essentials4.pdf; Suppes, G.J. and T. Storvick. 2007. *Sustainable Nuclear Power, Academic Press Sustainable World Series*. London: Elsevier; Zinkle, S.J. and G.S. Was. 2013. Materials challenges in nuclear energy. *Acta Mater.* 61 (3):735–758.

Note: MOX, mixed oxide of uranium and plutonium.

thermal reactors in that the energies of the neutrons in the reactor are reduced by collisions with the moderator to thermal energy in the range of 0–1 eV. This process—deemed *thermalization*—is necessary to increase the likelihood of the neutron being captured by U-235 and resulting in fission. If the neutron energy is *not* reduced and, instead, *fast neutrons* (>1 MeV) are allowed to persist, the reactor is termed a *fast reactor*. Fast reactors are discussed in Section 9.4.3 in the context of Generation IV reactors.

Thermal reactors include both *heavy water reactors* and *light water reactors* (meaning they use D_2O or H_2O). There are two primary types of light water reactors: the *pressurized water reactor* (PWR) and the *boiling water reactor* (BWR). They are appropriately named; in the PWR water is under pressure so that it is not allowed to boil; in the BWR water is under lower pressure and it is allowed to boil, generating steam to turn the steam turbine. The light water-cooled and water-moderated PWR is the most commonly used reactor design in operation today.

The PWR essentially consists of three interconnected operations: (1) the reactor vessel, (2) the steam generator, and (3) the steam turbine/generator complex (Figure 9.3). The heat from the nuclear reaction heats the water surrounding the reactor core. This hot, pressurized water acts as the heating fluid for the separate production of steam in the steam generator. It is pumped through to the steam generator, but the two supplies of water never intermix (see Figure 9.3)—two separate loops of water, one circulating from the reactor core and one circulating in the steam generator, are used. The produced steam goes to the turbine to generate electricity and the exhausted steam is condensed and pumped back to the steam generator.

The heart of the reactor vessel is the *reactor core* that contains the *fuel rod assemblies*. The type of fuel that is used depends on the reactor. Conventional pressurized light water reactors require 3–5% enriched U-235 pellets, but other fuels,

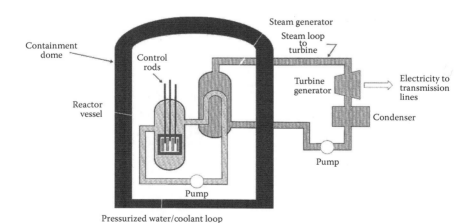

FIGURE 9.3 Schematic of a pressurized water reactor power plant. (Adapted from United States Nuclear Regulatory Commission. Pressurized Water Reactor. Last updated March 29, 2012. Available from http://www.nrc.gov/reading-rm/basic-ref/teachers/pwr-schematic.html.)

including naturally occurring uranium oxide or a mixture of uranium and pluto-nium oxides (MOX), can be used in certain cases (see Table 9.2). The MOX fuel is derived from the reprocessing of spent nuclear fuel (Section 9.3.3.2) and gener-ally cannot be exchanged 1:1 for reactor-grade fuel. Light water reactors require the enriched uranium because of the effectiveness of light water as the moderator. The hydrogen nucleus is a very efficient moderator (too efficient, in fact) in that it can capture a slow neutron to become deuterium, thus removing neutrons from the chain reaction. (The use of heavy water, a less effective moderator, allows for the use of natural uranium. Graphite, also, is a less effective moderator than water and is used when the production of Pu-239 is desired for nuclear weapons production: it both captures fewer thermal neutrons and requires more collisions with the neu-trons to reach thermal energies. As a result, the chances are increased that U-238 will capture a neutron and convert into Pu-239. However, graphite obviously cannot be used as a coolant.)

The fuel pellets are encased in a zirconium alloy (*zircaloy*) tube, chosen both because it is strong and because it has low neutron absorptivity. An array of fuel rods is bundled into the fuel rod assembly with some spaces left empty for insertion of control rods containing a neutron absorber, for example, boron-10, that can control the rate of reaction by capture of a neutron (see Equation 9.8).

$$_{5}^{10}\text{B} + {}_{0}^{1}n \xrightarrow{\text{neutron capture}} {}_{5}^{11}\text{B} \xrightarrow{\text{alpha decay}} {}_{3}^{7}\text{Li} + {}_{2}^{4}\text{He} \qquad (9.8)$$

Cadmium-113 is also an effective control element. The operation of the control rods by their insertion or withdrawal from the core effectively controls the rate of reaction (see Figure 9.4). The entire assembly is placed in the reactor core in a pool of water that plays the dual role of both coolant and moderator. The entire reactor vessel is then contained within a protective dome to prevent release of radioactive materials.

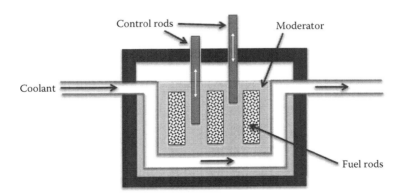

FIGURE 9.4 Schematic of control rods in reactor vessel. (Adapted from Los Alamos National Laboratory, http://www.lanl.gov/science/NSS/issue1_2011/story4full.shtml. Accessed June 25, 2013.)

9.2.3.2 Other Types of Nuclear Reactors

9.2.3.2.1 Heavy Water Reactors

Light water reactors are far and away the most common reactors in operation, but the pressurized heavy water reactor (also known as the CANDU reactor for Canadian deuterium uranium) should be noted because it uses heavy water in the role of coolant and moderator and, as a result, can make use of natural uranium fuel, bypassing the expense of uranium enrichment. Deuterium does not absorb neutrons as effectively as light water, so the nuclear reaction can be sustained without having to use a higher proportion of the U-235 in the fuel. (Graphite, too, is a moderator that can be used with natural uranium fuel.)

9.2.3.2.2 Breeder Reactors

The *fast breeder reactor* (FBR) is so named because more fissile material is formed than consumed, a feat that can be accomplished if the fragment nuclei are *also* fissile. The *breeding ratio* is the ratio of fissile material produced to the fissile material consumed; any reactor that achieves a *breeding ratio* of >1 is characterized as a breeder reactor. In a conventional light water reactor, Pu-239 is formed by random neutron capture, but not at a rate high enough to sustain the chain reaction (hence the fuel must be enriched in U-235). By using a uranium–plutonium oxide fuel mixture (usually ≈20% Pu-239) bombarded with fast neutrons, Pu-239 can undergo fission to produce more second-generation neutrons than any other fissile isotope. At the same time, U-238 present in the fuel can undergo neutron capture to transmute to more Pu-239 (recall Equation 9.7). Furthermore, the reactor core is surrounded by a "blanket" of natural or depleted uranium to capture any stray neutrons and create even more Pu-239.

That a breeder reactor requires fast neutrons means that no moderator is needed, although (obviously) a coolant still is. By using a coolant with a high mass number, thermalization is minimized. The coolant of choice in FBRs is liquid sodium, the use of which clearly presents safety issues of its own as sodium reacts violently with both air and water (Equations 9.9a through 9.9c).

$$Na^\circ + H_2O \rightarrow NaOH + H_2 \qquad (9.9a)$$

$$Na^\circ + O_2 \rightarrow \tfrac{1}{2} Na_2O \qquad (9.9b)$$

$$4Na^\circ + O_2 \rightarrow 2Na_2O \qquad (9.9c)$$

The BN-600 600 MW FBR in Russia had 27 sodium leaks between 1980 and 1997, resulting in 14 sodium fires (Cochran et al. 2010). Another serious safety issue is associated with the potential loss of coolant. In a fast-neutron reactor, the Pu-239 concentration is high enough that the chain reaction will continue (or may even increase) if the coolant is lost, possibly leading to a small nuclear explosion. There are several other reasons that the commercialization of FBRs has failed. A key reason for its development in the first place was the scarcity of uranium fuel—but this has not proven to be an issue. Breeder reactors are both very expensive to build and to operate and have a reputation for being unreliable, needing extended periods off-line for maintenance and repairs (primarily a result of using the liquid sodium coolant). Another major concern associated with breeder reactors is the proliferation of Pu-239 that can be used in nuclear weapons. Several FBRs have been built around the globe, but breeder reactors are, for all practical purposes, a dying breed (Cochran et al. 2010).

9.2.3.2.3 Advanced Pressurized Water Reactors

The nuclear power fleet is rapidly aging (the average age is currently thirty years) and many are being shut down or retired for good reasons: they have reached the end of their safe and reliable operating lifetime and electricity produced from natural gas is proving to be tough economic competition. However, this means that more and more electrical power generation is coming off-line. Newer Generation III/III+ reactors have been designed for greater safety. The Westinghouse AP1000 and the European Pressurized Reactor (EPR) are both examples of Generation III reactors that operate on essentially the same principles as a PWR but designed with the intention of greater safety, higher efficiency, less waste, and better economics. While the idea behind the EPR is to lower construction costs with a more compact design, the first EPR to be constructed (in Olikiluto, Finland) was to have opened in 2009, then 2012, and now (as of this writing) is delayed until 2016—all due to construction delays and greatly increased costs.

9.2.3.2.4 Thorium Reactors

Naturally occurring thorium, atomic mass number 90, exists entirely as Th-232 which is not a fissile material for thermal neutrons. It is, however, fertile: upon neutron capture it is converted into U-233 which *is* fissile (Equation 9.10).

$$^{232}_{90}\text{Th} + {}^{1}_{0}n \xrightarrow{\;neutron\,capture\;} {}^{233}_{90}\text{Th} \xrightarrow{\;beta\,decay\;} {}^{233}_{91}\text{Pa} + {}^{0}_{-1}\beta \xrightarrow{\;beta\,decay\;} {}^{233}_{92}\text{U} + {}^{0}_{-1}\beta$$

$$(9.10)$$

Thus, a thorium reactor requires a "seed" fissile material (U-233, U-235, or Pu-239) to initiate the production of neutrons, but once the product U-233 is formed, the reaction is off and running. This is the basis of the thorium–uranium fuel cycle.

There are many advantages to using thorium-232 as a nuclear fuel, including its abundance (it is 3–4 times more abundant than uranium), relative availability (it can be mined from the mineral monazite, $Sm_{0.2}Gd_{0.2}Th_{0.15}Ce_{0.15}Ca_{0.1}Nd_{0.1}(PO_4)_{0.9}$, as the phosphate or from thorite, $Th(SiO_4)$ as the silicate), it does not require enrichment, and there is no opportunity for nuclear proliferation of Pu-239. In fact, the waste generated from a thorium reactor, while still hazardous, is lower in volume and less radioactive compared to that from a conventional uranium reactor (Cooper et al. 2011). Given these advantages and the fact that the nuclear capabilities of thorium have been known since the 1950s, why did uranium become the fuel of choice for nuclear energy? It is the result of history: because thorium reactors do not provide a source of Pu-239 for nuclear weapons programs, they were set aside as nuclear energy developed under the shadow of the Cold War and the nuclear arms race. The uranium fuel cycle with its production of Pu-239 allowed for the development of a nuclear stockpile.

As a result of being rejected in favor of uranium, thorium does not have decades of research and development to facilitate its entry into commercialization. Nevertheless, progress on thorium-based nuclear energy is continuing. A heavy water BWR in Halden, Norway has recently begun a thorium fuel burn using pellets consisting of a thorium oxide ceramic matrix with approximately 10% plutonium oxide to drive the fission reaction (World Nuclear News 2013). And thorium energy proponents are encouraging continued development of the *liquid fluoride thorium reactor* (LFTR). This reactor is a variation of the *molten salt reactor* (Table 9.2) and uses a combination of two molten salts as the fuel: a core of $^{233}UF_4$ dissolved in a lithium–beryllium fluoride surrounded by a "blanket" of $^{232}ThF_4$ in the same solvent. As the $^{233}UF_4$ generates neutrons, the blanket of $^{232}ThF_4$ captures them to generate more U-233. This uranium fuel is converted into gaseous UF_6 and fed back into the core as fresh thorium fuel is supplied to the blanket (Jacoby 2009). Not only does the LFTR present the advantages noted above due to the use of thorium fuel, it also presents greater safety features, primarily because the fuel is molten. Should a catastrophic event occur, the escape of the molten fuel will quickly cease as the fuel solidifies upon escape from the high-temperature core. It is also run at low pressure, thus obviating the need for housing that can withstand high pressures and therefore reducing construction costs. The high temperature of a molten salt reactor also means that thermal efficiencies are increased (from about 35% to 50%) (Cooper et al. 2011; Jacoby 2009).

In addition to the reactors described above, some of which are in operation, some under construction, and some planned for construction, the next generation of nuclear reactors—Generation IV—has been evolving and is discussed in Section 9.4.3. However, progression to Generation IV is not likely for many years to come.

9.3   URANIUM PRODUCTION

The *nuclear fuel cycle* is the entire process of dealing with nuclear fuel from mining, to refining and enrichment, to use in the reactor, and then reprocessing or disposal of the depleted fuel. If the uranium and plutonium oxides are reprocessed and returned to a reactor as a fuel, the phrase *fuel cycle* is appropriate. If the material is used once and discarded, the "cycle" is known as a *once-through* cycle. Some of the more important aspects of the nuclear fuel cycle—focusing on uranium—are covered in the following sections.

9.3.1   URANIUM MINING

Uranium is a naturally occurring element that is obtained as its oxides from ores in a wide variety of deposits in Earth's crust. These ores primarily consist of *uraninite* (UO_2) and *pitchblende* ($U_2O_5 \cdot UO_3$). The generic uranium oxide that is ultimately used as the feedstock for uranium enrichment (*vide infra*) is U_3O_8, often referred to as UOX or *yellowcake*. Pure uranium is hard, silvery, extremely dense (1.6 times denser than lead) and, of course, radioactive. The two naturally occurring isotopes—U-235 and U-238—are both radioactive, but only U-235 is fissile. The concentration of U-235 in naturally occurring uranium is only 0.7%, the remainder being U-238.

A majority (roughly 64%) of the world's uranium production is from just three countries—Kazakhstan, Canada, and Australia (World Nuclear Association 2013b). Extraction of uranium-containing ores from Earth's crust is also energy intensive and environmentally disruptive. About 20% of uranium ore is mined from open pits and roughly 30% from underground excavations where the ore is crushed and the uranium oxides leached out by treatment with sulfuric acid (Equations 9.11a and 9.11b).

$$UO_3(s) + 2H^+(aq) \rightarrow UO_2^{2+}(aq) + H_2O \tag{9.11a}$$

$$UO_2^{2+}(aq) + 3SO_4^{2-}(aq) \rightarrow UO_2(SO_4)_3^{4-}(aq) \tag{9.11b}$$

A process that has grown steadily in the last few decades is *in situ leach mining* or *in situ recovery* (ISL or ISR) and, in fact, this method is now the most commonly used for uranium recovery. In ISL, the uranium oxide is dissolved directly from ore *in situ* using either an acid (typically sulfuric acid) or, when limestone or other carbonate minerals are present that would also dissolve in the acid, an alkali leaching solution. The uranium-laden solution is then pumped to the surface. This method, of course, requires careful monitoring to ensure that groundwater supplies are not contaminated (World Nuclear Association 2012). If the leachate is acidic, the dissolved $UO_2(SO_4)_3^{4-}$ can be extracted into an organic solvent by treatment with a lipophilic quaternary amine (R_3NH^+) that coordinates with the trisulfate, extracting it into an organic solvent for subsequent concentration and conversion to the oxide.

9.3.2 URANIUM ENRICHMENT

The very low concentration of U-235 (the fissile isotope) is problematic because most conventional nuclear reactors require a higher proportion of fissile material. Most commercial reactor fuel is enriched to between 3% and 5% U-235, a product known as *reactor-grade uranium*. (Nuclear weapons require uranium enriched to greater than 90% U-235 (World Nuclear Association 2013a).) How is the U-235 isotope enriched? The methods of enrichment must take advantage of the sole difference between the two isotopes: the fact that U-235 has 92 protons and 143 neutrons while U-238 has 92 protons and 146 neutrons. Even with a difference of only three neutrons, these two isotopes can be separated and the proportion of U-235 increased.

The first step in the enrichment process is the conversion of uranium oxide to uranium hexafluoride (UF_6), a gas at room temperature (Equations 9.12a and 9.12b).

$$UO_2(s) + 4HF(g) \rightarrow UF_4(s) + 2H_2O(g) \tag{9.12a}$$

$$UF_4(s) + F_2(g) \rightarrow UF_6(g) \tag{9.12b}$$

The $^{235}UF_6$ and $^{238}UF_6$ can then be separated by one of two main methods: effusion (often less accurately referred to as diffusion) or centrifugation. While the effusion method is rapidly becoming obsolete, it still accounted for processing of 25% of the enriched U-235 supply in 2010. The basis of enrichment by diffusion is the rate at which the two isotopic UF_6 gases diffuse through a thin membrane. Graham's law (Equation 9.13) quantifies this property; r_1 and r_2 are the diffusion rates for the two gases of masses M_1 and M_2:

$$\frac{r_1}{r_2} = \sqrt{\frac{M_2}{M_1}} \therefore \text{for uranium,} \quad \frac{r_{235}}{r_{238}} = \sqrt{\frac{352}{349}} = 1.004 \tag{9.13}$$

Thus, based on the miniscule difference in the mass of the two uranium isotopes, gaseous diffusion technology can enrich the uranium after several passes through the membrane.

The centrifugation method is projected to account for up to 93% of the enriched uranium supply by 2017 and is the more efficient process since centrifugal force is proportional to mass, not the square root of mass (World Nuclear Association 2013a). In this process, the mixture of UF_6 gaseous isotopes is spun in cylinders at 50–70,000 rpm and centrifugal forces separate the two, concentrating the slightly heavier $^{238}UF_6$ closer to the cylinder wall. The gas centrifuge technology consumes only about 5% as much electricity as the gaseous diffusion technology, but it should be noted that the enrichment process accounts for almost one-half the cost of nuclear fuel. The carbon footprint of nuclear power, although relatively small, can be mostly attributed to the enrichment process as the electricity it requires typically comes from coal-fired power plants.

New methods of uranium enrichment are under development and a "third-generation" laser-based process is nearing commercial implementation. In this process, a laser precisely tuned to ionize a U-235 atom (once again in a UF_6 feedstock) creates $[^{235}UF_6]^+$ which is then separated from $[^{238}UF_6]$ by ionic attraction to a negatively charged collector plate. For all of these separation methods, the next step in the nuclear fuel cycle is conversion of the enriched uranium hexafluoride gas to uranium dioxide (Equation 9.14):

$$UF_6(g) + 2H_2O(g) + H_2(g) \rightarrow UO_2(s) + 6HF(g) \tag{9.14}$$

The enriched uranium dioxide is then fabricated into sintered ceramic pellets that are then manufactured into fuel rods.

9.3.3 FUEL REPROCESSING AND WASTE HANDLING

Crucial to the concept of sustainable nuclear energy is a *closed fuel cycle*, that is, recycling the radioactive fuel and fission products to as full an extent as possible. The once-through mindset of conventional nuclear power cannot continue if we are to achieve any sense of sustainability with respect to nuclear power. That said, a closed fuel cycle requires extensive reprocessing of nuclear waste. The once-through nuclear fuel cycle is inherently unsustainable, but it is certainly legitimate to weigh the safety issues of excessive handling of radioactive waste (as in reprocessing) against the safety concerns of "disposing" of the waste for thousands of years.

9.3.3.1 Depleted Uranium

There are several sources of waste in the nuclear fuel cycle, from the initial mining to the treatment of spent fuel. For example, after naturally occurring uranium sources have been mined, crushed, and leached out of the ore, the remainder (the *tailings*) still contains radioactive uranium. There is also the spent fuel from the reactor: fuel rods are replaced on a regular basis, rotating out partially depleted fuel rods and replacing them with fresh fuel. Some of the U-235 undergoes neutron capture instead of fission and, ultimately, only about 75% of the U-235 is consumed in the fuel rod. Hence, the spent fuel contains uranium isotopes, Pu-239, other radioactive fission products, and minor actinides, as shown in Figure 9.5. The unfortunate fact that about 1% of the spent fuel consists of plutonium isotopes means that it can be a source of material for nuclear weapons. And although the *minor actinides* (neptunium-237, americium-241, and curium-244) make up only the tiniest amount of the waste, they are the most hazardous in terms of long-term radiotoxicity (see below) (Hudson et al. 2012).

Nuclear waste can be classified as *low-level* (LLW), *intermediate-level* (ILW), or *high-level nuclear waste* (HLW) depending upon the percent of radioactivity:

Type of Waste	Radioactivity (%)
Low-level	1
Intermediate-level	4
High-level	95

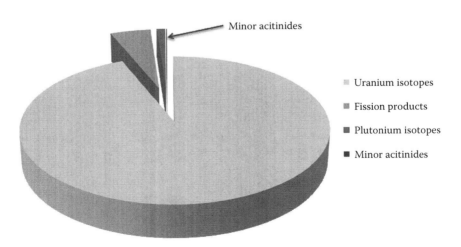

FIGURE 9.5 Typical composition of spent fuel from a light water reactor. (Fuel Cycle Stewardship in a Nuclear Renaissance. The Royal Society Science Policy Centre report 10/11. London, UK: Reproduced by permission of The Royal Society of Chemistry.)

High-level waste is hot both thermally (in terms of radiated heat from ongoing nuclear decay processes) and with respect to radioactivity and is stored in cooling ponds for an indefinite period of time as it decays (see below). The typical output for a 1-GW nuclear power plant is 200–350 m^3 of low- and intermediate-level waste per year and 10–20 m^3 of high-level waste (for comparison, the capacities of nuclear generating plants in the United States range from about 0.5 to 1.3 GW (U.S. Energy Information Administration 2011)). High-level waste that is not reprocessed (Section 9.3.3.2) undergoes *vitrification* in which it is encased in borosilicate glass and then sealed in casks for storage in geologic structures for thousands of years (the half-life for Np-237 is 2.2 million years). These storage areas are deep rock structures that are back-filled with some sort of impermeable material like clay. However, since the radioactivity level of high-level waste decreases significantly in the first few decades, the spent fuel is often "temporarily" stored in cooling ponds for several years (International Energy Agency 2007).

9.3.3.2 Reprocessing Technologies

Of particular interest for sustainability in nuclear energy is *reprocessing* of spent (or *depleted*) nuclear fuel. There are three main reasons to reprocess spent nuclear fuel: (1) to recycle the plutonium and uranium and reuse them in a nuclear reactor that can handle this type of *mixed oxide* fuel (MOX), thus increasing the efficiency of fuel consumption, (2) to decrease the volume of high-level waste, and (3) to separate out the minor actinides. As noted above, the minor actinides Np-237, Am-241, and Cm-244 are the most hazardous in terms of long-term radiotoxicity. Using the radiotoxicity of naturally occurring uranium as a baseline, it takes around 350,000 years for spent fuel to return to the same level as natural uranium ore. On the other hand, if the plutonium and uranium isotopes can be removed from the spent fuel waste stream, that same level of radiotoxicity can be reached in roughly 5000 years. If plutonium, uranium, and the minor actinides are removed, the radiotoxicity drops relatively quickly to match that of natural uranium in a relatively short period of 500 years (The Royal Society 2011). Furthermore, if the minor actinides can be separated and concentrated as their oxides, they can be converted in a separate nuclear process into less harmful nuclei, a process referred to as *partitioning and transmutation* (P & T). Thus, reprocessing of spent nuclear fuel is a very important goal—but it is replete with political repercussions due to the presence of Pu-239. Any reprocessing scheme could, potentially, separate out the Pu-239 for recovery and use in nuclear weapons. It is primarily for this reason that the United States no longer has any reprocessing facilities and does not reprocess spent nuclear fuel, unlike several other countries around the world. (For example, the French reprocessing technology is able to extract over 99% of both uranium and plutonium (Bodansky 1996).)

Reprocessing spent nuclear fuel is basically fractionating it in such a way as to have separate waste streams that can be reused, treated, or disposed of. An overview of spent nuclear fuel separation schemes is given in Figure 9.6. There are huge technological difficulties associated with these separations. Some of the waste components are volatile, some are neutron emitters, plutonium-239 presents a proliferation risk, and the minor actinides are tremendous radiotoxicity hazards. As can be imagined, the technologies that accomplish these separations are

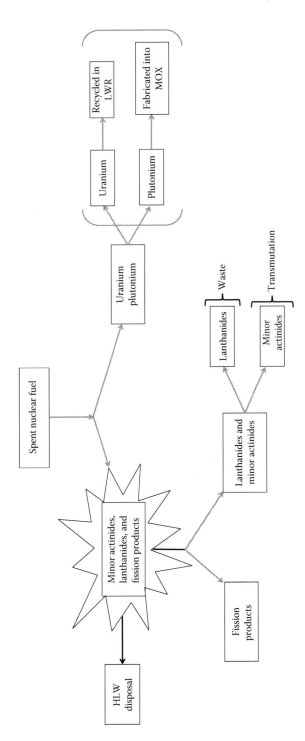

FIGURE 9.6   An overview of the partitioning of nuclear waste.

not trivial and it is far beyond the scope of this text to go through all of them. Instead, we will selectively focus on (1) the historical process, PUREX, which has been used for decades and (2) two small-scale processes for the separation of lanthanides from actinides (TALSPEAK and SANEX) and some recent research developments in this context.

PUREX stands for **p**lutonium **u**ranium **r**efining by **ex**traction and begins with the chopping and dissolution of the used reactor fuel in nitric acid to form a witch's brew of dissolved ions, including actinides (UO_2^{2+} and Pu^{4+}, among others, including the minor actinides) and lanthanides and various fission products (iodine-129, technetium-99, cesium-135, strontium-90, etc.). Some volatile elements (iodine, krypton, and xenon) are given off and captured during the dissolution process. The resulting acidic and highly radioactive solution is then extracted with a mixture of 30% tributyl phosphate (TBP, Figure 9.7) in a nonpolar organic solvent (kerosene or dodecane) to selectively extract the plutonium and uranium compounds into the organic phase as neutral complexes with tributyl phosphate $[UO_2(TBP)_2(NO_3)_2]$ and $[Pu(TBP)_2(NO_3)_4]$ (Whittaker et al. 2013). The remaining fission products (the lanthanides and the minor actinides) stay in the acidic aqueous phase (Whittaker et al. 2013). The subsequent separation of U from Pu may or may not be then undertaken depending upon the country's policy toward nuclear proliferation risk (WPFC Expert Group on Chemical Partitioning of the NEA Nuclear Science Committee 2012).

Improvements in reprocessing technology have developed over the years and continue to be an active area of investigation. Several reprocessing options have been developed to address the fundamental issues associated with recycling of spent nuclear fuel and are identified by a plethora of acronyms:

- SANEX—selective actinide extraction
- TRUEX—transuranic extraction (the transuranic elements are those of atomic number >92)
- DIAMEX—diamide extraction (removal of nonlanthanide fission products)
- TALSPEAK—trivalent actinide–lanthanide separation by phosphorus reagent extraction from aqueous komplexes

Others include UREX+, NUEX, COEX, GANEX, and so on. These processes basically vary in regard to exactly which chemicals are separated out: *all* of the minor actinides, only uranium, uranium plus plutonium, and so on.

The separation of lanthanides from actinides in the waste stream is necessary so that transmutation can be carried out to reduce the radiotoxicity of the waste stream. The high neutron absorption capability of some of the lanthanides makes transmutation less

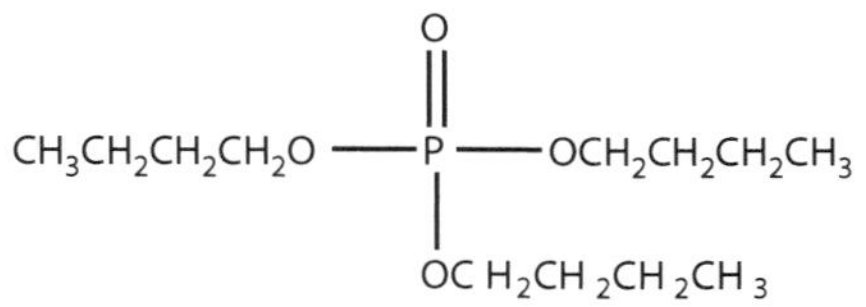

FIGURE 9.7 Tri-*n*-butyl phosphate, a key component of the PUREX process.

feasible if the lanthanides cannot be separated out (Hudson et al. 2012). Many of the processes listed above have been developed to address this challenge to remove enough of the actinides (≥99%) in high purity to outweigh the costs of this separation process.

The lanthanide–actinide separation is particularly difficult because of the similarity of the An(III) (actinide³⁺) and Ln(III) (lanthanide³⁺) behaviors. The trivalent actinides are very slightly more covalent (presumably due to the presence of 5f instead of 4f orbitals although the origin of this behavior is not well understood) and therefore interact more favorably with soft ligands like nitrogen and sulfur. The TALSPEAK process separates Ln(III) from An(III) as follows: a lipophilic ligand, bis-2-ethyl(hexyl) phosphoric acid (HDEHP, see Figure 9.5), is used to extract Ln(III) ions into an organic phase while diethylenetriamine-*N,N,N′,N″,N″*-pentaacetic acid (DTPA) complexes with the An(III) ions and retains them in the aqueous phase (Figure 9.8) (Braley et al. 2011). The TALSPEAK process is buffered with lactic acid and run at a pH of 3.5–3.6. The competition between the lipophilic extraction of the lanthanide ions and the aqueous retention of the actinide ions is revealed in the reaction and equilibrium coefficients for the aqueous phase (Equation 9.15) and organic phase (Equation 9.16). In this example L stands for the ligand DTPA and the acid, HA, is HDEHP with AHA representing the HDEHP dimer. An example of a $M(AHA)_{3(org)}$ complex for americium is given in Figure 9.9.

$$M^{3+} + H_3L^{2-} \rightleftarrows ML^{2-} + 3H^+ \therefore$$

$$K_{eq} = \frac{[ML^{2-}] \times [H^+]^3}{[M^{3+}]_{aq} \times [H_3L^{2-}]} \tag{9.15}$$

$$M^{3+}_{aq} + 3(HA)_{2(org)} \rightleftarrows M(AHA)_{3(org)} + 3H^+ \therefore$$

$$K_{eq} = \frac{[M(AHA)_3]_{org} \times [H^+]^{3+}_{aq}}{[M^{3+}]_{aq} \times [(HA)_2]^3_{org}} \tag{9.16}$$

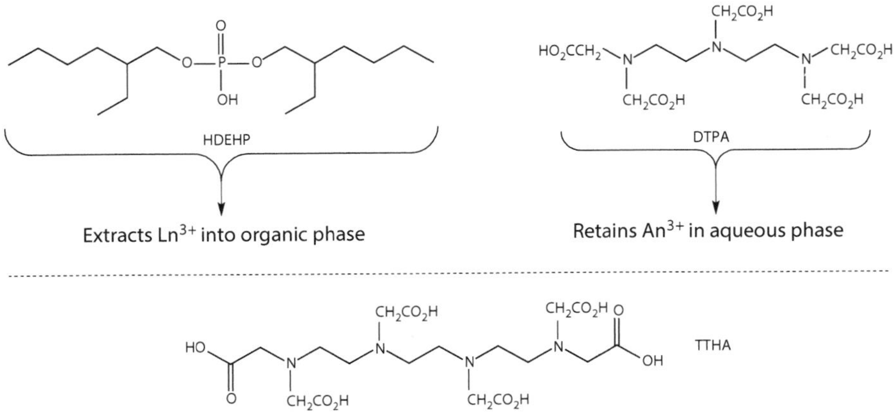

Extracts Ln³⁺ into organic phase   Retains An³⁺ in aqueous phase

FIGURE 9.8 The premise behind the TALSPEAK separation of actinides from lanthanides.

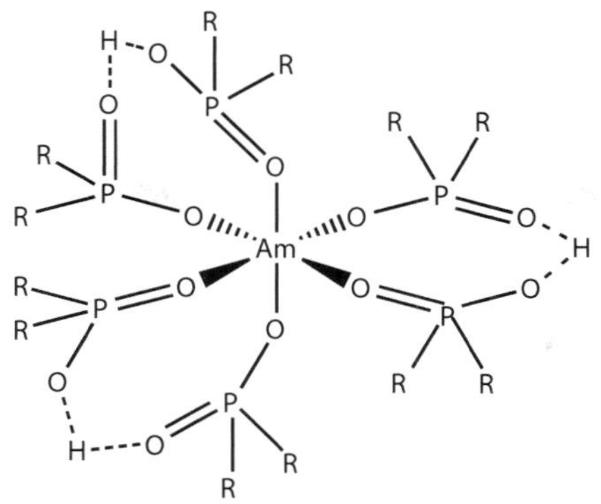

FIGURE 9.9 The Am(III) ion complexed with HDEHP dimers.

Combining these relationships gives the overall equilibrium for conventional TALSPEAK (Equation 9.17):

$$ML_{aq}^{2-} + 3(HA)_{2(org)} \rightleftarrows M(AHA)_{3(org)} + H_3L_{aq}^{2-} \qquad (9.17)$$

The TALSPEAK process, while effective, is highly complex and has room for improvement; thus, research to tailor the chemical behavior of the lipophilic phosphoric acid agent and the hydrophilic complexing agent is ongoing. For example, substitution of triethylenetetramine-N,N,N',N'',N''',N'''-hexaacetic acid (TTHA, Figure 9.8) for the pentaacetic acid reagent DTPA resulted in a slight improvement in the separation of the actinides from heavy lanthanides but poorer efficiency in separation of the lighter lanthanides (Braley et al. 2011).

The SANEX process is a related solvent extraction process that is implemented after the removal of Pu and U by PUREX and removal of the non-lanthanide fission products by the DIAMEX process. In the SANEX process, a bis-(1,2,4-triazine) ligand (e.g., 6,6'-bis(5,5,8,8-tetramethyl-5,6,7,8-tetrahydro-1,2,4-benzotriazin-3-yl)2,2'-bipyridine (thankfully abbreviated as CyMe$_4$-BTBP, Figure 9.10)) is used in conjunction with phase-transfer agent N,N'-dimethyl-N,N'-dioctyl-2,(2'-hexyloxyethyl) malonamide (DMDOHEMA, Figure 9.10) to selectively extract Ln(III) into an organic solvent. SANEX ligands must be able to work well in solutions of low pH for which the 1,2,4-triazines are particularly well suited due to their weak basicity. Further, they must be lipophilic and stable to both radiolysis and hydrolysis. To that end, the presence of the four methyl groups of CyMe$_4$-BTBP is not arbitrary; it was found that the benzyl positions must be blocked to prevent decomposition of this ligand. Research efforts to improve the selectivity and effectiveness of this ligand has led to the development of many different possibilities (Hudson et al. 2012) with the more rigid phenanthroline triazine CyMe$_4$-BTBPhen (Figure 9.10) being particularly effective. An ORTEP drawing of the europium(III) ion complexed with CyMe$_4$-BTBPhen is shown in Figure 9.11 (Hudson et al. 2012). The logical next step in further development of these reprocessing technologies is attachment of the appropriate complexing agents to a solid support to simplify the technical aspects of

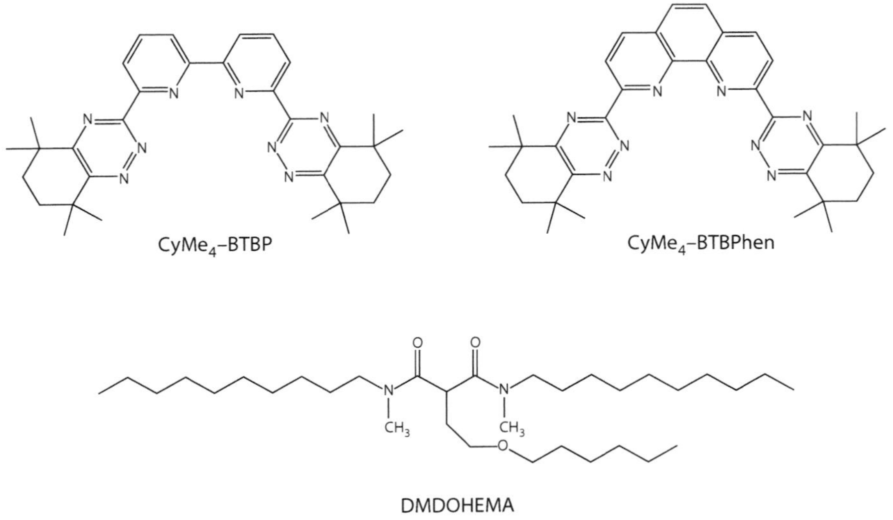

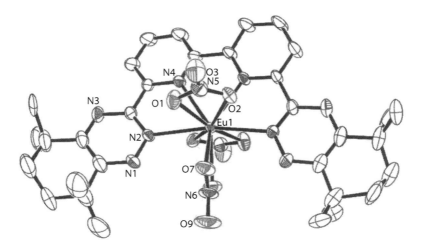

CyMe$_4$-BTBP   CyMe$_4$-BTBPhen

DMDOHEMA

FIGURE 9.10   Chemicals used in the SANEX process.

the separation, research that is already in progress (see, e.g., Raju and Subramanian (2007)). New developments in nuclear reactor technology (Generation IV, the topic of the next section) will require new methods of fuel reprocessing, such as *electrolytic reprocessing* (also known as *pyroprocessing*), a topic left to the interested reader for additional investigation.

FIGURE 9.11   (**See color insert.**) An ORTEP diagram of the europium(III) BTBPhen complex. (Reprinted with permission from Whittaker, D.M., T.L. Griffiths, M. Helliwell et al. 2013. Lanthanide speciation in potential SANEX and GANEX actinide/lanthanide separations using tetra-*N*-donor extractants. *Inorg. Chem.* 52 (7):3429–3444. Copyright 2013, American Chemical Society.)

9.4 FUTURE OF NUCLEAR ENERGY

9.4.1 GENERATION IV REACTORS

It should be apparent that an overwhelming concern with respect to nuclear power is the generation of waste. If only we could consume almost all of the troublesome nuclei, then we would not have to concern ourselves with hazardous and expensive reprocessing technologies. Generation IV reactors address this issue along with several other key aspects:

- Improved "burn-up" of nuclear fuel to result in less hazardous waste and a closed fuel cycle
- Improved safety features
- Transmutation of hazardous fission products
- Hydrogen cogeneration from excess process heat leading to improved efficiencies

In general, Generation IV reactors are *fast neutron* (or *fast-flux*) reactors: as opposed to their Generation II and III counterparts, the neutrons are *not* thermalized, so no moderator is needed. The fast neutrons mean that all of the transuranium elements undergo fission, resulting in a higher efficiency and lower radiotoxicity spent fuel. Enriched uranium is not needed and, in fact, a variety of fissile fuels can be used. However, these reactors will operate at much higher temperatures (up to 1000°C) and radiation levels. The demands that will be placed on the coolant are daunting and the coolant is a molten metal, molten salt, or high-pressure gas, depending on the reactor design (see Table 9.2 and recall the hazards associated with the use of a liquid sodium coolant described in Section 9.3.2.3). Yet the demands on the materials of the reactor core are even more formidable, with stress-corrosion cracking and radiation damage leading to hardening and embrittlement likely to be more extreme under the conditions of high temperature, high radiation flux, and high pressure (Zinkle and Was 2013).

Needless to say, these increased safety concerns require increased safety systems, which include things like a built-in "core catcher" should the reactor core suffer a meltdown; double containment walls for containment of radiation leaks; and the so-called passive safety systems. Passive safety systems rely on natural forces—gravity and entropy—to implement the needed safety action (e.g., in the event of a power failure, the control rods are designed to drop into the core under the pull of gravity, not requiring redundant power supplies). While these details make the Generation IV reactors more appealing than, say, a conventional reactor nearing the end of its operating lifetime, their contribution to electrical energy generation is undoubtedly many years in the future.

9.4.2 FUSION

And what about fusion? At this stage (and into the foreseeable future), it is far from commercially attainable, but there is enormous interest and activity in laying the groundwork for future fusion energy generation. A long sought-after goal, energy from nuclear fusion is attractive for its ability to maximize the power output while minimizing the generation of nuclear waste. The source of energy is the fusion of deuterium and tritium to form helium (Equation 9.18):

$$_1^2H + _1^3H \rightarrow {}_2^4He + {}_0^1n \ (+17.5 \ \text{Mev}) \tag{9.18}$$

The amount of energy released from this reaction is 3.5 MeV/nucleon (compared to 0.5 MeV/nucleon for a typical fission reaction (Irvine 2011)). There is no generation of carbon dioxide or other greenhouse gases and no possibility of an uncontrolled reaction. Furthermore, with the absence of Pu-239, the concern of nuclear proliferation is also absent.

Unfortunately, it takes Sun-like temperatures to initiate the fusion reaction (over 100 million degrees Celsius)! While this is easy enough to attain in the form of the hydrogen bomb, creating nuclear fusion in a controlled manner for generation of electricity is a severe technological challenge. In addition, the economic challenge is so great that to build a fusion reactor will require international cooperation. Such a cohort has been formed between India, the European Union, Russia, Japan, the United States, China, and Korea under the acronym ITER: the International Thermonuclear Experimental Reactor (http://www.iter.org/). The ITER project is in the process of building a prototype fusion reactor that plans to produce 500 MW output from 50 MW input. Construction began in southern France in 2010 with full operation targeted for early 2027.

The key to the ITER project is *magnetic confinement* of the fusion materials in a *tokamak* vessel (Figure 9.12). In a torus-shaped chamber, the deuterium and tritium nuclei will be heated under vacuum to temperatures *above 150 million degrees Celsius*, forming a gaseous mixture of positive ions and electrons (a *plasma*) in which fusion can occur. By the use of extremely strong magnetic fields, the plasma can be concentrated and held in the center of the torus so that no material touches the walls.

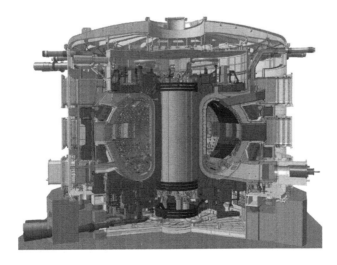

FIGURE 9.12   (**See color insert.**) The ITER tokomak fusion reactor. (From http://www.iter. org/doc/all/content/com/gallery/Media/7%20-%20Technical/In-cryostat%20Overview%20 130116.jpg)

If the plasma did make contact with the walls, the reaction would instantly cease and the material of the wall would be destroyed.

While fusion energy is appealing from the viewpoint of safe and sustainable nuclear power, this approach is astronomically challenging and expensive. Any sort of large-scale implementation of fusion energy is many decades away.

9.5   SUMMARY

It should be evident that research and development in the area of nuclear energy is ongoing even if the public perception of nuclear power is still largely negative. If there is one clear conclusion that can be made about the possible future use of nuclear energy as a sustainable energy source, it is "it is complicated."

- The capital costs associated with building, operating, and decommissioning a nuclear power plant are truly staggering and future investments are likely to be favorable only if a carbon tax is implemented on fossil fuel-generated energy.
- The issue of waste is far from settled, although the promise of Generation IV nuclear reactors and uranium–thorium fuel cycle provide some hope for safer nuclear energy generation without the formation of large volumes of waste that must be safeguarded for thousands of years. However, these technologies are many years from large-scale implementation.
- Sustainable nuclear power is achievable only if the fuel cycle is closed and the spent fuel reprocessed. Countries that do not allow reprocessing must look at the advantages and disadvantages of closing the fuel cycle from both a policy and a technological point of view.

OTHER RESOURCES

BOOKS

Bodansky, D. 1996. *Nuclear Energy: Principles, Practices, and Prospects.* Woodbury, NY: American Institute of Physics Press.
Irvine, M. 2011. *Nuclear Power. A Very Short Introduction.* New York: Oxford University Press.

ONLINE RESOURCES

International Atomic Energy Agency: http://www.iaea.org/.
International Collaboration on Nuclear Fusion: http://www.iter.org/.
United States Nuclear Regulatory Commission: http://www.nrc.gov/.
World Nuclear Association: http://www.world-nuclear.org/.

REFERENCES

Ansolabehere, S., J. Deutsch, M.J. Driscoll et al. 2003. *The Future of Nuclear Power, An Interdisciplinary MIT Study.* ISBN 0-615-12420-8. Cambridge, MA: Massachusetts Institute of Technology.
Armaroli, N. and V. Balzani. 2011. *Energy for a Sustainable World.* Weinheim, FRG: Wiley-VCH.

Bodansky, D. 1996. *Nuclear Energy: Principles, Practices, and Prospects*. Woodbury, NY: American Institute of Physics Press.

Braley, J.C., T.S. Grimes, and K.L. Nash. 2011. Alternatives to HDEHP and DTPA for simplified TALSPEAK separations. *Ind. Eng. Chem. Res.* 51 (2):629–638.

Char, N.L. and B.J. Csik. 1987. Nuclear power development: History and outlook. *IAEA Bull.* 3:19–23.

Cochran, T.B., H.A. Feiveson, W. Patterson, et al. 2010. *Fast Breeder Reactor Programs: History and Status*. Princeton, NJ: International Panel on Fissile Materials.

Cooper, N., D. Minakata, M. Begovic, et al. 2011. Should we consider using liquid fluoride thorium reactors for power generation? *Environ. Sci. Technol.* 45 (15):6237–6238.

Davis, S.J., K. Caldeira, and H.D. Matthews. 2010. Future CO_2 emissions and climate change from existing energy infrastructure. *Science* 329 (5997):1330–1333.

Fuel Cycle Stewardship in a Nuclear Renaissance. The Royal Society Science Policy Centre report 10/11. London, UK: Reproduced by permission of The Royal Society of Chemistry.

Hudson, M.J., L.M. Harwood, D.M. Laventine, et al. 2012. Use of soft heterocyclic N-donor ligands to separate actinides and lanthanides. *Inorg. Chem.* 52:3414–3428.

IEA/International Energy Agency. 2012. 2013. *Key World Energy Statistics*. Paris, France: International Energy Agency.

International Energy Agency. 2007. Nuclear Power. ETE04. International Energy Agency. http://www.iea.org/publications/freepublications/publication/essentials4.pdf

Irvine, M. 2011. *Nuclear Power. A Very Short Introduction*. New York: Oxford University Press.

Jacoby, M. 2009. Reintroducing thorium. *C. & E. News*, 48, November 16, 2009, 44–46.

Los Alamos National Laboratory, http://www.lanl.gov/science/NSS/issue1_2011/story4full.shtml. Accessed June 25, 2013.

National Aeronautics and Space Administration. 2013. *Solar Physics/the Solar Interior*. NASA 2011 [cited June 25, 2013]. Available from http://solarscience.msfc.nasa.gov/interior.shtml.

Patzek, T.W. and D. Pimentel. 2005. Thermodynamics of energy production from biomass. *CRC Crit. Rev. Plant. Sci.* 24 (5–6):327–364.

Raju, C.S.K. and M.S. Subramanian. 2007. A novel solid phase extraction method for separation of actinides and lanthanides from high acidic streams. *Sep. Purif. Technol.* 55:16–22.

Schnoor, J.L. 2013. Nuclear power: The last best option. *Env. Sci. Technol.* 47 (7):3019–3019.

Suppes, G.J. and T. Storvick. 2007. *Sustainable Nuclear Power, Academic Press Sustainable World Series*. London: Elsevier.

The Royal Society. 2011. Fuel Cycle Stewardship in a Nuclear Renaissance. The Royal Society Science Policy Centre report 10/11. London, UK: The Royal Society.

U.S. Energy Information Administration. 2013. *Nuclear Reactor Operational Status Tables*. U.S. Energy Information Administration 2011 [cited July 5, 2013]. Available from http://www.eia.gov/nuclear/reactors/stats_table1.html

United States Nuclear Regulatory Commission. Pressurized Water Reactor. Last updated March 29, 2012. Available from http://www.nrc.gov/reading-rm/basic-ref/teachers/pwr-schematic.html.

Whittaker, D.M., T.L. Griffiths, M. Helliwell et al. 2013. Lanthanide speciation in potential SANEX and GANEX actinide/lanthanide separations using tetra-N-donor extractants. *Inorg. Chem.* 52 (7):3429–3444.

World Nuclear Association. 2013. *In Situ Leach (ISL) Mining of Uranium*. World Nuclear Association 2012 [cited June 22, 2013]. Available from http://www.world-nuclear.org/info/Nuclear-Fuel-Cycle/Mining-of-Uranium/In-Situ-Leach-Mining-of-Uranium/-.UcYgfuuXKtc

World Nuclear Association. 2013. *Uranium Enrichment*. World Nuclear Association 2013a [cited June 22, 2013]. Available from http://www.world-nuclear.org/info/

Nuclear-Fuel-Cycle/Conversion-Enrichment-and-Fabrication/Uranium-Enrichment/-. UchTlT5ARal.

World Nuclear Association. 2013. *World Uranium Mining Production 2012*. World Nuclear Association, June 2013, 2013b [cited June 22, 2013]. Available from http://www.world-nuclear.org/info/Nuclear-Fuel-Cycle/Mining-of-Uranium/World-Uranium-Mining-Production/-.UchPSD5ARal.

World Nuclear News. 2013. *Thorium Test Begins*. World Nuclear Association 2013 [cited July 2, 2013]. Available from http://www.world-nuclear-news.org/ENF_Thorium_test_begins_2106131.html

WPFC Expert Group on Chemical Partitioning of the NEA Nuclear Science Committee. Spent Nuclear Fuel Reprocessing Flowsheet. NEA/NSC/WPFC/DOC(2012)15. 2012. OECD Nuclear Energy Agency, http://www.oecd-nea.org/science/docs/2012/nsc-wpfc-doc2012-15.pdf

Zinkle, S.J. and G.S. Was. 2013. Materials challenges in nuclear energy. *Acta Mater.* 61 (3):735–758.

10 Closing Remarks

> The question for humanity, then, is not whether humans and our civilizations will
> survive, but rather what kind of a planet we will inhabit.

Shellenberger and Nordhaus 2011

Science works. From initial empirical insights to theoretical explorations and finally to implemented designs we have managed to create a standard of living (for some) that was inconceivable a few decades ago. As John O'M. Bockis states in the foreword to the book *Future Energy*, "we have grown fat and happy on carbon." Process efficiencies have increased steadily; with continuing advancements in nanotechnology and analytical and computational methods. it is likely that they will continue to do so as the depth of our understanding of atomic level processes grows. We have a variety of energy options that could, potentially, reduce our dependence on carbon and begin to assuage the current assault on the environment. That is the good news. But as Bockis goes on to add "... the banquet is on its last course and there is really not much time left" (Letcher 2008).

The best solutions science has to offer cannot work without a brutally realistic perspective when it comes to sustainability. The Earth is, for all practical purposes, a closed system. In every endeavor—scientific or otherwise—our mindset must change to keep this perspective foremost. Other than energy from the Sun, there are essentially no additional material inputs on our planet. We are very much a part of this closed system and we must coexist with our outputs: what we do to the Earth we do to ourselves. Aboriginal cultures were keenly aware of this reality and worked and lived with the gifts and constraints of their environment; we would be wise to embrace their wisdom.

Given this reality, it is imperative that our careless use of resources be addressed. Waste abounds in our material world: in the Bakken fields of western North Dakota, the night sky is lit up with flares from "waste" gas—enough to heat one-half million homes per day (Manning 2013). As chemists, we generate waste in abundance and toss away carbon with abandon and added expense. Although most metals are recyclable, few are recycled to any great extent (Knowledge Transfer Network 2010). Waste heat, waste materials, waste water—we discard these resources at our own peril and recover them at great energetic and environmental costs. A serious focus on sustainability requires designing everything with recovery, recycling, and reuse in mind.

However, far and away the most draconian impact we are making on the planet is from our population growth. The ever-increasing number of humans shows absolutely no sign of waning (Figure 10.1) and is, as pointed out in the introduction to this book, an issue that has overwhelming consequences. In their book *Energy for a Sustainable World*, authors Armaroli and Balzani point out that to *maintain* the rate of our increasing energy consumption "we need to build *every day* about three

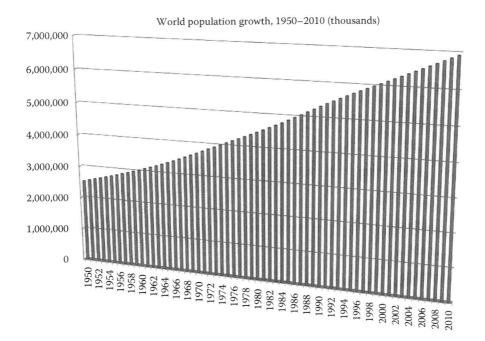

FIGURE 10.1 Global population growth, both sexes. (With permission from United Nations Department of Economic and Social Affairs, Population Division. 2013. World Population Prospects: The 2012 Revision (CD-ROM edition) 2013 [cited July 12, 2013]. Available from http://esa.un.org/unpd/wpp/Excel-Data/population.htm).

carbon-burning power plants, or two nuclear plants, or 10 km² of photovoltaic modules" (Armaroli and Balzani 2011). Our rate of population growth and the attendant rate of energy consumption is unsustainable no matter how much our efficiencies increase or what solutions scientists can provide. And of the more than 7 *billion* humans that now populate this planet, roughly 12% (the G8 nations of France, West Germany, Italy, Japan, the United Kingdom, the United States, Russia, and Canada) consume about one-half of the world's primary energy supply while the poorest 25% consume less than 3%, a disparity that is morally untenable (Armaroli and Balzani 2011).

We must transform how we use energy, how much, and where it comes from. It will require implementation of *all* of the approaches in this book—not just one or two—to begin to solve the problem of sustainable energy. But this is a social and cultural as well as technological problem. Ultimately, the problem of sustainable energy is immensely cross-cutting, requiring not only the input and ability of scientists, but also educators and ethicists, sociologists, politicians, and poets—all who can see, write, think, understand, communicate, and work together to face our conundrum. The issues associated with energy use and climate change provide the most interdisciplinary intersection of human problem solving.

Humans may not stop fighting wars, but as natural resources dwindle and climate change tightens its grip, we have an opportunity to recognize that all of us on the planet are engaged in the same struggle. Our challenges are a chance for us to come together, if we can keep them from driving us apart.

Blake 2013

REFERENCES

Armaroli, N. and V. Balzani. 2011. *Energy for a Sustainable World.* Weinheim, FRG: Wiley-VCH.

Blake, H.E. 2013. Preamble. *Orion*, May/June, 1.

Knowledge Transfer Network. 2013. *Minerals and Elements Review.* Chemistry Innovation Ltd. 2010 [cited May 8, 2013]. Available from http://www.chemistryinnovation.co.uk/stroadmap/files/dox/MineralsandElementspages.pdf.

Letcher, T.M. 2008. *Future Energy.* Oxford, UK: Elsevier.

Manning, R. 2013. Letter from Elkhorn Ranch. Bakken business. The price of North Dakota's fracking boom. *Harper's* March 2013, 29–38.

Shellenberger, M. and T. Nordhaus. 2011. Evolve. A case for modernization as the road to salvation. *Orion*, September/October. http://www.orionmagazine.org/index.php/articles/article/6402.

United Nations Department of Economic and Social Affairs, Population Division. 2013. World Population Prospects: The 2012 Revision (CD-ROM edition) 2013 [cited July 12, 2013]. Available from http://esa.un.org/unpd/wpp/Excel-Data/population.htm.

Appendix I: SI Units and Prefixes

Measured Quantity	SI Unit	Abbreviation
Length	meter	m
Mass	kilogram	kg
Time	second	s
Electric current	ampere	A
Temperature (thermodynamic)	kelvin	K
Amount of substance	mole	mol
Pressure	Pascal	Pa
Energy, work, quantity of heat	Joule	J
Electromotive force	Volt	V
Electrical conductance	siemens (A/V)	S
Electrical resistance	ohms (V/A)	Ω
Electrical charge	Coulomb	C

Prefix	Equivalent	Prefix	Equivalent
E (exa)	1×10^{18}	m (milli)	1×10^{-3}
P (peta)	1×10^{15}	μ (micro)	1×10^{-6}
T (tera)	1×10^{12}	n (nano)	1×10^{-9}
G (giga)	1×10^{9}	p (pico)	1×10^{-12}
M (mega)	1×10^{6}	f (femto)	1×10^{-15}
K (kilo)	1×10^{3}		

Source: Adapted from NIST Reference on Constants, Units, and Uncertainty. http://physics.nist.gov/cuu/Units/units.html. Accessed June 7, 2013.

Appendix II: Unit Conversions

A. Energy

	J	Erg	Cal[a]	eV	BTU	kWh	Quad	MToe
1 J =	1	10^7	0.239	6.24×10^{18}	9.48×10^{-4}	2.78×10^{-7}	9.48×10^{-19}	2.38×10^{-17}
1 Erg =	10^{-7}	1	2.39×10^{-8}	6.24×10^{11}	9.48×10^{-11}	2.78×10^{-14}	9.48×10^{-26}	2.38×10^{-24}
1 Cal[a] =	4.19	4.19×10^7	1	2.61×10^{19}	3.97×10^{-3}	1.162×10^{-6}	3.97×10^{-18}	1.59×10^{-19}
1 eV =	1.60×10^{-19}	1.0×10^{-12}	3.38×10^{-20}	1	1.52×10^{-22}	4.45×10^{-26}	1.52×10^{-37}	6.08×10^{-39}
1 BTU =	1.06×10^3	1.06×10^{10}	2.52×10^2	6.59×10^{21}	1	2.93×10^{-4}	1×10^{-15}	4×10^{-17}
1 kWh =	3.60×10^6	3.60×10^{13}	8.60×10^5	2.25×10^{25}	3.41×10^3	1	3.41×10^{-12}	8.57×10^{-11}
1 Quad =	1.0×10^{18}	1.0×10^{25}	2.52×10^{17}	6.59×10^{36}	1×10^{15}	2.93×10^{11}	1	25
1 MToe =	4.2×10^{16}	4.2×10^{23}	1.00×10^{16}	2.62×10^{35}	4×10^{13}	1.16×10^{10}	4×10^{-2}	1

[a] Thermochemical calorie.

B. Mass

	Tonne	Kilogram	Pound	Short ton	Long ton
1 Tonne (metric) =	1	1000	2205	1.10	0.98421
1 Kilogram =	1×10^{-3}	1	2.20	1.10×10^{-3}	9.84×10^{-4}
1 Pound (US) =	4.54×10^{-4}	0.454	1	0.0005	4.46×10^{-4}
1 Short ton (US) =	0.907	907	2000	1	0.893
1 Long ton (UK) =	1.02	1016	2240	1.12	1

C. Volume

	Gal (US)	Gal (UK)	Barrel	Cubic Foot	Liter	Cubic meter
1 U.S. Gallon =	1	0.833	2.38×10^{-2}	0.134	3.78	3.8×10^{-3}
1 U.K. Gallon =	1.20	1	2.86×10^{-2}	0.160	4.55	4.5×10^{-3}
1 Barrel (bbl) =	42.0	35.0	1	5.62	159.	0.159
1 Cubic foot (ft³) =	7.48	6.23	0.178	1	28.3	2.83×10^{-2}
1 Liter (L) =	0.264	0.220	6.29×10^{-3}	3.53×10^{-2}	1	1×10^{-3}
1 Cubic meter (m³) =	264	220.	6.29	35.3	1000	1

Source: Adapted from IEA/International Energy Agency. 2012 Key World Energy Statistics. 2013. Paris, France.

D. Pressure

	Bar	Torr	Pascal	Atmosphere	PSI
1 Bar =	1	7.52×10^{-4}	100,000	0.987	14.5
1 Torr =	1.33×10^{3}	1	1.33×10^{2}	1.32×10^{-3}	1.93×10^{-2}
1 Pascal (Pa) =	1×10^{-5}	7.50×10^{-3}	1	9.87×10^{-6}	1.45×10^{-4}
1 Atmosphere[a] (atm) =	1.01	760	1.01×10^{5}	1	14.7
1 Pounds/inch^2 (PSI) =	6.89×10^{-2}	51.7	6.89×10^{3}	0.0680	1

[a] Standard atmosphere.

E. Miscellaneous Conversions

$$1 \text{ Å} = 1 \times 10^{-8} \text{ cm}$$

$$E \text{ (eV)} = \frac{1239.87}{\lambda \text{ (nm)}}$$

Appendix III: Electricity: Units and Equations

Property	Unit	Definition/conversions	Other
Electromotive force	Volt (V)	$V = \dfrac{W}{A} = \dfrac{J}{A \cdot s} = \dfrac{N \cdot m}{A \cdot s} = \dfrac{J}{C}$	
Current	Ampere (A)	$A = \dfrac{V}{R} = \dfrac{C}{s}$	
Charge	Coulomb (C)	One coloumb is assigned to the amount of electric charge conveyed in one second by a current of one ampere	An alternate unit of charge is the Faraday; $1\ C = 1.036 \times 10^{-5}\ F$
Power	Watt (W)	$W = J/s$	
Conductance	Siemens (S)	$S = \dfrac{A}{V}$	

Appendix IV: Fossil Fuel Units and Abbreviations

	Oil		Natural gas
TOE (tonne oil equivalent)	The amount of energy released by burning 1 tonne of crude oil ($\approx$42 GJ). One tonne = 1000 kg	Cf or cu. ft.	Cubic feet
BBl	Barrel	Bcf	Billion cubic feet
b/d	Barrels per day	Bcm	Billion cubic meters
Mb/d	Thousand barrels/day	Tcf	Trillion cubic feet
MMbbl	Million barrels	Tcm	Trillion cubic meters

Appendix V: Important Constants

Avogadro's number	6.022×10^{23} mol^{-1}
Planck's constant (h)	6.626×10^{-34} J · s
Faraday's constant (F)	9.6485×10^4 C/mol
Speed of light (c)	2.998×10^8 m/s

Appendix VI: Acronyms

a-Si	Amorphous silicon
a-Si:H	Hydrogenated amorphous silicon
AB	Ammonia borane
AFEX	Ammonia fiber/freeze explosion
APS	Artificial photosynthesis
ATR	Attenuated total reflectance
BPA	Bisphenol A
BWR	Boiling water reactor
c-Si	Crystalline silicon
CANDU	CANada Deuterium Uranium reactor
CB	Conduction band
CC	Carbon capture
CCS	Carbon capture and sequestration *or* storage
CCU	Carbon capture and utilization
CGO	Ceria-gadolinium oxide
CHP	Combined heat and power
CIGS/CIGSe	Copper indium gallium selenide
CLC	Chemical looping combustion
CM	Carrier multiplication = EHPM
CNT	Carbon nanotube
CRP	Controlled radical polymerization
CSP	Concentrated solar power
CTL	Coal-to-liquid
CV	Cyclic voltammetry
CZTS, CZTSe	$Cu_2Zn(Sn_{1-x}Ge_x)S$, Se
DAFC	Direct alcohol fuel cell
DMC	Dimethyl carbonate
DMDOHEMA	N,N'-dimethyl-N,N'-dioctyl-2,(2'-hexyloxyethyl) malonamide
DMF	Dimethylformamide
DMFC	Direct methanol fuel cell
DOE	U.S. Department of Energy
DP	Degree of polymerization
DRIFTS	Diffuse reflectance infrared Fourier transform spectroscopy
DSC	Differential scanning calorimetry
DSSC	Dye-sensitized solar cell
DTPA	Diethylenetriamine-N,N,N',N'',N''-pentaacetic acid
DWCNT	Double-walled carbon nanotube
EDS	Energy-dispersive x-ray spectroscopy
EES	Electrical energy storage
E_g	Bandgap energy
EHP	Electron–hole pair
EHPM	Electron–hole pair multiplication
EIA	U.S. energy information administration
EMF	Electromotive force
EOR	Enhanced oil recovery

EPR	European pressurized reactor *or* evolutionary power reactor
EQE	External quantum efficiency
ESR	Steam reforming of ethanol
EV	Electric vehicle
FAE	Fatty acid ester
FAEE	Fatty acid ethyl ester
FAME	Fatty acid methyl ester
FBR	Fast breeder reactor
FF	Fill factor
FFA	Free fatty acids
FT	Fischer–Tropsch
FTO	Fluorine-doped tin oxide
GDL	Gas diffusion layer
GPC	Gel permeation chromatography
H4MPT	Tetrahydromethanopterin
HAWT	Horizontal axis wind turbine
HDEHP	Bis-2-ethyl(hexyl) phosphoric acid
HDO	Hydrodeoxygenation
HDPE	High-density polyethylene
HHV	Higher heating value
HWR	Heavy water reactor
IGCC	Integrated gasification combined cycle
IL	Ionic liquid
IPCE	Incident photon-to-current conversion efficiency
IQE	Internal quantum efficiency
ISL	*In situ* leaching
ISR	*In situ* recovery
ITER	International thermonuclear experimental reactor
ITO	Indium tin oxide = 0.9 In_2O_3/0.1 SnO_2
LCB	Lignocellulosic biomass
LDPE	Low-density polyethylene
LFR	Lead fast reactor
LFTR	Liquid fluoride thorium reactor
LHV	Lower heating value
LSGM	Lanthanum/Strontium/Gallium/Magnesium solid oxide
LWR	Light water reactor
MAS	Magic angle spinning
MEA	Membrane electrode assembly
MEC	Microbial electrolysis cell
MEG	Multiple exciton generation
MFC	Microbial fuel cell
MMD	Molecular mass distribution
MOF	Metal–organic framework
MSR	Molten salt reactor
MSW	Municipal solid waste
MWCNT	Multi-walled CNT
NAS	Sodium-sulfur battery
NETL	U.S. Department of Energy's National Energy Technology Laboratory
NRC	U.S. National Research Council
NREL	U.S. National Renewable Energy Laboratory

OPV	Organic photovoltaic
ORR	Oxygen reduction reaction
P3HT	Poly-3-hexylthiophene
PAN	Polyacrylonitrile
PAT	Polyalkylthiopene
PBI	Polybenzimidazole
PBMR	Pebble-bed modular reactor
PCE	Power conversion efficiency
PDI	Polydispersity index
PDI	Perylene-3,4:9,10-bis(dicarboximide)
PEM	Polymer electrolyte membrane
PEMFC	Polymer electrolyte membrane fuel cell OR proton exchange membrane FC
PES	Poly(ether) sulfone
PFSA	Perfluorinated sulfonic acid
PFSI	Perfluorinated sulfonic ionomer
PGM	Platinum group metals
PSC	Polymer solar cell
PSI	Photosystem I (photosynthesis)
PSII	Photosystem II (photosynthesis)
PTFE	Polytetrafluoroethylene
PUREX	Plutonium uranium extraction
PV	Photovoltaic
PVA	Polyvinyl alcohol
PVC	Polyvinyl chloride
PWR	Pressurized water reactor
QD	Quantum dot
QDSSC	Quantum dot-sensitized solar cell
QE	Quantum efficiency
QY	Quantum yield
RH	Relative humidity
SANEX	Selective actinide extraction
SEC	Size-exclusion chromatography
SEM	Scanning electron microscopy
SFR	Sodium fast reactor
SMR	Steam reforming of methane
SOFC	Solid oxide fuel cell
SPEEK	Sulfonated poly(ether) ether ketone
SPI	Sulfonated polyimide
SRE	Steam reforming of ethanol
SSZ	Scandium-stabilized zirconia
STEM	Scanning transmission electron microscopy
SWCNT	Single-walled carbon nanotube
TAG	Triacylglyceride
TALSPEAK	Trivalent actinide-lanthanide separation by phosphorus reagent extraction from aqueous komplexes
TCO	Transparent conducting oxide
TEM	Transmission electron microscopy
THF	Tetrahydrofuran
THGE-PE	1,1,1-Tris(p-hydroxyphenyl)ethane triglycidyl ether
TOF	Turnover frequency

TPP	Thiamine pyrophosphate
TRUEX	Transuranic extraction
UOX	Uranium oxide
VAWR	Vertical axis wind turbine
VB	Valence band
WGS	Water gas shift (reaction)
XAFS	X-ray absorption fine structure
XRD	X-ray diffraction
YSZ	Yttrium-stabilized zirconia

Index